Dried Fruit Products

Dehydration of fruits is a massive operation amounting to tens of billions of dollars' worth in the global market. The enormous variety of fruits grown around the world and the wide range of products made therefrom make this an attractive method for the development of novel and shelf-stable consumer products.

Dried Fruit Products offers a current approach linking the theory and practice of fruit drying, summarizing various techniques, their advantages and limitations, industrial applications, and simple design methods. Such dried fruit products as fruit pieces, fruit leathers, and fruit powders are dealt with in a way to inform their physical, chemical, sensory, and nutritional features, along with the characteristics of the process used to obtain them, such as drying method and drying equipment.

Key Features:

- Contains up-to-date information on fruit drying
- Presents a multi-perspective viewpoint of fruit drying
- Addresses both food science and chemical engineering aspects of fruit drying

Readers can gain knowledge on the various types of drying techniques and insightful thoughts on selecting the appropriate drying techniques for different fruit products.

Advances in Drying Science and Technology

Series Editor Arun S. Mujumdar, McGill University, Quebec, Canada

Drying of Biomass, Biosolids, and Coal: For Efficient Energy Supply and Environmental Benefits
Shusheng Pang, Sankar Bhattacharya, Junjie Yan

Drying and Roasting of Cocoa and Coffee
Ching Lik Hii and Flavio Meira Borem

Heat and Mass Transfer in Drying of Porous Media
Peng Xu, Agus P. Sasmito, and Arun S. Mujumdar

Freeze Drying of Pharmaceutical Products
Davide Fissore, Roberto Pisano, and Antonello Barresi

Frontiers in Spray Drying
Nan Fu, Jie Xiao, Meng Wai Woo, Xiao Dong Chen

Drying in the Dairy Industry
Cécile Le Floch-Fouere, Pierre Schuck, Gaëlle Tanguy, Luca Lanotte, Romain Jeantet

Spray Drying Encapsulation of Bioactive Materials
Seid Mahdi Jafari and Ali Rashidinejad

Flame Spray Drying: Equipment, Mechanism, and Perspectives
Mariia Sobulska and Ireneusz Zbicinski

Advanced Micro-Level Experimental Techniques for Food Drying and Processing Applications
Azharul Karim, Sabrina Fawzia and Mohammad Mahbubur Rahman

Mass Transfer Driven Evaporation of Capillary Porous Media
Rui Wu, Marc Prat

Particulate Drying: Techniques and Industry Applications
Jangam Vinayak, Chung-Lim Law, and Shivanand Shirkole

Drying and Valorisation of Food Processing Waste
Chien Hwa Chong, Rafeah Wahi, Chee Ming Choo, Shee Jia Chew, and Mackingsley Kushan Dassanayake

Drying of Herbs, Spices, and Medicinal Plants
Ching Lik Hii and Shivanand Shirkole

Drying of Aromatic Plant Material for Natural Perfumes
Viplav Hari Pise, Ramakant Harlalka, and Bhaskar Narayan Thorat

Dried Fruit Products
Felipe Richter Reis and Shivanand S. Shirkole

For more information about this series, please visit: www.routledge.com/Advances-in-Drying-Science-and-Technology/book-series/CRCADVSCITEC

Dried Fruit Products

Edited by

Felipe Richter Reis and Shivanand S. Shirkole

CRC Press
Taylor & Francis Group
Boca Raton London New York

CRC Press is an imprint of the
Taylor & Francis Group, an **informa** business

Cover image: Shutterstock

First edition published 2025
by CRC Press
2385 NW Executive Center Drive, Suite 320, Boca Raton FL 33431

and by CRC Press
4 Park Square, Milton Park, Abingdon, Oxon, OX14 4RN

CRC Press is an imprint of Taylor & Francis Group, LLC

Library of Congress Cataloging-in-Publication Data
Names: Richter Reis, Felipe, editor. | Shirkole, Shivanand S., editor.
Title: Dried fruit products / edited by Felipe Richter Reis and Shivanand
S. Shirkole.
Description: First edition. | Boca Raton, FL : CRC Press, 2024. | Series:
Advances in drying science and technology | Includes bibliographical
references and index.
Identifiers: LCCN 2023057701 (print) | LCCN 2023057702 (ebook) | ISBN
9781032495750 (hardback) | ISBN 9781032515809 (paperback) | ISBN
9781003402916 (ebook)
Subjects: LCSH: Cooking (Dried fruit) | Fruit--Drying. | LCGFT: Cookbooks.
Classification: LCC TX826.5 .D756 2024 (print) | LCC TX826.5 (ebook) |
DDC 641.4/4--dc23/eng/20240102
LC record available at https://lccn.loc.gov/2023057701
LC ebook record available at https://lccn.loc.gov/2023057702

ISBN: 978-1-032-49575-0 (hbk)
ISBN: 978-1-032-51580-9 (pbk)
ISBN: 978-1-003-40291-6 (ebk)

DOI: 10.1201/9781003402916

Typeset in Times New Roman
by KnowledgeWorks Global Ltd.

Dedication

We would like to dedicate this book to all the researchers who work in drying science and technology worldwide, especially to Professor Arun S. Mujumdar for his global leadership and vision in establishing drying R&D as an important field for academia and industry.

Contents

Advances in Drying Science and Technology

Series Editor: Dr. Arun S. Mujumdar

It is well known that the unit operation of drying is a highly energy-intensive operation encountered in diverse industrial sectors ranging from agricultural processing, to ceramics, chemicals, mineral processing, pulp and paper, pharmaceuticals, coal polymer, food, and forest products industries, as well as waste management. Drying also determines the quality of the final dried products. Meeting the need to make drying technologies sustainable and cost-effective via the application of modern scientific techniques is the goal of academic as well as industrial R&D activities around the world.

Drying is a truly multi- and interdisciplinary area. Over the last four decades, the scientific and technical literature on drying have seen exponential growth. The continuously rising interest in this field is also evident from the success of numerous international conferences devoted to drying science and technology.

The establishment of this new series of books, entitled Advances in Drying Science and Technology, is designed to provide authoritative and critical reviews and monographs focusing on current developments as well as future needs. It is expected that books in this series will be valuable to academic researchers as well as industry personnel involved in any aspect of drying and dewatering.

The series will also encompass themes and topics closely associated with drying operations, e.g., mechanical dewatering, energy savings in drying, environmental aspects, life-cycle analysis, techno-economics of drying, electro-technologies, control and safety aspects, and so on.

ABOUT THE SERIES EDITOR

Prof. Arun S. Mujumdar is recognized internationally for his extensive, original, innovative, and impactful research in drying science and sustainable technology. Indeed, he is often referred to as the "Drying Guru." He was the Founding Chair in 1978 of the International Drying Symposium (IDS) series and Editor-in-Chief of *Drying Technology: An International Journal* since 1988. The 4th enhanced edition of his *Handbook of Industrial Drying*, published by CRC Press, has just appeared. He is the recipient of numerous international awards, including honorary doctorates from Lodz Technical University, Poland, University of Lyon, France, and Western University, Ontario, Canada.

Please visit www.arunmujumdar.com for further details.

Preface

Dehydration of fruits is a massive operation amounting to tens of billions of dollars' worth in the global market. The enormous variety of fruits grown around the world and the wide range of products made therefrom make this an attractive method for the development of novel and shelf-stable consumer products. The high nutritional content of fruits makes them an important source of vitamins, minerals, and bioactive compounds for human health.

Fruits are highly perishable after harvesting due to their high moisture content along with the presence of sugars, which allows microbial degradation, especially by fungi. Drying fruits to water activity levels at which microbial activity cannot take place extends their shelf life so that they can be consumed off-season and shipped to regions where specific fruits cannot be grown. As will be noted by the readers, fruits can be dried in a wide assortment of dryers to produce dried products in several different physical forms.

This book is an outcome of the expertise and experience of an international team of researchers who have significantly contributed to the knowledge base on the drying of fruits. As this is indeed a vast subject, we have selected the following important aspects of broad interest: the global market for dried fruit products; dryer selection; drying characteristics; quality characteristics, including aroma and bioactive profile; storage aspects; emerging technologies; techno-economic evaluation; and future prospects.

This book contains chapters from expert authors in both industry and academia. This book is an important resource for those working in chemical processing, food processing, and agricultural processing, especially those focused on the drying of food and agricultural products.

Finally, we are truly grateful to the authors and reviewers for their time and efforts, which have enhanced the value and potential impact of this compilation. This book is aimed at academic researchers as well as scientists and technologists in the food industry. It can also be a useful reference for teaching courses in food science and technology. We believe that a significant growth in the production of various fruits around the world will help alleviate food insecurity while improving health and quality of life as well.

Felipe R. Reis, DSc
Federal Institute of Paraná, Campus Colombo, Brazil

Shivanand S. Shirkole, PhD
Institute of Chemical Technology Mumbai,
ICT-IOC Odisha Campus, Bhubaneswar, India

About the Editors

Felipe Richter Reis, D.Sc., is a Professor of Food Science at the Food Technology undergraduate course (B.Sc.), Federal Institute Paraná (IFPR), and at the Food Engineering graduate courses (M.Sc. and D.Sc.) offered at the Graduate Program in Food Engineering, Department of Chemical Engineering, Federal University of Paraná (UFPR), Brazil. All the above-mentioned courses received grades of excellence in a quality assessment performed by the Brazilian Ministry of Education in 2022. In both institutions, he teaches and advises students on research projects that emphasize the blanching and drying of fruits and vegetables.

Dr. Reis holds a B.Sc. degree in Food Engineering from the Federal University of Santa Catarina (UFSC), ranked 98 in Food Science and Technology in the Best Global Universities ranking. In addition, he holds an M.Sc. degree in Food Technology and a D.Sc. degree in Food Engineering, both from the Federal University of Paraná, ranked 220 in Food Science and Technology by the same ranking mentioned above. He has been a researcher in the drying area for more than ten years, which is the theme of his D.Sc. thesis entitled 'Vacuum drying of yacon: influence of process conditions on quality parameters and drying kinetics.' The results obtained in his thesis were published in the *Journal of Food Processing and Preservation* (Wiley, IF = 2.609) and in *Drying Technology* (Taylor and Francis, IF = 3.3). The objectives of his current research include studying the effect of pretreatments on drying and quality characteristics of dried fruit and vegetables. Additionally, he is continuously performing theoretical research to yield review papers, book chapters, and books published by world-class publishers. For example, a recent review that he authored was published in the renowned journal *Trends in Food Science and Technology* (Elsevier, IF = 15.3), from a publisher that also produced two book chapters for a Food Engineering series. With regard to edited and authored books, he has been contributing for almost a decade with world-class publishers, yielding unique manuscripts like 'Vacuum Drying for Extending Food Shelf-Life' (Springer Nature, 2014), 'New Perspectives on Food Blanching' (Springer Nature, 2017), 'Reports on the Processing of Exotic Fruits' (Springer Nature, 2019), and 'Volatile Compounds Formation in Specialty Beverages' (CRC Press, 2022). Dr. Reis acts as a reviewer of about 35 peer-reviewed journals, a task that he performs without any help from other people. Most of the papers that he reviews are submitted to world-class publishers like the American Chemical Society, Elsevier, Springer Nature, Taylor & Francis, and Wiley. Finally, he is the only South American member of the editorial board of the journals *Plant Foods for Human Nutrition* (Springer Nature, IF = 4.0) and *Drying Technology*.

Shivanand Shankarrao Shirkole is working as an Assistant Professor at the Department of Food Engineering and Technology, Institute of Chemical Technology Mumbai, ICT-IOC Campus, Bhubaneswar, India. He has pursued his PhD in 2020 from the Department of Food Process Engineering, National Institute of Technology, Rourkela, India. He has more than four years of industrial experience as a Plant

Engineer at Hyderabad, India-based Multi-Crop Seed Conditioning Plant. His broad areas of research are Low Moisture Food Safety and Thermal Processing of Food. Presently he is working on Sustainable Technologies for Food and Agri-products. He has been an Executive Committee Member of the Association of Food Scientists & Technologists (India), NIT Rourkela Chapter since May 2020. Also, he has been working for *Drying Technology – An International Journal* (Taylor & Francis) as Associate Editor since August 2021.

Contributors

Hannah Caroline Santos Araujo
Laboratory of Flavor and Chromatographic Analysis/ PROCTA
Federal University of Sergipe
São Cristóvão-SE, Brazil

Alessandro Benelli
University of Tuscia
Italy

Nikita S. Bhatkar
Institute of Chemical Technology Mumbai
ICT-IOC, Bhubaneswar
Odisha, India

Keshram Dulait
Institute of Chemical Technology Mumbai
ICT-IOC Bhubaneswar
Odisha, India

Pietro Ibba
University of Bozen
Italy

Mônica Silva de Jesus
Laboratory of Flavor and Chromatographic Analysis/ PROCTA
Federal University of Sergipe
São Cristóvão-SE, Brazil

Maria Terezinha Santos Leite Neta
Laboratory of Flavor and Chromatographic Analysis/ PROCTA
Federal University of Sergipe
São Cristóvão-SE, Brazil

Linlin Li
College of Food and Bioengineering
Henan University of Science and Technology
Luoyang, Henan, People's Republic of China

Sangeetha Macherla
Institute of Chemical Technology Mumbai,
ICT-IOC Bhubaneswar,
Odisha, India

Cinzia Mannozzi
University of Camerino
Italy

Riccardo Massantini
University of Tuscia
Italy

Maria Lucia Masson
Federal University of Paraná
Chemical Engineering Department
Graduate Program of Food Engineering
Curitiba, Brazil

Francesca Melini
Council for Agricultural Research and Economics
Italy

Anna Michalska-Ciechanowska
Department of Fruit, Vegetable and Plant Nutraceutical Technology, Faculty of Biotechnology and Food Science
Wrocław University of Environmental and Life Sciences
Wrocław, Poland

Roberto Moscetti
University of Tuscia
Italy

Narendra Narain
Laboratory of Flavor and Chromatographic Analysis/ PROCTA
Federal University of Sergipe
São Cristóvão-SE, Brazil

Viplav Hari Pise
Institute of Chemical Technology Mumbai
Maharashtra, India

Felipe Richter Reis
Federal Institute of Paraná,
Campus Colombo, Brazil

Rafael Donizete Dutra Sandes
Laboratory of Flavor and Chromatographic Analysis/ PROCTA
Federal University of Sergipe
São Cristóvão-SE, Brazil

Shivanand S. Shirkole
Institute of Chemical Technology Mumbai
ICT-IOC Bhubaneswar
Odisha, India

Parul Shukla
Institute of Chemical Technology Mumbai
ICT-IOC Bhubaneswar
Odisha, India

Savitha Srinivasan
Institute of Chemical Technology Mumbai
Maharashtra, India

Vimal
Institute of Chemical Technology Mumbai
ICT-IOC Bhubaneswar
Odisha, India

Min Zhang
School of Food Science and Technology
Jiangnan University
Wuxi, Jiangsu, People's Republic of China

1 Overview of the Current Global Market for Dried Fruit Products and General Aspects of Fruit Drying

Felipe Richter Reis[1,2]
[1]Federal Institute of Education, Science, and Technology of Paraná
Colombo, PR, Brazil
[2]Graduate Program of Food Engineering, Chemical Engineering Department, Federal University of Paraná
Curitiba, PR, Brazil

1.1 GENERAL ASPECTS ABOUT FRUITS AND DRIED FRUIT PRODUCTS

Initially, fruits are defined from all available points of views. From the botanical viewpoint, fruit is defined by a relevant plant physiology book as the final product of a matured ovary. This includes a wide range of structures, like a grain of cereal, a nut, and a peach (Grierson 2001). In addition, there is another classical definition of fruit, which is based on its usage by consumers. Fruit is, in this sense, a plant item that is generally eaten as dessert (Food and Agriculture Organization 1995). Figure 1.1 shows examples of fruits.

Dried fruit products (DFP) are obtained by removing water from whole fruits, pieces of fruit of different shapes and sizes, purees or juices. Advantages of DFP when compared to fresh fruits include intensified sweetness and flavor, decreased weight and volume and extended shelf-life. On the other hand, some drawbacks of fruit drying have been reported, such as nutrient loss, color degradation, turgidness loss, and shrinkage. Mild drying methods, like freeze-drying, and innovative drying methods, like conductive multi-flash drying and instant controlled pressure drop drying result in better preservation of DFP quality (Reis *et al.* 2022). Figure 1.2 shows examples of DFP.

Previous documents on fruit drying, such as books, book chapters, and review articles, have been published before, though not in such a comprehensive way. In this sense, is it relevant to emphasize the remarkable contribution of Prof. Arun S. Mujumdar to this field, who has edited interesting books on fruit drying which are not associated

DOI: 10.1201/9781003402916-1

FIGURE 1.1 Examples of fruits.

FIGURE 1.2 Examples of dried fruit products.

with any specific publishing company, such as: "Drying of Foods, Vegetables and Fruits" and "Processing and Drying of Tropical Fruits". A search on the Web of Science with the term "fruit drying– resulted in 940 review articles. When ordered by relevance, the first article from the list is the work of Fernandes *et al.* (2011). In such document, drying techniques, drying kinetics, and quality parameters of dried exotic tropical fruits were addressed. Other reviews on DFP include topics like health benefits (Kundu and Chun 2014; Chang *et al.* 2016; Alasalvar *et al.* 2023); product quality (Santos and Silva 2008; McSweeney and Seetharaman 2015; Omolola *et al.* 2017; Radojčin *et al.* 2021); specific drying techniques, like spray drying (Verma and Singh 2015; Shishir and Chen 2017), solar drying (Natarajan *et al.* 2022), microwave drying (Zhang *et al.* 2006), and infrared drying (Sossa *et al.* 2021); drying modelling

(Onwude *et al.* 2016); and general aspects of fruit drying (Sagar and Suresh Kumar 2010). Usually, these documents dealt with vegetables as well.

Processing of fruits into DFP has been put in place since prehistoric times as a preservation method since fruits originally present a shelf-life of only a few days due to their high value of water activity ($a_w > 0.97$). Nowadays, though, drying fruits is not only a subsistence activity anymore but also a profitable business on global level, as will be shown in the next section.

1.2 GLOBAL MARKET FOR DRIED FRUIT PRODUCTS

The trade for dried fruits reached $2.98 billion globally in the year 2021. The top exporter was Turkey, and the top importer was China (The Observatory of Economic Complexity 2023a). The fruits individually analyzed for their dried version by OEC are apples, apricots, grapes, and prunes, being dried grapes (raisins) the market leader with a world trade of $1.7 billion in 2021 (The Observatory of Economic Complexity 2023b). Figure 1.3 shows some raisins.

Companies from several countries lead the market of DFP, like Australia, Finland, Germany, Greece, India, United Arab Emirates, and the United States (Reportlinker.com 2023). In Massachusetts, for example, a company has been focused on innovative flavored, covered, and health-promoting DFP, such as dried cranberries flavored with cherry juice, covered with Greek yoghurt, and a probiotic fruit medley (Ocean Spray Cranberries 2023). A German company invested in fruit shapes, i.e., fruit gums with high fruit content and variable features, like customizable shape, vegan formulation, nutraceutical benefits, added probiotic cultures, or plant extracts (Paradise Fruits 2023). A company from Finland sells five species of whole berries which can be sweetened by liquid sugar, being available also as organic quality (Kiantama Oy 2023). Following the organic trend, a Greek company provides a specific link on its website that leads to its line of organic DFP (Geobres 2023). In Australia, one company has nothing less than 72 options of DFP in its portfolio, which includes several fruits processed into the following products: medley, diced,

FIGURE 1.3 Dried grapes (raisins).

rings, whole, salad, slices, chips, pitted, shredded, flakes, peel, and spears (Royal Nut Company 2023). Dates are the specialty of a company from the United Arab Emirates that, besides plain dates, sells products like chocolate-covered dates (AL FOAH 2023). Finally, in California, there are companies specialized in raisins (Lion Raisins 2023; National Raisin Company 2023).

1.3 GENERAL ASPECTS OF FRUIT DRYING

Drying of fruit follows the rules that are used for drying heat-sensitive materials, as fruit contains many thermolabile compounds. For example, fruit color is provided by pigments like carotenoids, anthocyanins, and chlorophyll, all of them considered as bioactive compounds. Then, obtaining high-quality DFP is not a simple task, usually requiring controlled process conditions, even though sun drying is still practiced in many regions. A recent review showed that high-quality DFP can be obtained by means of mild and innovative drying methods, which are usually preceded by innovative pre-treatments (Reis *et al.* 2022).

Some considerations could be made considering the DFP available in the market, as mentioned in the previous section. For example, dried fruits flavored with juice can be obtained by using pre-treatments consisting of immersion in fruit juices. In this sense, Kowalska *et al.* (2020) used chokeberry juice to provide flavor and antioxidant properties to apple chips. They found that an osmotic dehydration consisting of immersion of apple slices in a 50°Brix aqueous solution made of sucrose and chokeberry juice concentrate, followed by freeze-drying and/or microwave-vacuum drying, yielded apple chips with increased content of vitamin C and polyphenols, and increased antioxidant activity.

Another trending DFP, probiotic fruit snacks, can be made by incorporating probiotic cultures into dried fruits. One of the studies in this field was conducted by Galvão *et al.* (2020), who developed apple snacks incorporated with the probiotic microorganism *Bacillus coagulans*. They concluded that probiotic fruit snacks could be obtained by spraying a coating composed of hydroxyethyl cellulose, polyethylene glycol, and probiotic bacteria onto dried apple cubes ($a_w < 0.4$) obtained by fluidized bed drying.

Fruit gums are also on the rise, being a way to deliver not only a unique texture and flavor but also health-promoting compounds. Reis (2019) defined fruit gums, which he called structured fruits, as products obtained from fruit puree added with ingredients like glycerol, sucrose, and hydrocolloids, which are mixed, cooled, and finally dried. Figure 1.4 shows examples of fruit gums.

The demand for organic DFP is increasing as well, driven by consumers' health consciousness. According to the United States Department of Agriculture, organic produce must be grown on soil that had no prohibited substances applied for three years prior to harvest, which includes most synthetic fertilizers and pesticides. Organic produce must be used as ingredient if one wants to make organically processed foods, whose manufacturing requires additional care, like containing no artificial preservatives, colors, or flavors (United States Department of Agriculture 2023).

FIGURE 1.4 Fruit gums.

1.4 CONCLUSION

DFP are versatile, healthy, and shelf-stable options for nourishment and pleasantness of a wide range of consumers. A remarkable amount of DFP is traded worldwide, boosting economy in various countries. Innovative DFP are on the market, such as juice-infused, uniquely covered, probiotic, organic, and processed into natural gums. Academic research on this field has accompanied industry in such a way that studies dealing with these products were published in renowned journals, though not extensively. Future research should focus on developing and analyzing healthy and sustainable DFP while ensuring high sensory acceptability. The next chapters will present DFP from many perspectives like equipment, drying and quality characteristics, emerging technologies, and techno-economic analysis.

REFERENCES

AL FOAH, 'Our Products'. AL FOAH, 2023. https://alfoah.com/

Alasalvar, C., S. K. Chang, P. M. Kris-Etherton, V. K. Sullivan, K. S. Petersen, M. Guasch-Ferré, and D. J. A. Jenkins. 2023. Dried fruits: Bioactives, effects on gut microbiota, and possible health benefits—an update. *Nutrients* 15:1611. https://doi.org/10.3390/nu15071611

Chang, S. K., C. Alasalvar, and F. Shahidi. 2016. Review of dried fruits: Phytochemicals, antioxidant efficacies, and health benefits. *Journal of Functional Foods* 21:113–132. https://doi.org/10.1016/j.jff.2015.11.034

Fernandes, F. A. N., S. Rodrigues, C. L. Law, and A. S. Mujumdar. 2011. Drying of exotic tropical fruits: A comprehensive review. *Food and Bioprocess Technology* 4:163–185. https://doi.org/10.1007/s11947-010-0323-7

Food and Agriculture Organization, 'Fruit and Vegetable Processing'. Food and Agriculture Organization, 1995. https://www.fao.org/3/V5030E/V5030E00.htm

Galvão, A. M. M. T., S. Rodrigues, and F. A. N. Fernandes. 2020. Probiotic dried apple snacks: Development of probiotic coating and shelf-life studies. *Journal of Food Processing and Preservation* 44:e14974. https://doi.org/10.1111/jfpp.14974

Geobres, S. A., 'Products'. Geobres S.A., 2023. http://www.geobres.gr/products.asp

Grierson, W. 2001. Fruit Development, Maturation and Ripening. In *Handbook of Crop and Plant Physiology*, edited by M. Pessarakli, 143–160. Boca Raton, FL: CRC Press.

Kiantama Oy, ‘Sweetened Berries’. Kiantama Oy, 2023. https://kiantama.fi/en/product-category/sugar-infused-berries/

Kowalska, J., A. Marzec, E. Domian, S. Galus, A. Ciurzyńska, A. Lenart, and H. Kowalska. 2020. The use of antioxidant potential of chokeberry juice in creating pro-healthy dried apples by hybrid (convection-microwave-vacuum) method. *Molecules* 25:5680. https://doi.org/10.3390/molecules25235680

Kundu, J. K., and K. S. Chun. 2014. The promise of dried fruits in cancer chemoprevention. *Asian Pacific Journal of Cancer Prevention* 15:3343–3352. https://doi.org/10.7314/apjcp.2014.15.8.3343

Lion Raisins, ‘Products’. Lion Raisins, 2023. https://www.lionraisins.com/

McSweeney, M., and K. Seetharaman. 2015. State of polyphenols in the drying process of fruits and vegetables. *Critical Reviews in Food Science and Nutrition* 55:660–669. https://doi.org/10.1080/10408398.2012.670673

Natarajan, S. K., E. Elangovan, R. M. Elavarasan, A. Balaraman, and S. Sundaram. 2022. Review on solar dryers for drying fish, fruits, and vegetables. *Environmental Science and Pollution Research* 29:40478–40506. https://doi.org/10.1007/s11356-022-19714-w

National Raisin Company, ‘Ingredients’. National Raisin Company, 2023. https://www.nationalraisin.com/industrial-bulk-dried-fruit/

Ocean Spray Cranberries, ‘Products’. Ocean Spray Cranberries, 2023. https://www.oceanspray.com/Products

Omolola, A. O., A. I. O. Jideani, and P. F. Kapila. 2017. Quality properties of fruits as affected by drying operation. *Critical Reviews in Food Science and Nutrition* 57:95–108. https://doi.org/10.1080/10408398.2013.859563

Onwude, D. I., N. Hashim, R. B. Janius, N. M. Nawi, and K. Abdan. 2016. Modeling the thin-layer drying of fruits and vegetables: A review. *Comprehensive Reviews in Food Science and Food Safety* 15:599–618. https://doi.org/10.1111/1541-4337.12196

Paradise Fruits, ‘Solutions: Fruit Shapes’. Paradise Fruits, 2023. https://www.paradise-fruits.de/en/products/pure-fruit-shapes/

Radojčin, M., I. Pavkov, D. Bursać Kovačević, P. Putnik, A. Wiktor, Z. Stamenković, K. Kešelj, and A. Gere. 2021. Effect of selected drying methods and emerging drying intensification technologies on the quality of dried fruit: A review. *Processes* 9:132. https://doi.org/10.3390/pr9010132

Reis, F. R. 2019. *Reports on the Processing of Exotic Fruits*. Cham: Springer.

Reis, F. R., C. Marques, A. C. S. Moraes, and M. L. Masson. 2022. Trends in quality assessment and drying methods used for fruits and vegetables. *Food Control* 142:109254. https://doi.org/10.1016/j.foodcont.2022.109254

Reportlinker.com, ‘Dried Fruits Global Market Report 2023’. Reportlinker.com, 2023. https://www.reportlinker.com/p06284579/Dried-Fruits-Global-Market-Report.html

Royal Nut Company, ‘Dried Fruit’. Royal Nut Company, 2023. https://www.royalnutcompany.com.au/dried-fruit/

Sagar, V. R., and P. Suresh Kumar. 2010. Recent advances in drying and dehydration of fruits and vegetables: A review. *Journal of Food Science and Technology* 47:15–26. https://doi.org/10.1007/s13197-010-0010-8

Santos, P. H. S., and M. A. Silva. 2008. Retention of vitamin C in drying processes of fruits and vegetables—A review. *Drying Technology* 26:1421–1437. https://doi.org/10.1080/07373930802458911

Shishir, M. R. I., and W. Chen. 2017. Trends of spray drying: A critical review on drying of fruit and vegetable juices. *Trends in Food Science & Technology* 65:49–67. https://doi.org/10.1016/j.tifs.2017.05.006

Sossa, Z., J. Wilder, G. L. Orozco, M. L. M. García, O. M. Peña, and S. N. Sánchez. 2021. Infrared drying trends applied to fruit. *Frontiers in Sustainable Food Systems* 5. https://doi.org/10.3389/fsufs.2021.650690

The Observatory of Economic Complexity, 'Dried Fruits'. The Observatory of Economic Complexity, 2023a. https://oec.world/en/profile/hs/dried-fruits

The Observatory of Economic Complexity, 'Dried Fruits'. The Observatory of Economic Complexity, 2023b. https://oec.world/en/profile/hs/grapes-dried

United States Department of Agriculture, 'Organic 101: What the USDA Organic Label Means'. United States Department of Agriculture, 2023. https://www.usda.gov/media/blog/2012/03/22/organic-101-what-usda-organic-label-means

Verma, A., and S. V. Singh. 2015. Spray drying of fruit and vegetable juices—A review. *Critical Reviews in Food Science and Nutrition* 55:701–719 https://doi.org/10.1080/10408398.2012.672939

Zhang, M., J. Tang, A. S. Mujumdar, and S. Wang. 2006. Trends in microwave-related drying of fruits and vegetables. *Trends in Food Science & Technology* 17:524–534. https://doi.org/10.1016/j.tifs.2006.04.011

2 Guideline and Selection of Dryers for Different Fruit Products

Vimal, Nikita S. Bhatkar, and Shivanand S. Shirkole

2.1 INTRODUCTION

The moisture content present in most of the fruits is approximately in the range of 80–90%. The abundance of moisture and sugars in fruits makes them susceptible to the easy growth of microorganisms like bacteria and fungi. Most of the fruits also deteriorate as the quality and quantity of the harvested crops get reduced or lost before they reach to customer. The reasons behind these losses are poor handling practices, inadequate transport, lack of storage facilities (cold storage), lack of awareness, and inadequate infrastructure (Valenzuela 2023). The various types of enzymatic reactions take place if the fruits are not harvested at the right time and post-harvest operations are not done properly. Pectinase enzyme can result in softening of the fruits, while browning reactions can also take place due to polyphenol oxidases, whereas peroxidase enzyme can affect flavour, texture, colour, and nutrition (Tumuluru 2023). Various types of serious health problems can occur because of sensorial and nutritional quality degradation due to post-harvesting losses (Hosseinifarahi 2023). Post-harvest management can help to reduce post-harvest losses, economic losses, and food waste (Valenzuela 2023). Freezing, pickling, canning, packaging, irradiation, use of preservatives, curing, fermentation, drying, and hurdle technology are some of the methods by which food products can be preserved for longer periods of time (Tumuluru 2023). Drying, osmotic dehydration, freezing, canning, blanching, and use of preservatives are some of the major techniques that can be used for fruit preservation (Garcia and Barrett 2002).

Drying has been considered as the most common and simple method of preservation for a very long time. Drying reduces the free water and somewhat percentage of bound water in the fruit sample, which increases the concentration of sugars, acids, and other components present in fruits. This can help in the prevention of microbial growth and longer storage times. Drying or reduction of moisture content can be possible by thermal drying (with the help of dryers), osmotic dehydration (commonly using sugar solution), and mechanical dewatering (physical force application). The drying can make the fruits available whole year, converting the fresh into dried product while maintaining their quality, and decreasing the weight and volume of the end product, which makes the transportation easier (Calín-Sánchez 2020). Drying can be possible from sun drying, i.e., primitive drying technique to refractance window (RW) drying, i.e., fourth generation dryer. Various types of dryers are available in the market nowadays like solar dryer, hot air dryer, vacuum dryer, spray dryer, drum

DOI: 10.1201/9781003402916-2

dryer, freeze dryer, microwave dryer, and infrared dryer. The selection of the dryer becomes a difficult task for the industries, which type of dryer is best suited for their purpose. Due to lack of knowledge, most of the time industries choose wrong dryers, which do not meet their requirements in future, leading to losses of crores of money along with the loss of the commodities. Industries make some strategies or decisions, like choosing the same dryer that they bought last time or selecting the dryer that is generally used at most of the sites, or the industry-scaled version of the lab-scale process dryer or the dryer that was recommended by the agent or supplier, which can cause them to bear losses or even can shut down the companies (Kemp 1999).

Judicial steps in the selection of dryers based on the case-to-case basis of the feed are one such approach that can be followed to prevent such losses. Thus, in this book chapter, we will discuss the factors which we must consider while selecting the dryers for fruits, specifically so that the best use of investment and time can be made.

2.2 DIFFERENT FRUIT PRODUCTS AND THEIR IMPORTANT PARAMETERS

The selection of the dryer for different fruit products is highly dependent on the feed and its nature. In case of fruits and their related products, the most commonly involved feed nature is the whole fruit, cut forms, pulp, puree, and juices. The cut forms further can be classified as slices, cubes, quarters, halves, or any other cut forms. The waste from fruits processing industry is gaining huge importance due to the presence of nutrients and bioactives. These materials are nowadays used as feed for drying, depending on the end applications; they include the seeds, peels, or pomace of the fruits. The nature of the feed in each case is critical for deciding the dryer and further finalizing its characteristics. Table 2.1 gives examples of different

TABLE 2.1
Type of Dryer Based on the Nature of the Feed

Feed type	Fruit	Dryer	References
Whole Fruit	Banana, Strawberry, Blueberry, Cranberry, and Tart Cherries	Hot air drying	(Da Silva, e Silva et al. 2014, Nemzer, Vargas et al. 2018)
	Strawberry, Blueberry, Cranberry, and Tart Cherries	Freeze drying	(Nemzer, Vargas et al. 2018)
	Longan	Combined infrared and hot air	(Nuthong, Achariyaviriya et al. 2011)
	Blueberry	Tray dryer	(Chen and Martynenko 2013)
	Fig	Solar dryer	(El-Sebaii *et al.* 2002)
Cut Pieces	Apples	Fluidized bed dryer	(Kaleta, Górnicki et al. 2013)
	Mango	Freeze-drying	(Sogi *et al.* 2013)

(Continued)

TABLE 2.1 *(Continued)*
Type of Dryer Based on the Nature of the Feed

Feed type	Fruit	Dryer	References
	Apple	Microwave-assisted pulse-spouted vacuum drying	(Mothibe, Wang et al. 2014)
	Mango, strawberry	Infrared drying	(Sogi *et al.* 2013, Adak, Heybeli et al. 2017)
	Persimmon	Hot air dryer	(Giovagnoli-Vicuña, Moraga et al. 2017)
Slices	Dragon fruit, Kiwifruit	Microwave vacuum dryer	(Tian, Wu et al. 2015, Raj and Dash 2020)
	Tamarind (Dehulled)	Tray dryer	(Rajkumar, Deepa et al. 2023)
	Star Fruit	Convective tray dryer	(Dash, Gope et al. 2013)
	Strawberry, Apple	Electrohydrodynamic drying	(Paul, Astatkie et al. 2022)
	Mango, Apple	Refractance window	(Ochoa-Martínez, Quintero et al. 2012, Rajoriya, Shewale et al. 2019)
	Apple	Freeze drying	(Schulze, Hubbermann et al. 2014)
	Pineapple, Jujube, and Sapota Slices	Hot air drying	(Ramallo and Mascheroni 2012, Gupta and Patil 2014, Elmas, Varhan et al. 2019)
	Kiwifruit	Pulsed vacuum drying	(Liu, Wei et al. 2020)
	Kiwifruit	Radio frequency-vacuum drying	(Zhou, Xu et al. 2018)
	Pear	Ultrasound-assisted infrared drying	(Dujmić, Brnčić et al. 2013)
	Date	Microwave	(İZLİ 2016)
Pulp/Juice	Strawberry, Blueberry, Cranberry, and Tart Cherries	Refractance window drying	(Nemzer, Vargas et al. 2018)
	Persimmon, Dragon fruit, and Tamarind	Spray drying	(Tze, Han et al. 2012, Bhusari, Muzaffar et al. 2014, Du, Ge et al. 2014)
	Mango, Apple	Drum dryer	(Germer, Tonin et al. 2018)
	Orange	vacuum spray drying	(Islam, Kitamura et al. 2017)
Peels/pomace kernels	Orange	Microwave	(Kammoun Bejar, Kechaou et al. 2011)
	Apple	Freeze drying	(Henríquez, Almonacid et al. 2013)
	Apple Peel, Mango (peels and kernels)	Freeze dryer	(Henríquez, Almonacid et al. 2013, Sogi *et al.* 2013)
	Apple pomace	Continuous vacuum belt dryer	(Yan and Kerr 2013)
	Grape pomace	Electrohydrodynamic	(Martynenko and Kudra 2016)
	Mango (peels and kernals)	Infrared dryer	(Sogi *et al.* 2013)

types of dryers used for different types of feed for drying. This section discusses the different types of feeds involved in the drying operation of fruit products.

2.2.1 Whole Fruit

Drying of whole fruit is usually practiced for fruits such as grapes, dates, small berries, figs, and plums. Apart from these, there are a few varieties of apples, apricots, and pears that are dried as a whole. Here, the nature of the feed is such that it experiences a hindrance to the removal of water at the surface level. The drying of such feed is an energy-intensive operation, as the feed offers a huge resistance to the mass transfer involved. The increased amount of the energy involved in case of whole drying can be understood based on the difference in the drying time involved as compared to the feed of same nature but dried in cut forms. For instance, drying of cut figs takes around 12 hours as against the drying of whole figs, which takes 32 hours in total to reach the same equilibrium moisture content as reported by El-Sebaii (2002). This difference in time actually causes higher energy consumption throughout the entire process. During the drying of whole and pureed black chokeberry in freeze dryer, a 14.06% higher energy efficiency was reported for puree than whole fruit (Taskin 2020). The drying of the whole fruit can be accelerated to some extent by removing the outer peel; in this case, the nature of the feed offers a lesser resistance to the mass transfer. The effective diffusivity of the feed increases with the removal of peel, facilitating a higher drying rate and thus lower energy consumption (Xanthopoulos 2010).

Another common problem associated with the drying of whole fruit is the uneven drying, which can potentially lead to case hardening effect. The drying conditions need to be fixed in such a way that these problems are tackled. One of the benefits of drying whole fruits in dryers is the ease of handling the feed and the dried food due to the larger size compared to similar feed with cut portions. Adding to this, smaller cut portion requires a close or continuous monitoring system to avoid the over-drying of fruits, which is not the case in the drying of such feeds. The commonly used dryers for the drying of whole fruit are the open sun drying, solar, tunnel, tray, oven drying, and freeze dryers.

2.2.2 Cut Fruits

Partially dehydrated or semi-dried fruits refer to those where the drying of the fruit is usually carried out to a desired moisture content higher than the dried fruits. In this case of drying, the feed can be any kind of cut fruit of any shape or size. The different forms of the cut fruits can be slices, cubes, quarters, halves, and so on. The main problem with the drying of such feed is the enzymatic browning of cut pieces. The activity of the inherent enzymes usually leads to the browning of the fruits when exposed to oxygen; to overcome this issue, feed of this type is usually pretreated before drying. The pretreatment actually reduces the load of drying due to osmotic dehydration effect, or increment in the pore size of the feed due to pre-treatments such as ultrasonication or blanching. The dryers used in such feed will be subjected to a large amount of drip loss and left over pieces of dried fruit, which will eventually need cleaning and maintenance.

In case of osmotically dehydrated fruits that are one form of partially dehydrated fruits, the dryer is usually a tank filled with the osmotic solution of desired concentration of a given osmotic agent maintained at a certain temperature. The construction of the material is of critical importance in this case. The tank is in continuous contact with the osmotic solution, the material of the tank should be such that it is able to sustain the required pH, temperature, and resistant to any kind of chemical reaction with the osmotic solution. The equipment should be such that the temperature and concentration of the solution are continuously monitored and altered based on the requirements (Alemeida 2016).

2.2.3 Juice

Dried fruit powders are becoming more popular due to their convenience, flavour, nutritional value, versatility, and consumer demand for natural and clean label products. This has led to an increase in demand for these products in the market. Spray dryer, freeze dryer, tray dryer, infrared dryer, foam mat dryer, and RW dryer are some of the dryers that have been used in drying fruit powders. Spray dryer is the most preferred dryer which is used for preparation of fruit powders. During the preparation of fruit powders, there are numerous problems that can occur, like oxidation of some fruit during drying because of temperature, and exposure to air can lead to colour change, flavour, loss, and loss of nutritional quality like heat-sensitive vitamins, phytochemicals, enzymes, etc. If the fruit is not dried to proper moisture or inadequate amount of anticaking agents is added, it can result in formation of lumps after grinding. The fruits are rich in sugars (fructose and glucose) and acids, which can result in high hygroscopicity, thermoplasticity, and low glass transition temperature (T_g) that can lead to stickiness of powder to wall of dryer, which ultimately reduces the yield of powder (Muzaffar *et al.* 2015). Stickiness can be overcome by adding high molecular weight carrier agents like maltodextrin, gum arabic, and silicon dioxide, which increase the T_g and thus prevent stickiness and powder hygroscopicity. The feed is concentrated to 40–60% to form a stable powder and enable it to dry in spray dryer. The modification in the dryer design and working can also help to reduce the stickiness of fruit powders. This can be possible in several ways, like by creating low humidity and low temperature conditions, inserting cold air, cooling walls, and scrapping chamber walls. The surface of the collector wall in the dryer, if coated with coating material, can minimize stickiness at both low and high temperatures. As the height/length of the spray drier increase, it gives droplets more time to stay in contact with hot air, which results in complete drying. Roustapour *et al.* (2006) found that with an increase in the distance of collector from roof of dryer the final moisture content decreases. The speed of atomizer should be higher, which can lead to release of small-sized droplets, which results in more evaporation of moisture and quicker drying. For example, the feed with high sugar content in a co-current type of flow is more suitable in case of spray drying. This suggests that the selection of the dryer for the feed with high in sugars and acid should be such that the height of the spray dryer is higher. In the spray dryer, cooling of walls can be possible in between drying, which decreases the chances of thermoplastics to adhere on the wall, and there should be arrangement that can allow the dehumidified cold air to pass, which can help in removal of some of the sticky particles. The selection of hybrid dryers is nowadays in trend to reduce the drying time, increase efficiency, and deliver fruit powders with better product quality.

2.2.4 Pulp and Puree

For drying pulp, there are various methods reported like tray dryer, vacuum dryer, RW dryer, and freeze dryer. The feed in this case is similar to the feed of previous one with high in sugar and acids; however, in this case, the addition of carrier agents or bulk agents is not there. RW and freeze dryers are considered as best for drying fruit pulps as they retain more nutritional properties. RW can be an ideal dryer in this case based on the data available in literature. RW dryer is an emerging novel dryer considered as a fourth-generation dryer. The working of the equipment is dependent on refraction of radiations generated from the water when heat is supplied. The heat can be supplied via any mode of heat transfer, i.e., conduction, convection, and radiation. For getting better results, a combination of modes of heat transfer is possible. The Mylar film, which is a polyester film, is used most commonly for placing the pulpy medium and for the refractance phenomenon. The thickness of the Mylar film should typically be 0.25 mm for drying fruit pulps and puree. The drying rate depends upon the feed thickness. Lesser the thickness faster the drying as the water vapour has to travel shorter distance to escape. Thickness of the inlet can be decided on the basis of end product characteristics whether we require the product with least colour change, rehydration ratio, wettability, solubility, texture, moisture content, and water activity. The temperature of the water bath can also help in changing the drying time. The higher the temperature of the water bath, the higher the rate of drying (Shende and Datta 2019). Drying with the help of drum dryer is most employed industrial practice in this case. For drum drying, two hollow cylinders which are made up of cast iron or stainless steel (for holding and drying of sample), feeder (for feeding the sample upon drum), supporting frame (for holding the whole system), scrapper (for removing the dried layer), and auxiliaries are required. Radius of drums having a range of 0.25–3.0 m and length of 1–6 m are suitable. The steam temperature is usually of 200°C, which can heat the drum from inside. The thickness of the sample can be varied from 0.5 mm to 2 mm, thickness is directly proportional to drying time. The residence time depends upon the end product's characteristics and its utilization. It is worth noting that residence time is less and can cause severe damage to product characteristics if sample is held for a long time. For heat-sensitive food materials, vacuum drum drying can be a possible, which can help in achieving the similar drying coefficients. Some food materials contain higher amount of sugar, which is difficult to scrap the dried layer from a very high temperature of drum due to sample's high thermoplastic behaviour. So, to avoid this, the installation of some equipment near to scrapper, which can bring the temperature down, that can help in bringing the product temperature and changing the state from rubbery to glassy, which can facilitate easy removal of the dried layer for pulp and puree as a feed (Tang, Feng *et al.* 2003).

2.2.5 Peels and Seeds

The nature of the feed in case of fruit seed it is mainly constituted of carbohydrate, protein, lipids, and moisture content. The carbohydrate, protein, lipid, and moisture vary between 40–70%, 10–30%, 30–60%, and 5–15%, respectively, depending on the fruit, its variety, and maturity state. Thus, the nature of feed in case of seed can be high in protein, lipids, and carbohydrates. Along with that, it is important to understand that the end product in this case is usually used as the feed for extraction purposes (Panja 2018). This is because the drying operation is one of the prerequisites for the extraction of bioactives from the feed. Similar is the case with the peel, where the major constituents become the carbohydrates, mostly the dietary fibres, cellulose, and hemicellulose. The nature of the feed is high in moisture and fibrous in nature, tending to hold the moisture while drying. The most commonly used dryers in this case are tray, cabinet, freeze, solar, and microwave dryers.

2.3 DRYERS AVAILABLE WITH DIFFERENT OPERATIONAL CONSTRAINTS

The selection of dryers can be based on different selection criteria among them. Feeding nature is one of the main criteria, which was explained in the previous section. Figure 2.1 comprehensively enlists the different criteria for the selection of dryers for fruits and their products. The scale of operation (small or large), source of energy (solar, fuel, or electrically powered), batch continuous or semi-continuous mode of operation, solid, liquid, or paste as a feed material, single or multistage operation, movement of fruit products (stationary or moving), operating conditions, medium used for drying along with its direction of flow, drying rate, process time, desired characteristics of the end product, and the economic constraints are the factors affecting the selection of dryer. This section deals with the selection criteria based on the dryer characteristics and qualities desired for a fruit product.

2.3.1 Scale of Operation and Energy Source

The scale of operation for dryers can be either small or large. This particular criterion is dependent on the location or place where the drying is conducted. Small scale operations are prevalent at farm levels, community levels, or small food business operators. The capacity of these dryers can vary from ten to a few hundred kilograms. On the other hand, the large-scale capacity can vary from several hundred to thousands of kilograms. These types of dryers are mostly used by large food business operators, large scale fruit processing industries, and so on.

Another criterion for the selection of dryer is the source of energy used to power the entire process of drying. It can be using the solar energy, fuel-operated, or electrically operated dryers. Solar conduction, cabinet, tunnel dryers, and solar hybrid dryers utilize solar energy for the drying operations (Sharma *et al.* 2009). On an industrial scale, the fuel used for the drying of the fruit products contributes majorly

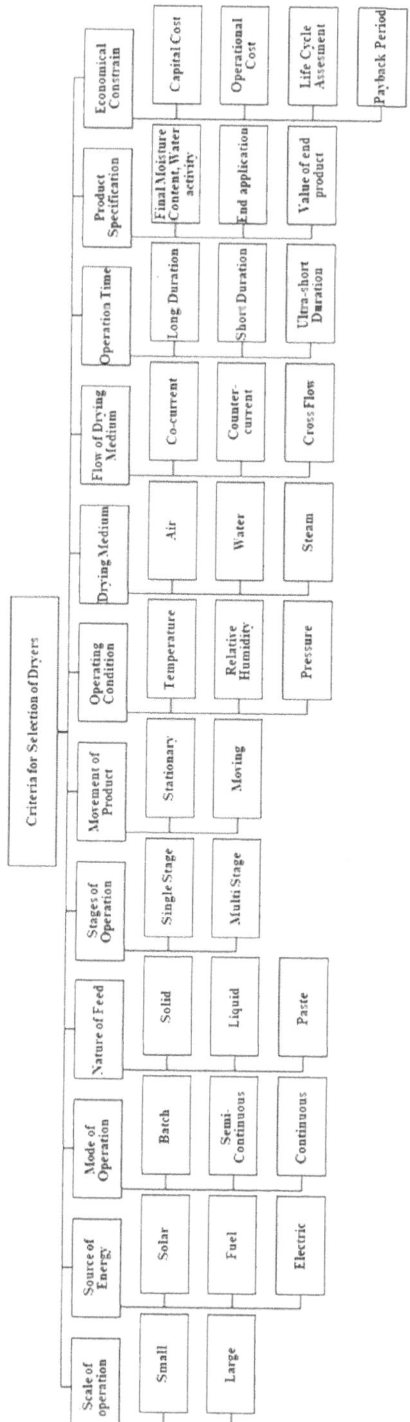

FIGURE 2.1 Criteria for selection of dryers.

to the operating cost of the dryer, thus very crucial for the selection of the dryer to be used (Bhandari 2015). Natural gas, coal, oil, and biogas are some of the options for the fuels. Direct fired dryers, rotary, fluidized, and spray dryers can operate on fuel-based dryers. Electricity as a source of energy to power the dryer is used on a large scale nowadays. All kinds of dryers can be powered by electricity to carry out the drying operation. The hybrid mode solar dryer is also designed to make the dryer functional during non-availability of the solar energy.

2.3.2 Mode of Operation and Stages of Operation

The mode of operation for drying can be categorized as batch process or continuous. The batch mode of dryer mostly includes tray, cabinet, vacuum, fluidized, and freeze dryers, whereas belt, conveyors, rotary, spray, drum, and tunnel dryers fall under the regime of continuous dryer. The batch mode is cost-effective to operate on a small scale and offers a better control over the process with lower capital cost; however, lower production rate, labour intensiveness, and lack of scalability are some of the drawbacks of batch operations. The continuous mode of operation provides an edge over the batch one in terms of scalability, production rate, less labour intensive, and provides a more consistency in the quality of the dried fruits. Higher initial investments, limited control over the drying operation, complexities in the operation, and lack of variability in the operations remain the challenges to be resolved. However, it is important to select the mode of operation on the basis of the nature of the feed along with the desired product characteristics. For drying of mango pulp, a continuous type of foam mat dryer was found to be better in terms of efficiency and the final product characteristics as compared to the cabinet type of dryer operated in a batch mode (Rajkumar *et al.* 2007).

Single and multistage drying refer to the drying of fruits in a single condition and more than one drying condition, respectively.

2.3.3 Selection Based on Heating Methods

The selection of the mode of heat transfer depends on different criteria, such as nature of feed, method of drying, capacity, desired quality of the end products required, and so on. The modes of transfer can be broadly classified as convective, conductive, and radiative modes of heat transfer.

Convective mode of heat transfer is the most dominant mode of heat transfer in dryers like hot air oven, tray dryer, cabinet dryer, and similar dryer. The Fluidized bed and rotary dryers are the drying methods where the convective mode of heating is most dominant. The mode of heat transfer is usually faster in this case. Such type of dryer could be helpful for the removal of initial moisture from the surface of the feeds. Fruits high in moisture are more suitable for this purpose as the convective mode of drying has a higher coefficient of heat and mass transfer, which enables faster drying; however, the literature on drying of fruits with such a selection criteria for a specific mode of heat transfer is limited and the selection is based on other criteria based on the desired characteristics of the final fruit products.

Conduction mode of heat transfer is a dominant heat transfer phenomenon for drying fruit products in drum dryers, to some extent in case of RW, and other conduction dryers such as solar conduction dryer. This mode of heat transfer is also evident when the food product is rested on the dryer for a longer period of time due to heating of the plate. The dryer where such mode is dominant can be highly detrimental to the product quality as the fruits are in direct contact with the hot surface. For instance, mango fruit powder dried using drum drying showed darker colour and quality losses (Caparino *et al.* 2012). Same study showed RW dryer (with conductive mode) to be as good as the original sample. However, the production rate of drum dryer is much higher than that of RW drying, and the scale of operation for latter is much lower than the former. The RW drying is more suitable for heat labile and highly nutritious fruit with high value and lower production rates, whereas drum drying is suitable for higher production rates and products where a higher quality of product is less desired.

Radiative mode of heat transfer is dominant method in dryers, where drying is mediated with the help of electromagnetic radiation. Microwave dryers, infrared dryers, and radiofrequency dryer are examples of these dryers. These kinds of dryers are known for their higher efficiency, uniform drying, and selective heating. The production rate offered by this mode of heating is higher than that offered by other two modes of heat transfer, owing to the lower drying time. However, optimization of the drying condition is required to get the desired product quality. The process needs close monitoring as the drying is quick and can be detrimental to the product quality.

2.3.4 Based on Operating Conditions

The main operating conditions in case of drying are the temperature and pressure of operation, which determine the drying time, rate, product quality, and characteristics. This is because the physiological conditions of temperature and pressure determine the chemical and physical changes taking place in the fruits and related products. The quality of the final product is dependent on the temperature of exposure and atmospheric conditions it is exposed to. The atmospheric conditions are dependent on the medium used for the drying purpose. A lower temperature ensures a minimal change to the fruit; however, a longer exposure can potentially lead to the loss of heat-labile content. It was observed that the ascorbic acid was reduced by increasing the drying temperature from 50 to 70°C. Thus, the dryers, such as vacuum, refractance dryer, sun dryer, or freeze dryer, are recommended in such cases. Similarly, when the drying is desired to be carried out under maintained atmospheric conditions like low pressure or controlled relative humidity, vacuum, or cabinet dryers are the options, respectively. Another highly important aspect while considering the dryer operating condition along with the cost of operation is the differences in the level of electricity consumption with different temperature and relative humidity combinations. At low temperature, the electricity consumption decreases with an increase in the relative humidity, and at higher temperature the electricity consumption increases with an increase in the relative humidity (Hnin *et al.* 2018).

2.3.5 Based on the Ease of Handling and Movement of the Fruit Material

Ease of handling of the dryer is a factor which should be considered while selecting a dryer in the long run. Fruits are delicate commodities which are easily prone to damage if not handled judicially. The post-harvest losses account for approximately 15–20%; further losses during the processing will lead to tremendous waste generation. Thus, an appropriate dryer needs to be selected to curb the handling losses. Dryers that offer the ease of handling fruits include conveyor dryer, fluidized bed dryer, rotary dryer, pneumatic dryer, and similar ones where the human interventions are limited and the product is conveyed without damage (Aghbashlo *et al.* 2014). The dryers such as drum dryers, spray dryers, and freeze dryers where operations issues, such as continuous provision to provide the required condition of drying, careful monitoring, and continuous inspection are desired, are difficult to handle. Involvement of labour while loading and unloading the sample from the dryer may also potentially lead to the losses and add labour charges. The dryers involving lower human interventions, along with the one offering ease of handling, need to be selected.

2.3.6 Economical Constraints

While carrying out any selection procedures, everything ends up at the financial status of the process; thus, the economic constraints are the essence of the guidelines for the selection of the dryer for fruit. The cost of drying operation involves fixed cost covering the cost of initial investment for the equipment, depreciation, insurance, installation, and setting up costs and variable cost involving cost of raw materials, energy consumption, utility costs, and labour costs. The cost of each part of the drying equipment is dependent on its capacity to be installed; for instance, a vacuum dryer with unheated agitator having capacity of 0.02–9 m^3 will have a fixed cost involved. Similarly, the material used to build the equipment will lead to different costs depending on the element used; for example, carbon steel parts cost less than the 304 stainless steel material used for fabrication (Bhandari 2015). However, the surface in contact with the food is recommended to have stainless steel 304 as per the guidelines; nevertheless, other parts, such as exhaust fans or other parts not in contact with the food, can be fabricated with alternative materials. Freeze dryers, fluidized bed dryers, vacuum dryers, rotary dryers, spray dryers, and superheated steam dryers involve a high initial capital cost, whereas hot air dryers, tray dryers, cabinet dryers, solar dryers, and drum dryers require a lower initial capital investment.

Operational costs can be reduced by reducing the utilities such as electricity and water consumption. The operating conditions play a vital role in this case, as explained earlier. Freeze dryers and high-temperature operating dryers show higher operational costs, while solar dryers, hot air dryers operating at low temperatures, and drum dryers are some examples of low cost operating dryers.

Certainly, the selection of dryers is based on the cost of the processing involved; however, the selection should be based on the end application of the products along with desired product characteristics. Further sections throw light on this aspect of selection.

2.4 PROPERTIES OR CHARACTERISTICS OF THE END PRODUCT REQUIRED

The product qualities are very crucial criteria when it comes to the drying of fruits and related products. The quality of dried fruit products can be further classified based on the moisture content, nutrient profile, and physical properties. Based on the moisture content, the dried fruit products can be classified as partially dehydrated and low moisture food products, the lower moisture content fruit products offer a higher shelf life and are thus desired. Similarly, lower water activity of the dried fruit product is an indicative of the microbial safety of the product. In case of nutritional profile of the food is judged based on the amount of bioactive compounds present in it. Generally, a higher phenolic content, higher flavonoids, vitamins, and higher antioxidant activity are desired for the nutritionally superior product. The physical characteristics are important to the dried products considering the consumers' preferences, minimal colour change with respect to the fresh product, higher rehydration ratio, and similar textural properties are some of the physical parameters to be satisfied. Table 2.2 gives a list of different types of dryers used in literature to achieve specific desired characteristics of the final products. This section discusses the product's characteristics in detail.

TABLE 2.2
Dryers Type Based on the Desired Characteristics of the Product

Desired Product Characteristics	Fruit	Dryer	References
Better Colour Retention	Sour Cherries, Raspberry	Freeze dryer	(Wojdyło, Figiel et al. 2014, Si, Chen et al. 2016)
	Strawberry, Cranberry, Cherry, and Kiwifruit	Refractance window drying	(Jafari, Azizi et al. 2016, Nemzer, Vargas et al. 2018)
	Mango	Vacuum drying	(Sehrawat, Nema et al. 2018)
	Chokeberry	Convection-vacuum-microwave	(Samoticha, Wojdyło et al. 2016)
	Pomegranate arils	Microwave dryer	(Horuz and Maskan 2015)
	Plum, Orange (Peel)	Microwave-vacuum dryer	(Michalska, Wojdyło et al. 2016, Bozkir 2020)
Preservation of Bioactive components			
Total Phenolic content (High)	Raspberry	Infrared radiation and microwave vacuum drying	(Si, Chen et al. 2016)
	Blackberry	Spray dryer	(Franceschinis *et al.* 2014)
	Plum	Convective Pre-drying and microwave finish-drying	(Michalska, Wojdyło et al. 2016)
	Mango (Peel), Strawberry, Blueberry, Cranberry, Cherry, and Chokeberry	Freeze dryer	(Sogi *et al.* 2013, Jafari, Azizi et al. 2016, Samoticha, Wojdyło et al. 2016)

(Continued)

TABLE 2.2 *(Continued)*
Dryers Type Based on the Desired Characteristics of the Product

Desired Product Characteristics	Fruit	Dryer	References
Total flavonoid content (High)	Raspberry	Infrared radiation and microwave vacuum drying	(Si, Chen et al. 2016)
	Plum	Convective Pre-drying and microwave finish-drying	(Michalska, Wojdyło et al. 2016)
Anthocyanin content (High)	Plum, Strawberry, Blueberry, Cranberry, Cherry, and Chokeberry	Freeze dryer	(Jafari, Azizi et al. 2016, Michalska, Wojdyło et al. 2016, Samoticha, Wojdyło et al. 2016)
Antioxidant activity (High)	Raspberry	Infrared drying	(Si, Chen et al. 2016)
	Mango (Peel), Chokeberry	Freeze dryer	(Sogi *et al.* 2013, Samoticha, Wojdyło et al. 2016)
	Blackberry	Spray dryer	(Franceschinis *et al.* 2014)
	Plum	Microwave-vacuum dryer	(Michalska, Wojdyło et al. 2016)
Physical Properties			
Rehydration ratio (High)	Mango	Low-pressure superheated steam drying	(Sehrawat, Nema et al. 2018)
	Kiwifruit	Refractance window drying	(Jafari, Azizi et al. 2016)
	Pomegranate arils	Microwave	(Horuz and Maskan 2015)
Water Activity/ Moisture Content (Low)	Chokeberry	Freeze dryer	(Samoticha, Wojdyło et al. 2016)
	Mango (Peel)	Vacuum dryer	(Sogi *et al.* 2013)
	Blackberry	Spray dryer	(Franceschinis, *et al.* 2014)
	Plum, Orange (Peel)	Microwave-vacuum dryer	(Michalska, Wojdyło et al. 2016, Bozkir 2020)
Bulk density (High)	Blackberry	Spray dryer	(Franceschinis, *et al.* 2014)
	Pomegranate arils	Hot air dryer	(Horuz and Maskan 2015)
	Plum	Microwave-vacuum dryer	(Michalska, Wojdyło et al. 2016)
	Orange (Peel)	Tray dryer	(Bozkir 2020)
	Mango (Peel)	Cabinet drying	(Sogi *et al.* 2013)
Solubility (High)	Blackberry	Spray dryer	(Franceschinis, *et al.* 2014)
	Plum, Mango (Peel)	Freeze dryer	(Michalska, Wojdyło et al. 2016)

2.4.1 Moisture Content of Final Product and Water Activity

The moisture content and water activity are the two governing factors that can affect physical properties (texture, colour, stickiness, density, etc.), shelf life, processing nature, and its utilization. To reduce the moisture content and water activity of fruit to optimum level is the main requirement for the usage of end product in a particular food application. Like for raisins, the moisture content requirement is approximately 14–16% (Sharma *et al.* 2013). If the moisture content is beyond 18%, it can lead to growth of moulds, and if the moisture content is below 11%, the raisin with undesirable flavour would be obtained. So, with this example, it is clearly understandable that drying food products is important, but drying up to a particular limit is even more important. With osmotic dehydration, the moisture content can be reduced by 50–70%, which also depends on several factors, including the initial moisture content of the food, the concentration of the osmotic solution, and the processing time. The reduction of moisture content to a certain level with the help of any dryer is possible just by continuous monitoring of the moisture content, followed by optimization of the entire process depending upon the desired condition.

2.4.2 Minimal Effect on the Nutritional Profile

After moisture content and water activity, the product's nutritional quality is equally important. The nutritional quality factor comes into play only if the end product usage is focused on nutritional and other quality parameters. Retention of essential nutrients (like bioactive components, vitamins, etc.) for health benefits (of these important nutrients) and consumer preference are the current needs of an ideal dryer. Losses of these essential elements of the dried fruits along with quality deterioration are some of the reasons which can occur if the nutrient loss happens during drying. Different factors during the drying which can affect the nutritional properties are type of dryer, temperature range, type of product and feed, drying time, pretreatment, etc. Franceschinis *et al.* (2014) observed that when the dryer was changed from spray dryer to freeze dryer, it was found that total phenolic content increased by 317 units, and it was also observed that when the carrier medium was changed from maltodextrin to maltodextrin-trehalose mixture can result in decrease in antiradical activity and total phenolic content of freeze-dried sample as well as spray dried sample. So, from here, it clearly depicts how important the type of process and selection of dryer are. The total phenolic content, total flavonoid content, antioxidant activity, and total anthocyanin content increase when the drying temperature of hot air dryer was increased from 50 to 70°C in case of cherry laurel fruit (Turkmen *et al.* 2020), while Yılmaz *et al.* (2017) reported the retention of ascorbic acid reduces when the drying temperature was increased from 50 to 70°C in case of vacuum as well as cabinet drying of 2 mm slices. This could be attributed to the higher the drying rate, which in turn reduces the time of exposure to high temperature. This suggests that, in case of heat labile nutrient preservation, dryers involving lower temperature operation should be selected (freeze drying, RW drying, and sun drying). Similar to temperature, the presence of oxygen in the air also influences the quality of the dried fruit, and thus, the dryer selection. Presence of vacuum has two major factors: reducing the temperature of evaporation and carrying out the drying in a reduced oxygen environment. The vacuum dryer shows better retention of ascorbic acid in dried pomegranate fruit leather. Thus, in such cases, vacuum dryer should be preferred.

2.4.3 Minimal Effect on the Physical Characteristics

Texture, colour, density, and aroma are some of the physical characteristics of any product, whether it is fresh or dried product. If these characteristics of the product are not up to mark, they can easily influence consumer decisions while buying the product. Hardness of the hot air convective drying blueberries decreases from 30 to 13 when drying temperature decreases from 90 to 60°C (Zielinska and Michalska 2016). Whereas the total colour change also increases in case of jujube fruit when temperature of convection dryer increases from 50 to 70°C (Wojdyło *et al.* 2019). When the dryer was changed from freeze dryer to spray dryer, the total colour change value also decreased from 2.23 to 1.84 in case of mango pulp (Zotarelli *et al.* 2022). Jafari *et al.* (2017) reported with increase in level of carrier agent (maltodextrin) and drying temperature of powder in case of pomegranate juice powder, the density of the powder decreases. The similar trend was observed in case of water solubility index when maltodextrin concentration increases by 25–30%. So, the effect on physical characteristics of the end product should be kept in mind when the type of dryer, drying rate, and drying temperature are decided for the process. These characteristics can affect the end product application; thus cost of the end product also varies. Here, it can be recommended that the dryer be chosen based on the end application and taking cost into consideration.

2.5 CHARACTERISTICS OF AN IDEAL DRYER FOR FRUIT PRODUCTS

An ideal dryer for drying fruit and related products would be one that satisfies all the characteristics provided in Figure 2.2. A dryer with lower capital and operating costs is most desirable considering the initial investment and production cost of the products. A dryer is more flexible when the operation can be carried out in both small and large quantities. This is because fruits are natural commodities, and the harvest quantity of the fruit is always variable. Other factors, such as weather, affect the production quantity. Higher efficiency of the dryer is a desirable characteristic an operator asks for. Higher efficiency in terms of energy and product throughput will not only ensure economical savings in long run but also help to build a sustainable environment, which is another characteristic of an ideal dryer. A better index to represent the characteristics of an ideal dryer is a dryer with a low life cycle assessment score, which is an indicative score for lower environmental impact considering all the life stages of a dryer.

Considering the fruit as a feed material, most important objectives are to achieve a higher production rate, carry out drying in shorter duration, maximize retention of the nutritional status of the fruit product, preservation of the bioactive components, and a higher resemblance to the original sample on drying. Considering this, an ideal

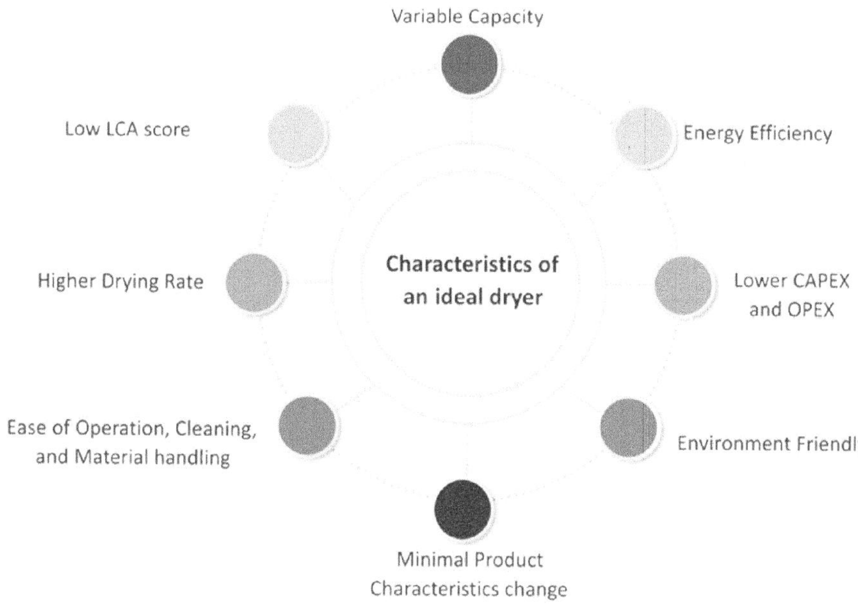

FIGURE 2.2 Available dryer based on the nature of the feed.

dryer would impart a higher drying rate, minimal effect on the nutritional and bioactive components with ease of operation and lower maintenance inputs.

2.5.1 General Steps for the Selection of a Dryer for Fruit Products

The general steps for the selection of the dryer for the fruit-related product will begin with the nature of the feed to dry in the dryer (Figure 2.3). The nature of the feed can be solid, liquid, or paste like based on the end product required or the fruit selected.

For instance, for mango fruit, it can be drying of cut mango pieces (solid), mango juice (liquid), or mango pulp (paste). Figure 2.4 shows the various types of dryers to be used for different types of feeds. Further, the mode of operation needs to be finalized depending on the required product flow rate. If a higher output rate is desired then a continuous mode of operation is recommended or else batch mode dryer with lower feed input can suffice for the requirements. Depending on the feed input rate, the scale of operation can be decided that is based on the availability of the feed, as in case of seasonally available fruits that can be operated in a batch mode. Another benefit of operating in batch mode is that the mode is quite versatile and can be easily modified for other fruit available in next season. The continuous mode of operation is more suitable for fruit with yearlong supply.

The further selection criteria should be based on the application for the end product from the drying operation. More or less, the applications could be categorized as feed for the extraction of bioactives, ingredient for functional food, and as a bulking agent in some food or direct consumption. As previously mentioned, the fruit peels, seeds, and pomace, though considered as waste, are abundant sources of

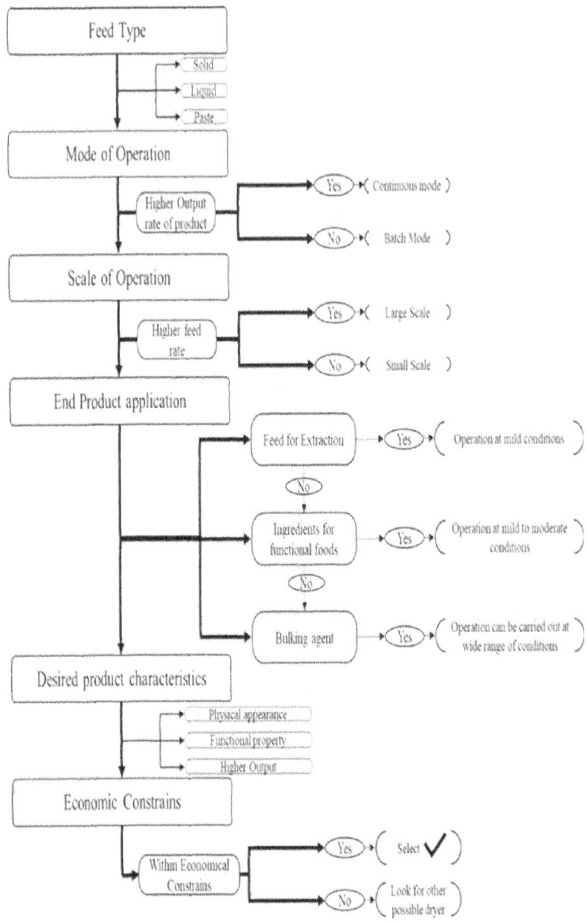

FIGURE 2.3 Flow chart for selection of dryer for fruit product.

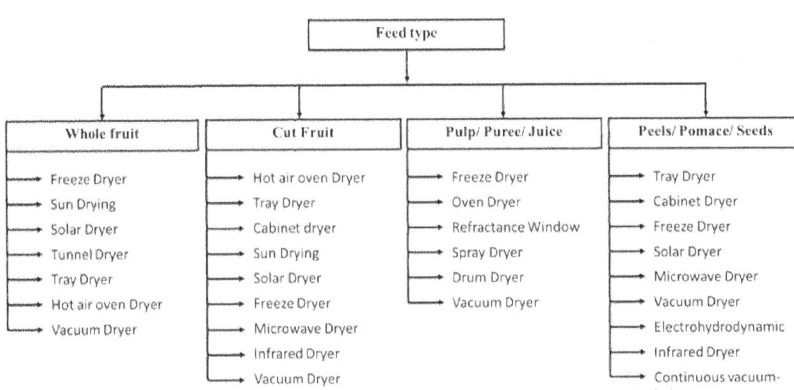

FIGURE 2.4 Characteristics of an ideal dryer.

nutraceutical agents; these can easily be used to extract the bioactives; however, drying is usually a prerequisite in extraction operation to improve the yield. The retention of bioactive components during the drying operation is the main objective in this case of end product application; thus, drying using sophisticated dryer with minimal effect on the bioactive compound should be chosen. In case of drying mango seeds and peels for preservation of antioxidant activity, the freeze drying was found to give the best results for retention of carotenoids, phenolic content, and antioxidant activity (Sogi *et al.* 2013). The freeze drying method is quite expensive considering both the capital and operational costs, but the end application of this product would also yield a high-value product, so freeze drying would be suitable in this case. In the second category of the end application, the product could be used as an ingredient for formulation of functional food; in this case, the cost of production is a constraint thus another suitable method is required. Continuing the same example, drying both peels and seeds of mango in cabinet dryer retains substantial amount of the bioactives; the quality is less superior to freeze dryer method, but it can still be applied to get the benefit over cost (Sogi *et al.* 2013). Similarly, next category is where the end application of dried form of fruit is mere consumption or as a bulking agent in some other product. For instance, mango pieces required to be added as an ingredient can be dried by the hot air drying method or any method that is economically feasible. Further, the characteristics of the product can be considered such as stickiness in case of powder dried. Here, methods such as spray dryer with longer heights can be used (Muzaffar *et al.* 2015). Also, another characteristic where retention of colour is the aim one can choose both freeze drying and RW drying. The selection criteria should be based on the economic constraints, as in the former example, RW dryer would be more preferable considering both cost and colour retention.

2.6 CONCLUDING REMARKS

Fruits are one of the largest agricultural commodities dried at an industrial level. The inappropriate selection of dryer can lead to huge quantum of losses, both to the industry as well as be fatal to the environment. Selection of an appropriate dryer is the only solution to this issue. The selection criteria are dependent on the feed type for the drying operation, the capacity, and mode of operation, characteristics of the products desired, and the end application of the dried form. Finally, the economical aspect of the process is the one deciding the entire selection process.

REFERENCES

Adak, N., Heybeli, N., & Ertekin, C. (2017). "Infrared drying of strawberry." Food chemistry 219: 109–116. https://doi.org/10.1016/j.foodchem.2016.09.103

Aghbashlo, M., et al. (2014). "Measurement techniques to monitor and control fluidization quality in fluidized bed dryers: A review." Drying Technology **32**(9): 1005–1051.

Alemeida, J., et al. (2016). "Alternative prototype of dehydrator osmotic with control automatic of level and temperature solution of sucrose." Revista De Engenharia e Tecnologia **8**(2): 152–160.

Bhandari, B. (2015). Handbook of Industrial Drying, Edited by A. S. Mujumdar: CRC Press: Boca Raton, FL. ISBN: 978-1-4665-9665-8, Taylor & Francis.

Bhusari, S. N., Muzaffar, K., & Kumar, P. (2014)."Effect of carrier agents on physical and microstructural properties of spray dried tamarind pulp powder." Powder technology 266: 354–364. https://doi.org/10.1016/j.powtec.2014.06.038

Calín-Sánchez, Á., et al. (2020). "Comparison of traditional and novel drying techniques and its effect on quality of fruits, vegetables and aromatic herbs." Foods **9**(9): 1261.

Caparino, O., et al. (2012). "Effect of drying methods on the physical properties and microstructures of mango (Philippine 'Carabao'var.) Powder." Journal of Food Engineering **111**(1): 135–148.

Chen, Yougui, and Alex Martynenko. (2013) "Computer vision for real-time measurements of shrinkage and color changes in blueberry convective drying." Drying Technology 31(10): 1114–1123.

Du, J., Ge, Z., Xu, Z., Zou, B., Zhang, Y., & Li, C. (2014). "Comparison of the efficiency of five different drying carriers on the spray drying of persimmon pulp powders." Drying Technology 32(10): 1157–1166. https://doi.org/10.1080/07373937.2014.886259

El-Sebaii, A., et al. (2002). "Experimental investigation of an indirect type natural convection solar dryer." Energy Conversion Management **43**(16): 2251–2266.

Franceschinis, L., et al. (2014). "Physical and functional properties of blackberry freeze-and spray-dried powders." Drying Technology **32**(2): 197–207.

Garcia, E. and D. M. Barrett (2002). "Preservative treatments for fresh-cut fruits and vegetables." Fresh-Cut Fruits Vegetables: 267–304.

Germer, S. P. M., Tonin, I. P., de Aguirre, J. M., Alvim, I. D., & Ferrari, C. C. (2018)."Influence of process variables on the drum drying of mango pulp." Drying Technology 36(12) 1488–1500. https://doi.org/10.1080/07373937.2017.1410707

Giovagnoli-Vicuña, Claudia, et al. (2017) "Quality assessment and mathematical modeling of hot-air convective drying of persimmon (Diospyros kaki L.) fruit." International journal of food engineering 13(7): 20160333.

Gupta, S. V., & Patil, N. (2014). "Convective drying of osmo-dehydrated sapota slice." International Journal of Agricultural and Food Science 5(3): 219–226.

Hnin, K. K., et al. (2018). "Emerging food drying technologies with energy-saving characteristics: A review." *Drying Technology.* 37(12): 1465–1480.

Erhan H., & Maskan, M. (2015). "Hot air and microwave drying of pomegranate (Punica granatum L.) arils." Journal of Food Science and Technology, 52: 285–293. https://doi.org/10.1007/s13197-013-1032-9

Hosseinifarahi, M., et al. (2023). "Preservation of postharvest quality and bioactive compounds in sweet cherry using carboxymethyl cellulose films enriched with Thymus vulgaris L. Essential oil." Journal of Food Measurement Characterization: 1–11.

Islam, M. Z., Kitamura, Y., Kokawa, M., Monalisa, K.,Tsai., & Miyamura, S. (2017). "Effects of micro wet milling and vacuum spray drying on the physicochemical and antioxidant properties of orange (Citrus unshiu) juice with pulp powder." Food and bioproducts processing 101: 132–144. https://doi.org/10.1016/j.fbp.2016.11.002

Jafari, S. M., Azizi, D., Mirzaei, H., & Dehnad, D. (2016). Comparing quality characteristics of oven-dried and Refractance Window-dried kiwifruits. Journal of Food Processing and Preservation, 40(3): 362–372. https://doi.org/10.1111/jfpp.12613

Jafari, S. M., et al. (2017). "Influence of spray drying on water solubility index, apparent density, and anthocyanin content of pomegranate juice powder." Powder Technology **311**: 59–65.

Kammoun, B. A, Kechaou, N., & Mihoubi, N., B. (2011). "Effect of microwave treatment on physical and functional properties of orange (Citrus sinensis) peel and leaves." Journal of Food Processing & Technology 2(2): 109.

Henríquez, M., Almonacid, S., Lutz, M., Simpson, R., & Valdenegro, M. (2013). "Comparison of three drying processes to obtain an apple peel food ingredient." CyTA-Journal of Food 11(2): 127–135. https://doi.org/10.1080/19476337.2012.703693

Kemp, I. C. (1999). "Progress in dryer selection techniques." Drying Technology **17**(7–8): 1667–1680.

Mariani, Viviana Cocco, et al. (2014). "Hot-air drying characteristics of soybeans and influence of temperature and velocity on kinetic parameters." Journal of Food Process Engineering 37(6): 619–627. https://doi.org/10.1111/jfpe.12118

Martynenko, A., & Kudra, T. (2016). Electrohydrodynamic (EHD) drying of grape pomace. Japan Journal of Food Engineering, 17(4): 123–129. https://doi.org/10.11301/jsfe.17.123

Mothibe, K. J., Zhang, M., Mujumdar, A. S., Wang, Y. C., & Cheng, X. "Effects of ultrasound and microwave pretreatments of apple before spouted bed drying on rate of dehydration and physical properties." Drying Technology 32(15): 1848–1856. https://doi.org/10.1080/07373937.2014.952381

Muzaffar, K., et al. (2015). "Stickiness problem associated with spray drying of sugar and acid rich foods: A mini review." Journal of Nutrition Food Sciences 5(S12): 1.

Nemzer, Boris, et al. (2018). "Phytochemical and physical properties of blueberries, tart cherries, strawberries, and cranberries as affected by different drying methods." Food Chemistry 262: 242–250. https://doi.org/10.1016/j.foodchem.2018.04.047

Nuthong, Paradorn, et al. (2011). "Kinetics and modeling of whole longan with combined infrared and hot air." Journal of Food Engineering 102(3): 233–239. https://doi.org/10.1016/j.jfoodeng.2010.08.024

Ochoa-Martínez, C. I., Quintero, P. T., Ayala, A. A., & Ortiz, M. J. (2012). Drying characteristics of mango slices using the Refractance Window™ technique. Journal of Food engineering, 109(1): 69–75. https://doi.org/10.1016/j.jfoodeng.2011.09.032

Panja, P. (2018). "Green extraction methods of food polyphenols from vegetable materials." Current Opinion in Food Science **23**: 173–182.

Raj, GVS Bhagya, and Kshirod K. Dash. (2020). "Microwave vacuum drying of dragon fruit slice: Artificial neural network modelling, genetic algorithm optimization, and kinetics study." Computers and Electronics in Agriculture 178: 105814. https://doi.org/10.1016/j.compag.2020.105814

Rajkumar, P., et al. (2007). "Drying characteristics of foamed Alphonso mango pulp in a continuous type foam mat dryer." Journal of Food Engineering **79**(4): 1452–1459.

Rajkumar, P., J. Deepa, and C. Indu Rani. (2023). "Mathematical Modeling on Drying of Tamarind Fruits." https://doi.org/10.56042/jsir.v82i06.1913

Rajoriya, D., Shewale, S. R., & Hebbar, H. U. (2019). Refractance window drying of apple slices: Mass transfer phenomena and quality parameters. Food and Bioprocess Technology, 12: 1646–1658. https://doi.org/10.1007/s11947-019-02334-7

Ramallo, L. A., & Mascheroni, R. H. (2012). "Quality evaluation of pineapple fruit during drying process." Food and bioproducts processing, 90(2): 275–283. https://doi.org/10.1016/j.fbp.2011.06.001

Roustapour, O., et al. (2006). "An experimental investigation of lime juice drying in a pilot plant spray dryer." Drying Technology **24**(2): 181–188.

Samoticha, J., Wojdyło, A., & Lech, K. (2016). The influence of different the drying methods on chemical composition and antioxidant activity in chokeberries. LWT-Food Science and Technology, 66: 484–489. https://doi.org/10.1016/j.lwt.2015.10.073

Schulze, B., Hubbermann, E. M., & Schwarz, K. (2014). Stability of quercetin derivatives in vacuum impregnated apple slices after drying (microwave vacuum drying, air drying, freeze drying) and storage. LWT-Food Science and Technology, 57(1): 426–433. https://doi.org/10.1016/j.lwt.2013.11.021

Sharma, A., et al. (2009). "Solar-energy drying systems: A review." Renewable Sustainable Energy Reviews **13**(6–7): 1185–1210.

Sharma, A. K., et al. (2013). "Raisin quality: the deciding factors." *National Research centre for Grapes, Pune, India.*

Shende, D. and A. K. Datta (2019). "Refractance window drying of fruits and vegetables: A review." Journal of the Science of Food Agriculture **99**(4): 1449–1456.

Sehrawat, R., Nema, P. K., & Kaur, B. P. (2018). Quality evaluation and drying characteristics of mango cubes dried using low-pressure superheated steam, vacuum and hot air drying methods. Lwt, 92: 548–555. https://doi.org/10.1016/j.lwt.2018.03.012

Si, X., Chen, Q., Bi, J., Wu, X., Yi, J., Zhou, L., & Li, Z. (2016). Comparison of different drying methods on the physical properties, bioactive compounds and antioxidant activity of raspberry powders. Journal of the Science of Food and Agriculture, 96(6): 2055–2062. https://doi.org/10.1002/jsfa.7317

Sogi, D. S., et al. (2013). "Total phenolics, antioxidant activity, and functional properties of 'Tommy Atkins' mango peel and kernel as affected by drying methods." Food Chemistry **141**(3): 2649–2655.

Tang, J., et al. (2003). "Drum drying." Encyclopedia of Agricultural, Food, Biological Engineering. 211–214.

Taskin, O. (2020). "Evaluation of freeze drying for whole, half cut and puree black chokeberry (Aronia melanocarpa L.)." Heat Mass Transfer **56**: 2503–2513.

Tian, Yuting, et al. (2015). "Drying Characteristics and Processing Parameters for Microwave-Vacuum Drying of Kiwifruit (A ctinidia deliciosa) Slices." Journal of Food Processing and Preservation 39(6): 2620–2629. https://doi.org/10.1111/jfpp.12512

Tumuluru, J. S. (2023). "Introductory Chapter: Food Processing, Preservation, and Packaging–A Brief Overview." *Food Processing Packaging Technologies-Recent Advances.*

Turkmen, F., et al. (2020). "Effects of different drying methods and temperature on the drying behavior and quality attributes of cherry laurel fruit." Processes **8**(7): 761.

Tze, N. L, Han, C. P., Yusof, Y. A., Ling, C. N., Talib, R. A., Taip, F. S., & Aziz, M. G. (2012). "Physicochemical and nutritional properties of spray-dried pitaya fruit powder as natural colorant." Food Science and Biotechnology 2: 675–682. https://doi.org/10.1007/s10068-012-0088-z

Valenzuela, J. L. (2023). "Advances in postharvest preservation and quality of fruits and vegetables." Foods **12**(9): 1830.

Wojdyło, A., et al. (2019). "Influence of different drying techniques on phenolic compounds, antioxidant capacity and colour of Ziziphus jujube mill. Fruits." Molecules **24**(13): 2361.

Xanthopoulos, G., et al. (2010). "Study of the drying behaviour in peeled and unpeeled whole figs." Journal of Food Engineering **97**(3): 419–424.

Yan, H., & Kerr, W. L. (2013). Total phenolics content, anthocyanins, and dietary fiber content of apple pomace powders produced by vacuum-belt drying. Journal of the Science of Food and Agriculture, 93(6): 1499–1504. https://doi.org/10.1002/jsfa.5925

Yılmaz, F. M., et al. (2017). "The effects of drying conditions on moisture transfer and quality of pomegranate fruit leather (pestil)." Journal of the Saudi Society of Agricultural Sciences **16**(1): 33–40.

Zhou, X., Ruzhen, X., Beihua, Z., Shaopei P., Qianqian L., Ramaswamy, H. S., & Shaojin, W. (2018). "Radio frequency-vacuum drying of kiwifruits: Kinetics, uniformity, and product quality." Food and Bioprocess Technology 11: 2094–2109. https://doi.org/10.1007/s11947-018-2169-3

Zielinska, M. and A. Michalska (2016). "Microwave-assisted drying of blueberry (Vaccinium corymbosum L.) fruits: Drying kinetics, polyphenols, anthocyanins, antioxidant capacity, colour and texture." Food Chemistry **212**: 671–680.

Zotarelli, M. F., et al. (2022). "Rehydration of mango powders produced by cast-tape drying, freeze drying, and spray drying." Drying Technology **40**(1): 175–187.

Wojdyło, A., Figiel, A., Lech, K., Nowicka, P., & Oszmiański, J. (2014). Effect of convective and vacuum–microwave drying on the bioactive compounds, color, and antioxidant capacity of sour cherries. Food and Bioprocess Technology, 7: 829–841. https://doi.org/10.1007/s11947-013-1130-8

3 Drying Characteristics of Dried Fruit Products

Linlin Li[1] *and Min Zhang*[2]
[1]Henan University of Science and Technology, Luoyang, China
[2]Jiangnan University, Wuxi, China

3.1 INTRODUCTION

Drying characteristics refer to the behavior and properties of a material during the process of drying. The drying characteristics of a material can vary depending on its composition, physical properties, as well as the drying method and pretreatments used. Studying drying characteristics is of great importance in food industries and their applications.

Understanding the drying characteristics of a material allows for the optimization of drying processes. By studying factors such as drying rate, drying time, and heat transfer, industries can develop efficient and cost-effective drying methods. This optimization can lead to energy savings, reduced production time, and improved overall process efficiency. In addition, drying characteristics directly affect the quality and shelf life of fruit products. Improper drying can lead to undesirable changes in texture, color, flavor, and nutritional composition. By studying the drying characteristics, manufacturers can determine the appropriate drying conditions to preserve the product's quality, appearance, and taste. This is particularly important in industries such as food processing, pharmaceuticals, and textiles. Also, drying characteristics provide valuable data for process control and predictability. By monitoring and understanding the behavior of a material during drying, industries can establish control parameters and develop models for predicting drying times and outcomes. This enables better planning, scheduling, and resource management. Furthermore, knowledge of drying characteristics influences the design and development of drying equipment. Understanding how different materials respond to drying enables the design of equipment that optimally addresses specific drying requirements. Improved drying equipment can enhance productivity, reduce energy consumption, and minimize product losses.

Overall, understanding the drying characteristics of materials allows for efficient, controlled, and high-quality drying processes, leading to improved product attributes, reduced costs, and enhanced customer satisfaction.

3.2 BASICS OF DRYING CHARACTERISTICS

The basis of drying characteristics, including drying kinetics and drying mathematical models, is the basis for understanding the principles of the drying process, clarifying the mechanism of action of the influencing factors, and thus optimizing the drying process and improving drying efficiency and product quality.

DOI: 10.1201/9781003402916-3

3.2.1 Drying Kinetics

The drying kinetics is used to describe the combined macroscopic and microscopic mechanisms of heat and mass transfer during drying, and it is affected by drying conditions, types of dryers and the characteristics of the materials to be dried. Studying drying kinetics is a means to choose appropriate drying methods and to optimize and control the drying processes. Some key aspects related to drying kinetics include drying curve and drying time, drying rate, drying temperature curve, drying efficiency, and drying consumption.

3.2.1.1 Drying Curve

The drying curve is the relationship between the moisture content of the material and the drying time. It shows the variation of the moisture content of the material over time and provides important information on the drying process. By observing the slope of the drying curve, one can understand the variation in drying rate. A steeper curve indicates a higher drying rate, while a gentler curve indicates a lower drying rate. This information is crucial for assessing the efficiency and performance of drying equipment or drying techniques. In addition, the drying curve provides insights into the equilibrium moisture content of the material during the drying process. As the curve gradually levels off, it indicates that the material has reached a state of drying equilibrium. This helps determine the criteria for the end of drying and evaluate the suitability of the drying process. Furthermore, the drying curve reflects the drying phase; it usually shows different phases, such as the initial phase, the fast drying phase, and the final phase. The changes in these phases reflect the behavior of the material under different drying conditions. By analyzing the shape and slope of the curve, it is possible to understand the drying characteristics of the material and the transformation during the drying process.

> When plotting a drying curve, the moisture content of the material is usually expressed in two ways, namely moisture content on a dry basis and moisture content on a wet basis. The moisture content on a wet basis is calculated on the basis of the mass of the material, while the moisture content on a dry basis is calculated on the basis of the solid dry matter in the material. In the various calculations for drying and moisture conditioning, it is easier to use the moisture content on a dry basis. The equilibrium moisture content (EMC) plays an important role in the drying process. It is the moisture content of a wet solid in equilibrium with air of given humidity and temperature.

Table 3.1 lists the equilibrium moisture content (EMC) for some foodstuffs. As can be seen from the table, the EMC of a material varies with the type of material. For most materials, the effect of temperature on the EMC is minimal, and the EMC depends only on the relative air humidity.

Drying time, which should be strictly expressed as the total drying time, refers to the duration required for the material to reach a specific moisture content or the endpoint of drying. It depends on the initial moisture content, drying conditions, and the moisture diffusion properties of the material. The determination of drying time is crucial for effective management and control of the drying process. Shorter drying times can improve production efficiency, reduce energy consumption, and ensure

TABLE 3.1
The EMC for Some Foodstuffs

	Relative Air Humidity/%								
Materials	**10**	**20**	**30**	**40**	**50**	**60**	**70**	**80**	**90**
Flour	2.20	3.90	5.05	6.90	8.50	10.08	12.60	15.80	19.00
White Bread	1.00	2.00	3.10	4.60	6.50	8.50	11.40	13.90	18.90
Biscuit	2.10	2.80	3.30	3.50	5.00	6.50	8.30	10.90	14.90
Starch	2.20	3.80	5.20	6.40	7.40	8.30	9.20	10.60	12.70
Apple	-	-	5.00	-	11.00	8.00	25.00	40.00	60.00
Wheat	-	-	9.30	-	-	13.00	-	-	24.00
Rye	6.00	8.40	9.50	12.00	12.50	14.00	16.00	19.50	26.00
Oat	4.60	7.00	8.60	10.00	11.60	13.60	15.00	18.00	22.50
Barley	6.00	8.50	9.60	10.60	12.00	14.00	16.00	20.00	29.00

product quality. However, excessively long drying times can lead to energy waste, extended production cycles, and instability in product quality.

To determine the appropriate drying time, experiments and tests are typically conducted. Reaching the EMC is one way of determining the endpoint of drying. Another method is that, by testing for different drying times, changes in the moisture content of the material can be observed in order to determine when the desired drying level has been reached. This ideal drying level can usually be considered to be the safe moisture content of the material. Additionally, preliminary estimates of drying time can be made based on experience and previous drying data. It is important to note that the accuracy and reliability of drying time depend on the specific circumstances and the drying method employed. Therefore, close monitoring and adjustment during the drying process are necessary to ensure that the drying time meets product requirements and ensures the quality and stability of the material.

3.2.1.2 Drying Rate Curve

The drying rate is the rate at which moisture is removed from a material. It can be influenced by factors such as temperature, airflow, humidity, and the moisture gradient between the material and its surroundings.

By convention, the drying rate (*DR*), is defined as:

$$DR = -\frac{M_S}{A}\frac{dX}{dt} \tag{3.1}$$

under constant drying conditions. Here, *DR* ($kg{\cdot}m^{-2}{\cdot}h^{-1}$) is the rate of water evaporation, *A* is the evaporation area (that may be different from heat transfer area), *Ms* is the mass of bone dry solid, *X* is the dry-basis moisture content, and *t* is time. If *A* is not known, then the drying rate may be expressed in kg water evaporated per hour.

The drying rate curve is a graphical representation describing the rate of change of the moisture content of the material during the drying process. It shows the drying

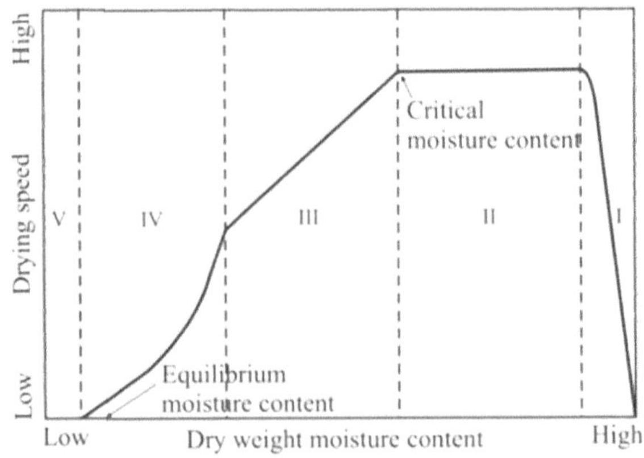

FIGURE 3.1 Drying rate curves: A typical drying curve for food products. (From Babu et al., *Renewable & sustainable energy reviews*, 2018, 90: 536–556. With permission.)

rate as a function of moisture content on a dry basis and reveals the different stages and rate changes in the drying process (Figure 3.1).

Figure 3.1 shows a typical drying rate curve for food products. The first stage, preheat period, is where sensible heat is transferred to the product and the contained moisture. This is the heating up of the product from the inlet condition to the process condition, which enables the subsequent processes to take place. The rate of evaporation increases dramatically during this period with mostly free moisture being removed. A high initial drying rate is detrimental to the drying process of products that can peel, crack, or shrink at high temperatures, as it encapsulates moisture within the product and prevents it from moving from the inside of the product to the outside. In the event of cracking or shrinkage, it may also reduce the quality of the product and distort some materials. During the second stage, constant rate period (II), free moisture persists on the surfaces, and the rate of evaporation alters very little as the moisture content reduces. During this period, drying rates are high, and higher inlet air temperatures than in subsequent drying stages can be used without detrimental effects on the product. There is a gradual and relatively small increase in the product temperature during this period. This period is not seen in many foodstuffs drying as the free moisture on the surface of the material is quickly removed. The constant rate drying period ends at critical moisture content. Then falling rate drying period(s) (III and IV) start and prevail till it reaches drying endpoint. During falling rate, drying rate decreases because the moisture inside the material is slowly transported to the surface by a gradual increase in material temperature before it is evaporated from the surface. Under this condition, diffusion and capillarity effects fully control such reduced drying rates[1]. The quantity of moisture removed during the falling rate period is small, but the time taken is considerably high and thus becomes the controlling factor in the overall drying process. Hence, the falling rate period prediction has a very important role and effect on the total drying time; it is dependent on particle size, spread thickness, and external conditions like temperature, relative humidity, and flow velocity rate of air[2]. When drying reaches EMC, there is no increased exchange of moisture between the material to be dried and the surrounding air. Drying operation is stopped at this point. It would not be possible to dry materials below equilibrium moisture content unless high temperature and dehumidified air are used. Therefore, final moisture content is proportionately controlled by contact approach of drying air temperature and its relative humidity before leaving the dryer.

3.2.1.3 Drying Temperature Curve

The drying temperature curve illustrates the relationship between material temperature (or drying temperature) and drying time during the drying process. It provides valuable information about the temperature profile and its impact on the moisture removal from the material being dried.

There are two basic temperatures: the dry bulb temperature and the wet bulb temperature. The dry bulb temperature is the temperature of the vapor-gas mixture and is usually determined by immersing a thermometer into the mixture. In process applications, it is usually the process control set point for the dryer. It is also referred to as the "temperature". The wet bulb temperature is the steady temperature reached by a small amount of liquid evaporating into a large amount of rapidly moving unsaturated vapor-gas mixture. It is measured by passing the gas rapidly past a thermometer bulb kept wet by a saturated wick and shielded from the effects of radiation.

The temperature curve for the product increases from the feed temperature (most commonly ambient temperatures) to approximately the wet bulb temperature during the constant rate period and then to almost dry bulb temperatures as it nears the end of the drying cycle. During the constant rate period, the product temperature remains reasonably constant due to the effects of evaporative cooling. That is, when evaporating water with a hot stream of gas, the water film at the surface of the solid will assume the wet bulb temperature of the drying gas, and the evaporative process can be considered to be adiabatic (a process in which the system is changing without the transfer of heat to or from the environment). As the drying process proceeds through the falling rate period, the moisture content progressively reduces and the rate of moisture removal decreases markedly. The product temperature starts to increase more rapidly because the effects of evaporative cooling are reduced.

3.2.1.4 Effective Moisture Diffusivity and Activation Energy

The assessment of the effective moisture diffusivity (D_{eff}) is part of the drying kinetics study and is used as a measure of the degree of moisture transport. The determination of the effective diffusion coefficient of water is based on the following assumptions: the primary mass transfer mechanism is diffusive in nature; the external resistance to mass transfer is negligible compared to the internal resistance; the initial water content is homogeneous throughout the sample; and the diffusion coefficient is constant and not a function of moisture concentration. D_{eff} can be calculated according to the following equation:

$$MR = \frac{X_t - X_e}{X_0 - X_e} = \frac{8}{\pi^2}\sum_{n=0}^{n=\infty}\frac{1}{(2n+1)^2}\exp\left[-\frac{(2n+1)^2\pi^2 D_{eff}t}{4L^2}\right] \tag{3.2}$$

where MR is the moisture ratio, X_t is the moisture content at any time t, X_e and X_0 are equilibrium and initial moisture contents, D_{eff} is the effective moisture diffusivity and L is the half-thickness of the slices for drying from both top and bottom sides and t is the drying time.

For the drying process, activation energy (E_a) is an important parameter to evaluate the required energy to remove moisture from inside to the outside of the drying product. E_a describes the effect of temperature on the D_{eff} during the drying process. The lower the E_a value, the more sensitive the D_{eff} is to temperature changes during the drying process. Therefore, for materials with low E_a, increasing the drying temperature can increase the drying rate more effectively. The E_a can be determined by an Arrhenius equation (Equation 3.3), which can be linearized by plotting ln D_{eff} versus T^{-1}.

$$D_{eff} = D_{\infty} \exp\frac{-E_a}{RT} \tag{3.3}$$

where D_{∞} is the pre-exponential factor (m^2s^{-1}), R is the universal gas constant, 8.314 $J{\cdot}K^{-1}mol^{-1}$, and T is the absolute temperature (K).

3.2.1.5 Drying Efficiency

Drying efficiency refers to the effectiveness and productivity of the drying process in removing moisture from a material or substance. It is a measure of how well a drying system or method can achieve the desired level of dryness in a given time-frame while utilizing resources effectively. Drying efficiency, i.e., the thermal efficiency of a dryer is the most important parameter used to analyze the performance of a dryer. It is the ratio between the heat utilized and the total heat consumed in the drying process. To determine the drying efficiency of the dryer, Equation 3.4 was used[3]:

$$\eta = \frac{\lambda W\left(M_i - M_f\right)}{Pt\left(100 - M_f\right)} \tag{3.4}$$

where η is the drying efficiency, λ is the latent heat of vaporization of water, W is the sample's initial weight, M_i is the initial moisture content, and M_f is the final moisture content. P is the power of the dryer, and t is the drying time. Because in each experiment for same samples, λ, W, M_i, and M_f were constant, drying efficiency of the dryer would depend on the performance of the heat source that is, Pt term.

High drying efficiency implies that a significant amount of moisture is removed from the material in a shorter period, using minimal energy and resources. It involves achieving the desired dryness level while minimizing drying time, energy consumption, and costs. A large number of studies have used shorter drying times or higher drying rates and lower energy consumption as indicators of high drying efficiency when evaluating drying efficiency.

3.2.1.6 Drying Energy Consumption

Drying processes are one of the most energy-intensive unit operations. Drying has been reported to account for anywhere from 12 to 20% of the energy consumption in the industrial sector. For example, drying accounts for around 60% of the total energy needed for production of cereals. This amount of energy consumption is considerable compared with average energy consumption figures in tillage (16%), planting and protection (12%), harvest (6%), and transport (6%)[4].

The high energy consumption in the drying of food products also results in significant carbon emissions. According to studies, greenhouse gases such as carbon dioxide emitted during the drying process have a non-negligible impact on climate change and environmental pollution. This poses a challenge to the achievement of global carbon emission targets. In order to achieve the goal of reducing carbon emissions and improving energy efficiency, countries have set energy-saving and emission-reduction targets. The food drying industry needs to respond positively to these targets by improving technology and management practices to reduce energy consumption and reduce carbon emissions.

Researchers and related companies in the field of food drying are actively exploring new food drying technologies and equipment to improve drying efficiency and energy efficiency; in addition to technological innovation, researchers are also optimizing operating parameters and control strategies during the food drying process. Through precise monitoring and control, energy wastage can be reduced and drying efficiency improved. With the development of renewable energy sources, some research is focusing on how renewable energy can be applied to food drying processes to reduce reliance on conventional energy sources.

3.2.2 Drying Mathematical Models

Food drying kinetics is a complex problem that requires a simplified representation to predict various drying characteristics and optimize drying parameters. Drying kinetic models are important in determining ideal drying conditions, which are important parameters for equipment design, optimization, and product quality improvement. Two scales are used in the mathematical model: a macroscopic scale (0.001–1 m) model and a microscopic scale (10^{-6}–10^{-3}m) model. A multiscale model is a model that incorporates both of these features. The microscopic scale model provides more accurate predictions of the transport processes and deformations of food materials throughout the drying process compared to the macroscopic scale model, which predicts at the cellular level. Process design, optimization, energy integration, and control are all achieved through the use of effective models. Therefore, the mathematical models and their applications in predicting food drying kinetics should be continuously developed and improved. Thin-layer drying is a widely used method for determining the drying kinetics of fruits and vegetables. The classification of thin-layer drying mathematical models is shown in Figure 3.2. Most of the drying models are macro-scale models, which are used to predict the drying behavior. These models can be categorized as empirical models, semi-empirical or semi-theoretical models, and theoretical models.

3.2.2.1 Empirical Models

Drying mathematical models are an important means of describing the drying behavior of fruit. It is difficult to fit the entire drying process exactly with linear

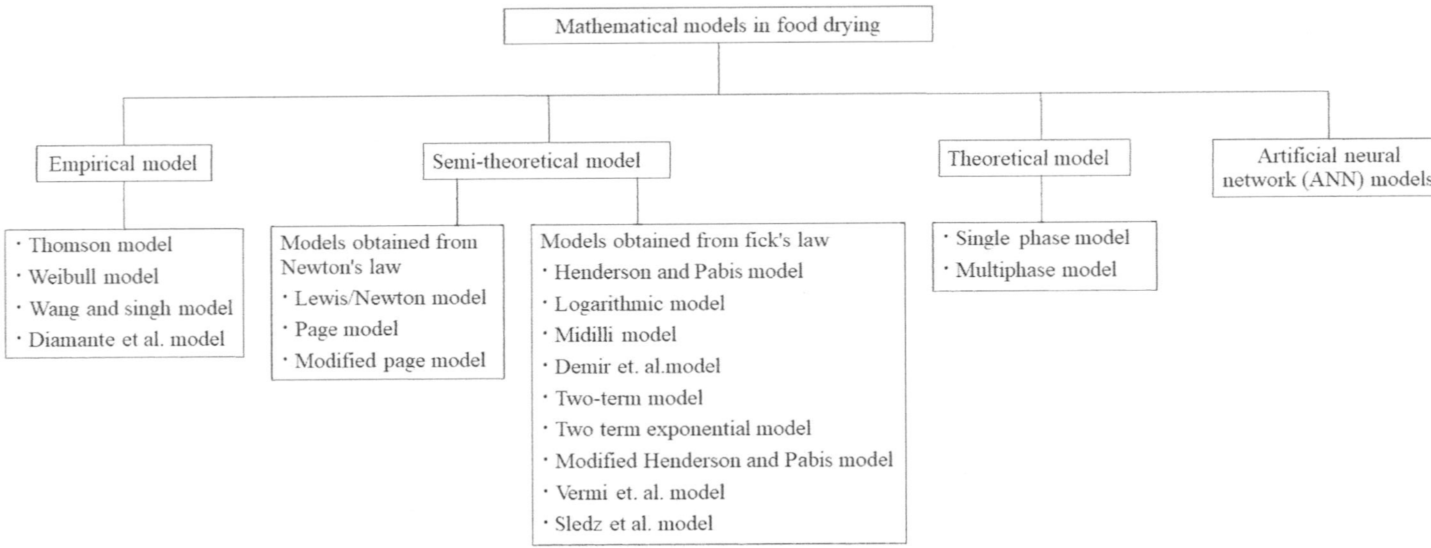

FIGURE 3.2 Classification of mathematical models in food drying.

equations. Therefore, traditional empirical models are often used to simulate drying curves to characterize the drying process and predict drying results.

The moisture content data at different drying conditions were converted into moisture ratio (MR), which is defined as:

$$MR = \frac{X_t - X_e}{X_0 - X_e} \tag{3.5}$$

where X_t, X_0, and X_e are the dry basis moisture content of the material at time t, the initial moisture content (d.b.) and the final equilibrium moisture content (d.b.), respectively. And then the curve fitting calculations of MR with drying time were carried out for the drying models proposed by various researchers.

Empirical models provide a direct relationship between average moisture content and drying time and are fitted to the experimental data by direct least square method. These models heavily rely on experimental conditions, and they do not follow the theoretical basis of the drying process, i.e., the kinetic relationship between the rate constant and the moisture concentration. Consequently, they may provide inaccurate parameter values. Moreover, these models lack a physical explanation and are solely derived from experimental data. These limitations are significant when applying empirical models in thin-layer drying. Table 3.2 lists four empirical models suitable for thin-layer drying kinetics of fruits. The Weibull model is one of the most appropriate empirical models widely used in the literatures.

3.2.2.2 Theoretical Models

Theoretical models are based on Fick's second law of diffusion. It considers both the external and internal resistance to moisture transfer during drying. This resistance is related to the geometry of the material, the mass diffusivity, and the electrical conductivity of the material[10]. The resistance can be estimated from the Lumped element model describing mass transfer[11]. The model is presented in the following Equations 3.5 and 3.6:

$$\frac{\delta M}{\delta t} = \nabla^2 K_1 \tag{3.6}$$

$$\frac{\delta T}{\delta t} = \nabla^2 K_{12} \tag{3.7}$$

TABLE 3.2
Empirical Models Used for Thin-Layer Drying Kinetics of Fruits

Model	Model Equation	Suitability	References
Thomson model	$t = a\ \ln MR + b\ \ln MR$	Blueberries	5
Weibull model	$MR = a - b\ \exp(-kt^n)$	Persimmon	6
Wang and Singh model	$MR = 1 + b\ t + a\ t^2$	Banana	7
Diamante et al. model	$\ln(-\ln MR) = a + b\ \ln t + c\ \ln t^2$	Kiwi fruit and apricot	8, 9

where K_1 is the effective diffusivity (D) and K_{12} is the thermal diffusivity (α). For the constant values of α and D, Equations 3.5 and 3.6 can further be presented as:

$$\frac{\delta M}{\delta t} = D\left[\frac{\delta^2 M}{\delta x^2} + \frac{b_1}{x}\frac{\delta M}{\delta x}\right] \tag{3.8}$$

$$\frac{\delta T}{\delta t} = \alpha\left[\frac{\delta^2 T}{\delta x^2} + \frac{b_1}{x}\frac{\delta T}{\delta x}\right] \tag{3.9}$$

The Lumped element model only considers the influence of time on the dependent variable and does not take into account variations in product temperature. It assumes that the dry air temperature is uniformly distributed within the product. The model incorporates assumptions from the Luikov equation[12], which neglects the pressure variable and assumes constant temperature.

Subsequently, the solution of Fick's second diffusion law was used as a theoretical model for thin-layer drying of foodstuffs[13].

3.2.2.3 Semi-theoretical Models

Semi-theoretical models are based on the diffusion theory by simplifying Fick's second law and are valid within the experimental range of temperature, relative humidity, air velocity, and moisture content. Semi-theoretical models can also be derived from Newton's law of cooling:

$$\mathrm{MR} = \exp(-kt) \tag{3.10}$$

where k represents the drying coefficient.

The semi-theoretical models give some understanding of the transport processes during the drying process, which have been used for describing drying kinetics of fruits. The various semi-theoretical drying model types, equations, and applicable fruits are shown in Table 3.3.

The Lewis, Page, and Modified Page models are derived from Newton's law of cooling. The other models listed in Table 3.3 are derived from Fick's second law of diffusion. These models cover (i) the exponential model and its simplified form, (ii) the two-exponential model and its modified form, and (iii) the three-exponential model and its simplified form. Factors that determine the applicability of these models include drying temperature, drying airspeed, material thickness, initial moisture content, and relative humidity[11]. Furthermore, under these conditions, it can be noted that the complexity of the model can be attributed to the number of constants. The number of constants varies between 1 (Newton model) and 6 (modified Henderson and Pabis model) (see Table 3.3). However, the selection of the most appropriate model for describing the drying behavior of fruits and vegetables does not depend on the number of constants.

With so many models available, it is important to choose a model that both captures the mechanism of moisture diffusion during drying and fits the moisture ratio curve well to characterize the drying kinetics of the fruit. The selection of a suitable model is based on the statistical parameters of the model fit. These statistical

TABLE 3.3
Semi-theoretical Mathematical Models Used for Fruits Drying

Model Name	Model Equation	References
Lewis/Newton model	$MR = \exp(-kt)$	14
Page model	$MR = \exp(-kt^n)$	15
Modified page model	$MR = \exp(-(kt)^n)$	16
Henderson and Pabis model	$MR = a\exp(-kt)$	17
Modified Henderson and Pabis model	$MR = a\exp(-kt) + b\exp(-k_0t) + c\exp(-k_1t)$	18
Two-term model	$MR = a\exp(-k_0t) + b\exp(-k_1t)$	19
Two term exponential model	$MR = a\exp(-k_0t) + (1-a)\exp(-k_1at)$	20
Logarithmic model	$MR = a\exp(-kt) + c$	21
Midilli model	$MR = a\exp(-kt) + bt$	22
Demir et al. model	$MR = a\exp(-kt)^n + b$	23
Vermi et al. model	$MR = a\exp(-kt) + (1-a)\exp(-k_0t)$	24
Sledz etal. model	$MR = \dfrac{b\exp(-kt)}{1+a\exp(k_it)}$	25

k, k_i, k_0, k_1 – drying rates (1/min); a, b, n – model parameters, t – time (s).

parameters include correlation coefficient (R or r), coefficient of determination (R^2 or r^2), reduced chi-square (x^2), sum square error (SSE), root mean square error ($RMSE$), mean relative error root square ($RRMS$), modelling efficiency (EF), mean percent error (MPE), and mean bias error (MBE). The higher the R and R^2 values for a particular model, the better the model is for predicting the drying behavior of fruits and vegetables. Similarly, the lower the x^2, SSE, $RMSE$, $RRMS$, EF, MPE, and MBE values of a particular model, the better the model is for predicting the drying kinetics of a particular product.

3.2.2.4 Artificial Neural Network (ANN) Models

Food drying is inherently a complex process. Continuous innovations in drying technology increase the difficulty of predicting the drying kinetics of dynamic and complex drying processes, which may sometimes be impossible to explain using traditional models. In recent years, the application of ANN models to various drying processes of fruit and vegetable foods has been extensively investigated.

An ANN is a mathematical model that mimics the structure and function of the biological nervous system, enabling learning and prediction by simulating the connections and information transfer between neurons. The basic structure of an ANN model, as shown in Figure 3.3, consists of an input layer, one or more hidden layers, and an output layer. The neurons in the input layer represent the model's input parameters, such as drying temperature, airspeed, drying time, and other drying process parameters. The output layer provides the predicted results, such as moisture content, shrinkage rate, and nutrient degradation rate, after drying. The hidden layers perform intermediate calculations between the input and output layers. In an ANN

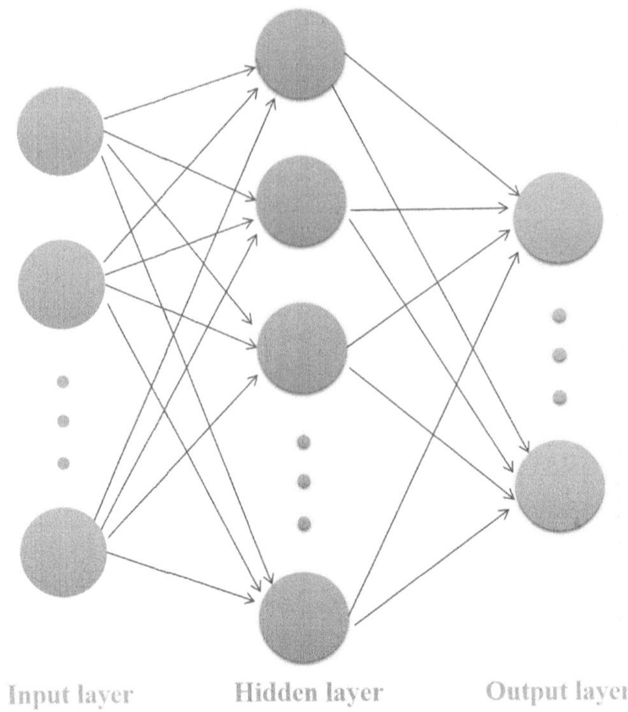

FIGURE 3.3 Illustration of an artificial neural network with one hidden layer.

model, each neuron in every layer is fully connected to all neurons in the adjacent layers. The neurons in the next layer are computed as the sum of all neurons in the previous layer.

As a regression tool, ANN models have been widely used in fruit drying performance prediction. An ANN model was developed by Şahin and Öztürk[26], with drying air temperature, wind velocity, air humidity, solar radiation, and drying time selected as input data and moisture ratio as the output layer. A hidden layer was mentioned as it was considered to be the most suitable architecture for drying process prediction, with 10 neurons defined in a hidden layer by trial and error method. The developed ANN model provided better predictions than 11 conventional drying models for predicting the experimental MR of different treatments and fresh figs under open sun drying. Similar predictive ANN models have been employed in various studies to predict the moisture content of different fruits during drying processes. For instance, researchers have reported the use of similar ANN models to predict the moisture content of Cantaloupe during microwave drying[27], as well as the moisture content of apples dried by convective and microwave drying methods[28]. Additionally, ANN models have been utilized to predict the time-dependent moisture changes of fruits such as bananas and pineapples[29,30].

If the processing time is used as an input parameter, a well-trained regression model can be used to study the drying kinetics, including moisture content, degradation of nutrients, and other time-dependent characteristics. The investigation of these kinetics can aid in designing appropriate drying processes and achieving desired stopping criteria, such as maintaining a certain level of nutrients or color in the dried product. ANN models can also be used for sensitivity analysis of input variables, which helps to rank the importance of the input variables and enables optimization of drying parameters[31]; when combined with real-time measurement techniques (e.g., low-field nuclear magnetic resonance), ANN models can be used to dynamically monitor and control the drying process[32,33].

3.3 FACTORS AFFECTING DRYING CHARACTERISTICS

There are many factors affecting the drying characteristics of fruit, the dominant ones being the type and characteristics of the fruit being dried, the pretreatment method before drying and the drying technology.

3.3.1 Fruit Materials

The drying characteristics of different fruit materials differ significantly, even if the same drying technology is used under the same drying process parameters. For example, under hot air drying conditions at 65°C, pomegranates took 240 min to dry, while lychees took 22 h[34]. This huge difference is due to the great differences in the chemical composition and physical properties of the different fruits.

For the same type of fruit, there are certain differences in chemical composition when the varieties are different or at different levels of ripeness, which will lead to differences in drying characteristics. The drying profile characteristics of date fruits at the fully ripe stage of Khalas and Ruzeiz varieties were studied using two solar dryers: direct solar dryer and indirect solar dryer, respectively. Results showed that Khalas date fruits are more dehydrated than Ruzeiz ones for the same period of time[35]. In addition, the drying characteristics of the same variety of fruit vary considerably depending on the ripeness of the fruit. Wang et al.[36] studied pulsed vacuum drying of Thompson seedless grapes at different ripeness levels, which resulted in significant differences in the chemical composition of seedless grapes, with total phenol concentration, soluble solids, and pH increasing significantly and titratable acid content decreasing significantly with increasing ripeness. The drying characteristics of seedless grapes differed significantly between ripeness levels, with the drying rate and effective moisture diffusion rate increasing significantly with increasing ripeness. The main reason for this is that the increase in ripeness leads to a significant increase in the free water content of the grapes and a decrease in the distribution of bound water, which is easily dehydrated during the drying process.

The shape and size of the fruit are also factors that affect its drying characteristics. Generally, fruits to be dried are cut into smaller pieces or slices to speed up the transfer of heat and mass. Shapes with a larger surface area provide more surface contact with the heating medium (air) and have more surface to allow moisture to escape; smaller particles or thinner layers reduce the distance for heat to reach the center of the food and also for moisture in the center of the food to reach the surface and escape. For a given volume, the drying rate of apple cubes was higher than that of slices because the exposed surface area was higher[37]. Shi et al.[38] found that the drying rate increased as the size of the berries decreased. This may be due to the fact that minimizing the size of the product reduces the distance of water diffusion, which leads to faster drying. In addition, the edge characteristics of the different shapes also influence the drying characteristics of the fruit. The sharp edges of the samples were the main cause of the weak airflow over the samples, which resulted in reduced contact between the drying medium and the samples. The hot air drying characteristics of three apples of the same volume and different geometric plate shapes, including square, circle, and triangle, at 110°C, 115°C, and 120°C showed that the drying times of triangle shape samples were 15% and 8% shorter than those of square shape samples at 110°C and 115°C, respectively, and 8% and 18% shorter than those of circle shape samples at 115°C and 120°C, respectively[39]. Triangle-shape samples with the smallest sharp edges had the greatest contact with the drying air and therefore the shortest drying times.

The characteristics of the fruit's skin state are also related to the drying characteristics. Berry fruits such as blueberries have a waxy layer on their surface, which acts as a barrier to water migration within the fruit, resulting in low drying rates and long drying times. In order to improve the drying characteristics of blueberries and their quality, certain treatments are often used to change the structure of the skin. For example, CO_2 laser perforation, peel grinding, or scoring treatments. Studies have shown that peeling, perforating, and scoring treatments can reduce freeze-drying times by 38.5%, 23.1%, and 15.4%, respectively, compared to untreated groups[40].

3.3.2 Pretreatments

In fruit drying processing, there are issues such as high energy consumption, long processing time, and significant loss of nutritional components. These problems are primarily addressed through innovative drying techniques and the development of drying equipment. Additionally, a common approach to addressing these issues is through pretreatments before drying.

Pretreatment techniques before drying involve the use of physical, chemical, biological, and other technical methods to treat the materials. This helps to accelerate the drying rate of the materials, improve the appearance and nutritional quality of the products, and extend their shelf life. Currently, commonly used pretreatment techniques for fruit drying include blanching, freeze-thaw, osmosis, chemical agents, ultrasound, high-pressure processing, high-pressure pulsed electric field, and carbon dioxide immersion treatment.

3.3.2.1 Blanching

Thermal blanching is widely used prior to drying. It involves subjecting fresh raw materials, either sliced or unsliced, to high-temperature hot water, boiling water, or atmospheric pressure steam. The main objective of blanching is to inactivate the enzymes involved in the spoilage of raw food and to reduce the microbiological load of the product to improve its preservation. In addition, blanching at high temperatures softens the tissue structure of fruits and vegetables, improves cell permeability, and thus increases the drying rate.

Hot water blanching is a common pretreatment method prior to drying, which involves immersing the fresh product in hot water at a constant temperature of 70 to 100°C for a few minutes[41]. It helps to speed up drying by altering the physical properties of the sample, such as the permeability of the cell membrane, dissociating waxes from the product surface, and forming fine cracks. It was reported that Danamrongrak et al.[42] rinsed bananas in boiling water for 3 min prior to heat pump drying, resulting in an 8% increase in drying rate and a 106% increase in moisture diffusivity. Doymaz et al.[43] soaked quince in hot water at 80°C for 1 min before conducting hot air drying, resulting in a reduction of drying time by over 14%. However, it is worth noting that the appropriate blanching temperature is important. Ando et al.[44] reported that a lower drying rate of carrots blanched at 60°C during air-drying than that of fresh samples was observed.

In order to minimize the loss of nutrients, especially water-soluble nutrients and solid substances dissolved in hot water, and reduce wastewater generation; a steam blanching system has been developed as a substitute for hot water blanching. Compared to hot water blanching, the leaching effect of steam blanching can be considered negligible, thus allowing for the retention of a significant portion of minerals and water-soluble components. Steam blanching disrupts the cellular structure, leading to tissue collapse and increased water freedom, which steam blanching disrupts the cellular structure, leading to tissue collapse and increased water freeing, which were intensified by prolonging the blanching time. Sun et al.[45] investigated the effect of steam blanching at different times on the far-infrared drying kinetics of broccoli. Steam blanching increased the drying rate and reduced the drying time by 9.26–48.15% compared to untreated samples.

Superheated steam impact blanching (SSIB), also known as high humidity hot air impact blanching (HHAIB), is a heat treatment technology developed in recent years that combines the advantages of superheated steam blanching and impact blanching technology. It was developed as an alternative to blanching in order to overcome the loss of nutrients during the pretreatment process of traditional hot water and steam blanching technology. HHAIB has high enthalpy, heat transfer rate, and energy efficiency. At a velocity of 14.4 m/s, temperature of 135°C, and relative humidity of 35%, the heat transfer coefficient of HHAIB is approximately 1403 W/(m^2·K), which is about 12 times higher than the heat transfer coefficient of hot air impingement at the same temperature and velocity[46]. Bai et al.[47] investigated the drying characteristics of seedless grapes blanched by HHAIB at different temperatures (90, 100, 110, and 120°C) and durations (30, 60, 90, and 120 s) for subsequent air-drying at temperatures ranging from 55 to 70°C. The drying time was decreased about 42% when the blanching time increased from 30 to 120 s. With the HHAIB retention time kept constant at 90 s, the drying curves for the four temperatures almost overlapped. As the blanching temperature increased from 90 to 120°C, the drying time reduced by approximately 11%. These results indicated that the blanching time has a greater impact on the drying time compared to the temperature. Under the experimental conditions, the grape samples did not exhibit a constant rate drying period and, like most food products, only a decreasing rate drying period. The lack of a constant rate period may be due to the inability to provide a constant supply of water for a considerable period of time during the initial stages of drying the product[48]. In addition, increasing the blanching time and blanching increased D_{eff}.

There are also many new blanching pretreatment methods, such as microwave blanching, ohmic blanching, and ultrasonic combined heat blanching. With the rapid development of the drying industry, these new blanching technologies will continue to be applied in the fruit drying industry.

3.3.2.2 Freeze-thaw

Freeze-thaw treatment is the process of freezing materials at low temperatures and then thawing them at high temperatures. The freeze-thaw treatment affects the drying characteristics and nutritional quality of the material by changing the form of moisture and solute distribution in fruits and vegetables. The formation of ice crystals during the freezing process leads to an increase in intercellular gaps and structural expansion. However, during the thawing process, the growth of ice crystals damages the cell walls and cell membranes of fruits, reducing the resistance to internal moisture diffusion and thereby enhancing the drying rate. When comparing the effects of blanching, freeze-thaw and vacuum-assisted osmotic pretreatments on the hot air drying characteristics of apples, the results showed that the freeze-thaw treated apple slices exhibited the fastest drying rate[49].

Freezing is a crucial step in the freeze-thaw treatment, and the freezing method has a significant impact on the drying characteristics of fruits subjected to freeze-thaw treatment. Edidiong Joseph Bassey et al.[50] studied the effects of freeze-thaw pretreatments using slow freezing (SFZ), fast freezing (FFZ), and ultra-fast freezing (UFZ) on the drying characteristics of red dragon fruit slices during mid-infrared (MIR) and near-infrared (NIR) drying. The results showed that the pretreatments reduced drying time, and enhanced drying rate and effective moisture diffusion rate, with these effects depending on the freezing method. During the MIR drying process, the pretreatments resulted in a reduction of drying time by 27.27%, 13.63%, and 27.27% for samples subjected to slow freezing (SFZS), fast freezing (FFZS), and ultra-fast freezing (UFZS), respectively. In the NIR drying process, the reductions were 25.00%, 16.66%, and 16.67% for SFZS, FFZS, and UFZS, respectively. These results may be attributed to the formation of ice crystals during the freezing pretreatment, which potentially led to cracks on the cell walls of the samples, facilitating rapid moisture removal and thereby reducing drying time relatively. Additionally, the observed reduction in drying time for SFZS and UFZS during MIR drying and for SFZS during NIR drying could be associated with the size of ice crystals formed and freeze fracturing during these pretreatments. Moreover, the pretreated samples presented higher D_{eff} in both MIR and NIR drying compared to the control group, which was attributed to the disruption of the cell structure due to the freeze-thaw pretreatment.

3.3.2.3 Osmotic Dehydration

Osmotic dehydration is one of the most extensive pretreatments carried out prior to drying to reduce energy consumption and improve food quality. It involves the immersion of the material in a hypertonic solution (mainly sugar or salt) for several hours with agitation. During osmotic pretreatment, the plant cell structure acts as a semi-permeable membrane, and a counter-flow transfer of mass occurs: solutes flow into the product, while water is transferred from the inside to the hypertonic solution[51]. The driving force for the removal of water from food materials is the osmotic pressure difference between the food material and the hypertonic solution.

Osmotic dehydration removes 10–70% of water from fruits and vegetables at ambient temperature without phase change, which provides an alternative method to reduce drying time and mitigate the thermal degradation of bioactive compounds observed during typical drying processes[51]. For example, Rodrigues & Fernandes[52] found that osmotic dehydration pretreatment using sucrose and mannitol reduced drying time to 6.3 h compared to un-pretreated samples.

The process conditions of the osmotic dehydration treatment, such as processing temperature, solution concentration, and solution-to-solid mass ratio, have an impact on the mass transfer and drying characteristics. An increase in solution concentration generally leads to an increase in the water mass transfer coefficient between the material and the solution, resulting in an increase in the osmotic pressure gradient. Ruiz-Lopez et al.[53] reported that pretreatment of chayote with 10% and 25% NaCl solutions (w/w) resulted in initial water content reductions of 2.85 and 5.79 kg/kg (d.b.), respectively, and drying time reductions of 48.0% and 62.1%, respectively. Similar results were reported by Oliveira et al.[54] and Brochier et al.[55] for the dehydration of yacon fruits. In terms of the solution-to-solid mass ratio, a low value may cause excessive dilution of the osmotic solution, resulting in a local reduction in osmotic driving force during the process, thereby reducing the degree of dehydration and consequently the subsequent drying time[56]. In terms of the temperature parameter for osmotic dehydration, an increase in temperature leads to a significant increase in the effective diffusion coefficient in the process[54]. This is because an increase in temperature will generally reduce the viscosity of the solution, thereby increasing the permeability of the membrane and thus reducing the external resistance to mass transfer. In general, the higher the degree of osmotic dehydration, the shorter the time required for subsequent drying of the fruit. However, osmotic dehydration also has a detrimental effect on the drying rate of some products due to increased resistance to water flux caused by shrinkage and solute uptake[57].

In conclusion, the effect of osmotic dehydration on the drying characteristics of fruit is mainly influenced by the degree of pretreatment dehydration and the physical properties of the dehydrated food material. In addition, due to the slow mass transfer efficiency of single osmotic dehydration treatment, ultrasound, vacuum, and high-voltage pulsed electric field techniques are often combined with osmotic dehydration. Numerous studies have shown that osmotic dehydration treatment in combination with the above techniques can significantly reduce drying times[58,59].

3.3.2.4 Chemical Reagent Treatment

The principle of chemical reagent treatment is to dissolve the waxy layer on the surface of fruits and vegetables and to change the structure of the skin, thus reducing the resistance to evaporation of water from the surface. Some of the reagents can also be used to inhibit polyphenol oxidase, prevent enzymatic browning, and inhibit microbial growth and antioxidant activity. Commonly used chemicals include ethanol, potassium carbonate, sodium hydroxide, citric acid, ethyl oleate, and sodium metabisulphite (Table 3.4).

TABLE 3.4
The Commonly Used Reagents for Drying Pretreatment of Fruits and Their Principles

Reagent	Principle	Suitability	References
Ethanol	It has a lower boiling point than water and is easily volatile. When drying, ethanol is rapidly vaporized by heat and the volume expands, which facilitates the expansion of the internal pore structure of the material and promotes the migration of water.	Banana	68
Potassium carbonate	Thinning the stratum corneum or making it intermittent distribution, destroys epidermal cells, thereby increasing the rate of water loss during drying.	Golden berry	69
Sodium hydroxide	Disrupts the surface wax layer and tissue fiber structure, thus promoting internal diffusion of water and surface evaporation.	Berry, Grape	38, 70
Ethyl oleate	The alkaline ethyl oleate dissolves the surface waxy layer structure and destroys the epidermal cells, thus promoting the evaporation of water and increasing the drying rate.	Plum	62
Sodium metabisulphite	The formation of sulphite is used to reduce the coloring matter formed by the oxidation of the material, causing it to fade; it can inhibit oxidation enzymes and prevent the material from decaying and deteriorating.	Nectarines	71
Citric acid	Inhibits oxidase activity, reduces enzymatic browning, and protects color. Mostly used as an acidulant and antioxidant.	Apple	72

The outer skin of fruits such as berries is covered by hydrophobic waxes, which act as a barrier to drying, resulting in a low rate of water evaporation[60]. As the wax coating consists mainly of oleanolic acid, treatment with alkaline solutions of ethyl oleate, sodium hydroxide, and potassium carbonate can dissolve the wax layer or disrupt the microstructure of the epidermal wax layer or even disrupt intracellular binding through the de-esterification of pectin[61]. This in turn enhances the permeability of skin to water, promotes water diffusion, and increases the rate of dehydration. Doymaz[62] reported that immersion of plums in K_2CO_3 solution (5%, v/v) and ethyl oleate (2%, v/v) for 1 minute at room temperature improved the hot air drying rate. The drying time of the pretreated samples was 29.4% shorter than that of the untreated samples. Tarhan[63] compared the effect of treatments with 4% ethyl oleate, 1% KOH, 1% NaOH, and water, at 23°C or 60°C for 1 min on the solar drying of plums, and found that those treated by 1% KOH or 1% NaOH at 60°C were the fastest ones reaching to the final weight loss percentage.

Acid reagents (Citric acid, Sodium metabisulphite, etc.) pretreatments are often used to improve product quality by inactivating enzymes, improving pigment stability, and altering the texture of produce, with little effect on drying characteristics.

As ethanol has a lower boiling point than water and is easily volatile, it rapidly vaporizes when heated during drying and the volume expansion has a role in expanding the internal pore structure of the material and promoting water migration. Zubernik et al.[25] studied the effect of pretreatment with ethanol solution on the convective drying kinetics of apples and found that the drying time was reduced to 13.4% after immersion in ethanol solution. Ten commonly used mathematical models were also used to evaluate the drying kinetics of apples. The most suitable model to describe the drying kinetics was found to be the model proposed by Sledz et al.[64] Many studies have shown that ethanol treatment can increase the efficiency of the drying process and minimize its energy consumption and costs[65–67].

3.3.2.5 Ultrasound

Ultrasound is a type of sound wave with a frequency greater than 16 kHz, which is outside the range of the human ear. Depending on the frequency, ultrasound can be divided into high and low frequencies. Low-frequency ultrasound (20–100 kHz) can be used as a pretreatment technique for fruit drying. The mechanism of ultrasound has three aspects[73,74]: (1) Sponge effect: the ultrasound action causes the material to shrink and swell continuously, forming a sponge-like structure. When the force generated by this structural effect is greater than the surface tension that holds the water inside the capillary, the water is easily transferred out through the tiny tubes. (2) Cavitation: within the ultrasonic pressure field, the formation, growth, and violent collapse of cavitation bubbles and the resulting series of physicochemical effects help to remove moisture that is tightly bound to the material. (3) Other effects: ultrasound promotes the formation of micro-channels, reduces the thickness of the heat transfer surface layer, and increases convective mass transfer efficiency.

Ultrasound has been shown to speed up drying by altering the microstructure of plant tissues. For example, ultrasonic pretreatment reduced drying time by 31% compared to untreated apple cubes, and the porosity of ultrasonically treated dried products was 9–14% higher than untreated samples, demonstrating that ultrasound affects drying characteristics by changing the pore structure of the material[75]. Tao & Sun[76] showed that within a certain range, drying kinetics increases with increasing sound intensity and ultrasonic efficiency is closely related to sound intensity.

3.3.2.6 Pulsed Electric Field

Pulsed electric field (PEF) is a non-thermal technique that involves the application of short duration high voltage field electrical pulses to food placed between two electrodes, usually in the range of 15–80 kV/cm. The mechanism of action is that the applied electric field increases the potential difference between the inside and outside of the membrane, and when the potential difference is higher than the critical value of the cell membrane, the cell membrane ruptures, thereby increasing the permeability of the cell membrane[77]. Yu et al.[78] showed that PEF could destroy the surface wax layer and improve the efficiency of sugar penetration and dehydration of blueberries, reducing the vacuum drying time by more than 50%. Parniakov et al.[79] also found that PEF could improve the freeze-drying efficiency of apple slices and

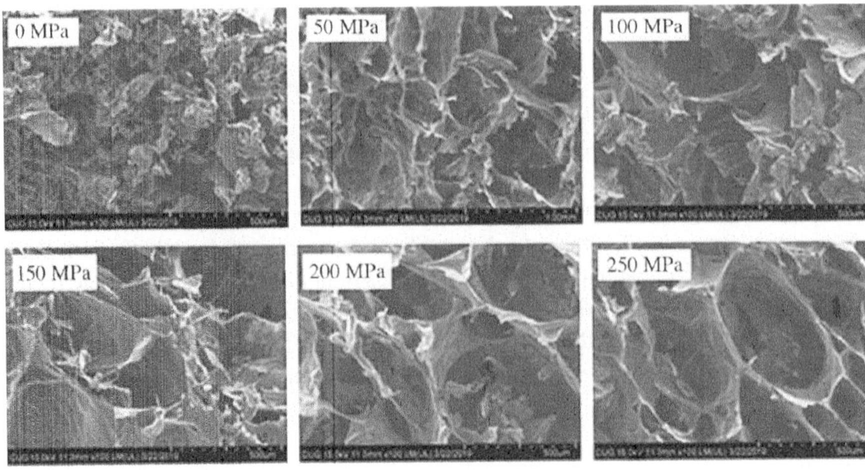

FIGURE 3.4 Effect of HHP on microstructure of strawberry slices. (From Zhang et al., *Food Chemistry*, 2020, 331: 127203. With permission.)

increase their porosity and rehydration. PEF can disrupt the cell membrane structure and increase cell permeability at low temperatures, increasing the rate of sugar penetration and dehydration of fruits and vegetables and reducing drying time[80].

3.3.2.7 High Hydrostatic Pressure

High Hydrostatic Pressure (HHP) technology is an innovative and emerging pressure-based technology. It involves applying a high-pressure shock wave (ranging from 100 to 800 MPa) through water to the material to achieve the desired residence time and temperature[81]. When fruits are pretreated with HHP, the high pressure-induced cell permeability promotes diffusion and provides higher drying rates[82]. The effect of HHP (50, 100, 150, 200, and 250 MPa) pretreatment on the vacuum freeze-drying of strawberry slices was investigated by Zhang et al.[83] HHP significantly increased the moisture mobility of strawberry slices, resulting in a 9–24% reduction in drying time. Microscopic images (Figure 3.4) showed the formation of microscopic holes or channels in the matrix and the breakdown of tissue structure by HHP. The accelerating effect of HHP pretreatment on the drying of bananas, plums, and carambola has been widely reported[84–86].

3.3.3 Drying Parameters

3.3.3.1 Temperature

The effect of drying temperature as the main operating parameter for most drying techniques, such as hot air drying, solar drying, oven drying, fluid bed drying, heat pump drying, infrared radiation drying, etc., on the drying characteristics of fruit drying is significant. The use of higher drying temperatures leads to higher drying rates, shorter drying times, and higher effective moisture diffusion rates. A study conducted by Ayala-Aponte et al.[87] reported that shorter drying times were required to dry yellow dragon fruit slices under hot air drying at high temperatures (70°C). A similar trend was observed for dragon fruit slices that had been permeably dehydrated prior to hot air drying. The D_{eff} doubled when the dragon fruit was dried at high temperatures. Chen et al.[88] also found that the D_{eff} values of date slices dried by medium wave infrared radiation drying were more than two times higher at 90°C than at 60°C. This is mainly due to the fact that the higher the temperature, the faster the water molecules in the date slices evaporate and the correspondingly higher D_{eff} values.

3.3.3.2 Air Velocity

In general, the air velocity affects the final moisture content of the sample, which can be explained by the fact that an increase in air velocity produces an increase in the convection coefficient, which increases the heat transferred to the sample, thus accelerating drying. However, other authors suggest that an air velocity of 1.5 m/s is sufficient to eliminate the resistance created by the food surface, thus promoting moisture diffusion[89,90]. Duarte et al.[91] showed that when the air velocity was increased from 2.4 to 2.7 m/s, the drying time was reduced and the moisture diffusion coefficient increased. However, when the air velocity was increased to 3.0 m/s, the moisture diffusion coefficient decreased.

3.3.3.3 Microwave Power

A higher microwave power density will result in shorter drying times. Approximate constant rate drying periods and falling rate drying periods were found in the microwave intermittent drying of apple slices at different microwave power densities (2.2, 3.5, 4.8, and 6.1 W/g). The drying time decreased with increasing microwave power density; when the microwave power density increased from 2.2 to 6.1 W/g, the D_{eff} increased more than threefold[92]. Similar observations were reported in the literature for microwave drying of black mulberry[93]. When the microwave power is increased from 90 to 360 W, the average drying time is reduced by 5.25 times. At high microwave power heating, mass transfer within the sample is faster as more heat is generated within the sample, creating a large vapor pressure difference between the center and the surface of the product due to the nature of microwave volumetric heating.

3.3.3.4 Combined Mode of Heat Input

As single drying mostly has certain drawbacks, combined drying technology has become a popular choice in order to achieve high-quality and efficient drying. This involves a combination of heat inputs, which may include infrared, convection, conduction, and microwave for drying. The addition of microwaves to conventional hot air drying can reduce drying time by up to 50% compared to using a single convection mode[94]. In contact ultrasound-assisted hot air drying, the application of contact ultrasound leads to a significant acceleration of internal mass transfer and the application of higher ultrasound power leads to faster drying rates. The increase in contact ultrasound power can reduce the energy consumption for hot air drying by up to 34.60%[95].

3.4 FACTORS AFFECTING DRYING EFFICIENCY

3.4.1 Product Parameters

The effect of product parameters on drying efficiency is less well reported. Only Aydin et al.[96] are known to have compared the drying efficiency of different types of fruit, such as apples, bananas, and grapes, using a new modular solar dryer, with drying efficiencies of 38.1%, 43.84%, and 42.12%, respectively. Taghinezhad et al.[97] showed that Kiwi thickness had a significant effect on drying efficiency, with low thickness leading to high drying efficiency. More studies have been conducted to evaluate the effect of product parameters on drying characteristics through drying time and drying rate.

3.4.2 Process Parameters

3.4.2.1 Temperature

In general, drying efficiency increases with increasing drying temperature, as a rise in the internal thermal gradient of the product will lead to a reduction in drying time. A study by Hazervazifeh et al.[98] confirmed this finding. The drying efficiency of apples increased by 18.05% when the hot air drying temperature was increased from 40 to 70°C. Similar observations were reported by Taghinezhad et al.[97] where the drying efficiency of hybrid hot air-infrared drying of kiwifruit increased with increasing temperature from 50 to 70°C.

3.4.2.2 Air Velocity

The study of Hazervazifeh et al.[98] showed that drying efficiency decreased with increasing air velocity, which may be related to the cooling effect of the airflow on the product. Although the increase in air flow rate resulted in a reduction in drying time, the drying time did not decrease proportionally.

3.4.2.3 Microwave Power

The increase in microwave power levels causes an increase in drying rates and a reduction in drying times, which are key reasons for improving drying efficiency. Many studies have confirmed that the drying efficiency was improved by increased microwave power[98,99]. However, a negative correlation between microwave power and drying efficiency was found in ultrasound-assisted microwave vacuum drying of persimmon fruit[100]. Authors suggested that as the microwave power increased, heat and mass transfer resistance were created. This resulted in the formation of a thicker surface, which was not conducive to the elimination of bound water from the center of the perspex slices. Evaporation must have taken longer to reach the outer surface of the persimmon slices. This may be the reason for the negative effect of microwave power on the MVD drying efficiency of persimmon slices.

3.4.2.4 Others

In the drying process under vacuum, the effect of the vacuum level on the drying efficiency cannot be ignored. As the vacuum level in the drying chamber increases, the drying efficiency also increases. This conclusion was confirmed in studies of microwave vacuum drying of dragon fruit slices and persimmons. The drying efficiency increases when the pressure in the drying chamber is reduced, this is because the pressure difference between the sample chamber and the surrounding atmosphere allows moisture to be rapidly removed from the sample.

In ultrasound-assisted fruit drying processes, it has been found that extending the ultrasonic pretreatment time increases the drying efficiency[97]. This is because, by increasing the ultrasonic pretreatment time, the variation in product texture increases, no hard layers are formed during the drying of this pretreated product, and the product dries faster. Dash et al.[100] showed that the increase in ultrasonic power negatively affected the drying efficiency, possibly due to the mechanism of rapid alternating compression and rapid expansion of the material forming larger pores that absorb moisture and increase the moisture content of the sample.

3.5 FUTURE PERSPECTIVES

Drying characteristics are fundamental data involved in research related to the field of drying. While extensive studies have been conducted, there are still several issues that are worth further exploration:

1. The guidance significance of drying characteristics for optimizing drying processes and enhancing product quality has not been fully realized, particularly regarding the transformation from fixed-parameter drying modes to more optimal dynamic drying modes based on the characteristics of the drying process. Further investigation is needed to better adjust drying parameters to meet the requirements of different materials and process conditions.
2. Due to the wide variety of drying techniques and fruit and vegetable types involved in fruit and vegetable drying, achieving comparability of drying characteristics between different drying techniques and produce categories is a challenge. It is important to establish standardized testing methods and evaluation criteria that can be referenced by both the industry and research community to reduce redundant research efforts.
3. Drying models play a significant role in optimizing drying processes and designing drying equipment. However, current research primarily focuses on describing drying kinetics, and the role of these models in guiding process optimization is not yet prominent. Exploring how to fully leverage drying models in guiding process optimization requires further consideration. This may involve optimizing model parameters, improving the accuracy and stability of the models, and integrating models with advanced optimization algorithms.
4. Although artificial neural network (ANN) models have been introduced in the study of drying kinetics models, there is still limited work in this area. Further research is needed to fully exploit the advantages of ANN and its potential in assisting drying process optimization.
5. The significant promoting effect of pretreatment techniques on fruit and vegetable drying has been demonstrated. However, there is a lack of

classification and summary of different fruit types based on their physical or chemical characteristics to determine which pretreatment techniques are suitable for maximizing the drying promotion effect for specific fruit characteristics.

6. The research on drying efficiency is relatively limited, and it deserves more attention in future studies.

3.6 CONCLUDING REMARKS

Drying characteristics are the behavior and properties of a material during the drying process. An in-depth understanding of how materials change during the drying process can lead to efficient, controlled, and high-quality drying processes, resulting in improved product properties, reduced costs, and increased consumer satisfaction. A concise overview of the basics of drying characteristics and affecting factors is given. The relevant parameters for evaluating drying characteristics, drying kinetics, and mathematical models reported in literatures were involved. Three main factors, i.e., the characteristics of the fruit material, the drying process parameters, and the pretreatment method employed, influence the drying characteristics. However, the research on drying efficiency is very limited, especially the effect of product parameters on drying efficiency.

REFERENCES

1. Babu, A.K., G. Kumaresan, V.A.A. Raj, and R. Velraj, Review of leaf drying: Mechanism and influencing parameters, drying methods, nutrient preservation, and mathematical models. *Renewable & Sustainable Energy Reviews*, 2018, 90: 536–556.
2. Shinde, A., S. Das, and A.K. Datta, Quality improvement of orthodox and CTC tea and performance enhancement by hybrid hot air-radio frequency (RF) dryer. *Journal of Food Engineering*, 2013, 116(2): 444–449.
3. Singh, K.K., Development of a small capacity dryer for vegetables. *Journal of Food Engineering*, 1994, 21(1): 19–30.
4. Motevali, A., S. Minaei, M.H. Khoshtaghaza, and H. Amirnejat, Comparison of energy consumption and specific energy requirements of different methods for drying mushroom slices. *Energy*, 2011, 36(11): 6433–6441.
5. Shi, J., Z. Pan, T.H. McHugh, D. Wood, E. Hirschberg, and D. Olson, Drying and quality characteristics of fresh and sugar-infused blueberries dried with infrared radiation heating. *LWT-Food Science and Technology*, 2008, 41(10): 1962–1972.
6. Doymaz, I., Evaluation of some thin-layer drying models of persimmon slices (Diospyros kaki L.). *Energy Conversion and Management*, 2012, 56: 199–205.
7. Fadhel, M.I., R.A. Abdo, B.F. Yousif, A. Zaharim, and K. Sopian. *Thin-layer drying characteristics of banana slices in a force convection indirect solar drying*. in *Recent Researches in Energy and Environment – 6th IASME/WSEAS International Conference on Energy and Environment, EE'11*. 2011. World Scientific and Engineering Academy and Society.
8. Diamante, L., M. Durand, G. Savage, and L. Vanhanen, Effect of temperature on the drying characteristics, colour and ascorbic acid content of green and gold kiwifruits. *International Food Research Journal*, 2010, 17: 441–451.
9. Diamante, L.M., R. Ihns, G.P. Savage, and L. Vanhanen, Short communication: A new mathematical model for thin layer drying of fruits. *International Journal of Food Science & Technology*, 2010, 45(9): 1956–1962.

10. Cihan, A., and M.C. Ece, Liquid diffusion model for intermittent drying of rough rice. *Journal of Food Engineering*, 2001, 49(4): 327–331.
11. Erbay, Z., and F. Icier, A review of thin layer drying of foods: Theory, modeling, and experimental results. *Critical Reviews in Food Science and Nutrition*, 2010, 50(5): 441–464.
12. Luikov, A.V., Systems of differential equations of heat and mass transfer in capillary-porous bodies (review). *International Journal of Heat and Mass Transfer*, 1975, 18(1): 1–14.
13. Kucuk, H., A. Midilli, A. Kilic, and I. Dincer, A review on thin-layer drying-curve equations. *Drying Technology*, 2014, 32(7): 757–773.
14. El-Beltagy, A., G.R. Gamea, and A.H.A. Essa, Solar drying characteristics of strawberry. *Journal of Food Engineering*, 2007, 78(2): 456–464.
15. Akoy, E., Experimental characterization and modeling of thin-layer drying of mango slices. *International Food Research Journal*, 2014, 21(5): 1911–1917.
16. Vega, A., E. Uribe, R. Lemus, and M. Miranda, Hot-air drying characteristics of Aloe vera (Aloe barbadensis Miller) and influence of temperature on kinetic parameters. *LWT-Food Science and Technology*, 2007, 40(10): 1698–1707.
17. Meisami-asl, E., S. Rafiee, A. Keyhani, and A. Tabatabaeefar, Determination of suitable thin layer drying curve model for apple slices (variety-Golab). *Plant Omics*, 2010, 3(3): 103–108.
18. Zenoozian, M., H. Feng, S. Razavi, F. Shahidi, and H. Pourreza, Image analysis and dynamic modeling of thin-layer drying of osmotically dehydrated pumpkin. *Journal of Food Processing and Preservation*, 2008, 32: 88–102.
19. Sacilik, K., Effect of drying methods on thin-layer drying characteristics of hull-less seed pumpkin (Cucurbita pepo L.). *Journal of Food Engineering*, 2007, 79(1): 23–30.
20. Ergün, K., G. Çalışkan, and S.N. Dirim, Determination of the drying and rehydration kinetics of freeze dried kiwi (Actinidia deliciosa) slices. *Heat and Mass Transfer*, 2016, 52(12): 2697–2705.
21. Routray, W., and K. Rayaguru, Mathematical modeling of thin layer drying kinetics of stone apple slices. *International Food Research Journal*, 2012, 19(4): 1503–1510.
22. Ayadi, M., S.B. Mabrouk, I. Zouari, and A. Bellagi, Kinetic study of the convective drying of spearmint. *Journal of the Saudi Society of Agricultural Sciences*, 2014, 13(1): 1–7.
23. Demir, V., T. Gunhan, and A.K. Yagcioglu, Mathematical modelling of convection drying of green table olives. *Biosystems Engineering*, 2007, 98(1): 47–53.
24. Akpinar, E.K., Determination of suitable thin layer drying curve model for some vegetables and fruits. *Journal of Food Engineering*, 2006, 73(1): 75–84.
25. Zubernik, J., M. Dadan, J. Cichowska, and D. Witrowa-Rajchert, The impact of the pretreatment in ethanol solution on the drying kinetics and selected properties of convective dried apples. *International Journal of Food Engineering*, 2020, 16(1–2): 20180338.
26. Sahin, U., and H.K. Ozturk, Comparison between artificial neural network model and mathematical models for drying kinetics of osmotically dehydrated and fresh figs under open sun drying. *Journal of Food Process Engineering*, 2018, 41(5): e12804.
27. Zhu, G., G.S.V. Raghavan, W. Xu, Y. Pei, and Z. Li, Online machine vision-based modeling during cantaloupe microwave drying utilizing extreme learning machine and artificial neural network. *Foods*, 2023, 12(7): 1372.
28. Sharabiani, V.R., M. Kaveh, R. Abdi, M. Szymanek, and W. Tanas, Estimation of moisture ratio for apple drying by convective and microwave methods using artificial neural network modeling. *Scientific Reports*, 2021, 11(1): 1–12.
29. Sabzevari, M., N. Behroozi-Khazaei, and H. Darvishi, Real-time evaluation of artificial neural network-developed model of banana slice kinetics in microwave-hot air dryer. *Journal of Food Process Engineering*, 2021, 44(9): 1–14.

30. Sarkar, T., M. Salauddin, S.K. Hazra, and R. Chakraborty, Artificial neural network modelling approach of drying kinetics evolution for hot air oven, microwave, microwave convective and freeze dried pineapple. *SN Applied Sciences*, 2020, 2(9): 1–8.
31. Jahanbakhshi, A., M. Kaveh, E. Taghinezhad, and V.R. Sharabiani, Assessment of kinetics, effective moisture diffusivity, specific energy consumption, shrinkage, and color in the pistachio kernel drying process in microwave drying with ultrasonic pretreatment. *Journal of Food Processing and Preservation*, 2020, 44(6): 1–15.
32. Sun, Y.A., M. Zhang, A.S. Mujumdar, and D.X. Yu, Pulse-spouted microwave freeze drying of raspberry: Control of moisture using ANN model aided by LF-NMR. *Journal of Food Engineering*, 2021, 292: 110354.
33. Liu, W.C., M. Zhang, B. Bhandari, and D.X. Yu, A novel combination of LF-NMR and NIR to intelligent control in pulse-spouted microwave freeze drying of blueberry. *LWT-Food Science and Technology*, 2021, 137: 110455.
34. Song, C.F., Z. Cui, G.Y. Jin, A.S. Mujumdar, and J.F. Yu, Effects of four different drying methods on the quality characteristics of peeled Litchis (Litchi chinensis Sonn.). *Drying Technology*, 2015, 33(5): 583–590.
35. Al-Amri, A.M., M.A. Ismail, Y.A. Al Hassan, and E.A. Almuhanna, Effect of solar drying on I: Some physico-chemical properties of fruits of two date palm (Phoenix dactylifera L.) varieties. *Solar Energy*, 2021, 218: 425–434.
36. Wang, J., W.-S. Mu, X.-M. Fang, A.S. Mujumdar, X.-H. Yang, L.-Y. Xue, et al., Pulsed vacuum drying of Thompson seedless grape: Effects of berry ripeness on physicochemical properties and drying characteristic. *Food and Bioprocess Technology*, 2017, 106: 117–126.
37. Kaleta, A., and K. Gornicki, Evaluation of drying models of apple (var. McIntosh) dried in a convective dryer. *International Journal of Food Science and Technology*, 2010, 45(5): 891–898.
38. Shi, J., Z. Pan, T.H. McHugh, D. Wood, Y. Zhu, R.J. Avena-Bustillos, et al., Effect of berry size and sodium hydroxide pretreatment on the drying characteristics of blueberries under infrared radiation heating. *Journal of Food Science*, 2008, 73(6): 259–265.
39. Kian-Pour, N., and S. Karatas, Impact of different geometric shapes on drying kinetics and textural characteristics of apples at temperatures above 100 degrees C. *Heat and Mass Transfer*, 2019, 55(12): 3721–3732.
40. Li, Y., J. Bi, Y. Ma, S. Feng, Y. Guo, and J. Yi, Effects of punching and cutting epidermis pretreatment on freeze drying characteristics and qualities of blueberry. *Journal of Food Science and Technology (China)*, 2023, 41(2): 175–186.
41. Guida, V., G. Ferrari, G. Pataro, A. Chambery, A. Di Maro, and A. Parente, The effects of ohmic and conventional blanching on the nutritional, bioactive compounds and quality parameters of artichoke heads. *LWT-Food Science and Technology*, 2013, 53(2): 569–579.
42. Dandamrongrak, R., R. Mason, and G. Young, The effect of pretreatments on the drying rate and quality of dried bananas. *International Journal of Food Science and Technology*, 2003, 38(8): 877–882.
43. Doymaz, I., H. Demir, and A. Yildirim, Drying of quince slices: Effect of pretreatments on drying and rehydration characteristics. *Chemical Engineering Communications*, 2015, 202(10): 1271–1279.
44. Ando, Y., Y. Maeda, K. Mizutani, N. Wakatsuki, S. Hagiwara, and H. Nabetani, Impact of blanching and freeze-thaw pretreatment on drying rate of carrot roots in relation to changes in cell membrane function and cell wall structure. *LWT-Food Science and Technology*, 2016, 71: 40–46.
45. Sun, Z., L. Deng, T. Dai, M. Chen, R. Liang, W. Liu, et al., Steam blanching strengthened far-infrared drying of broccoli: Effects on drying kinetics, microstructure, moisture migration, and quality attributes. *Scientia Horticulturae*, 2023, 317: 112040.

46. Du, Z., Z. Gao, and S. Zhang, Research on convective heat transfer coefficient with air jet impinging. *Transactions of the Chinese Society of Agricultural Engineering*, 2006, 22(14): 1–4.
47. Bai, J.-W., D.-W. Sun, H.-W. Xiao, A.S. Mujumdar, and Z.-J. Gao, Novel high-humidity hot air impingement blanching (HHAIB) pretreatment enhances drying kinetics and color attributes of seedless grapes. *Innovative Food Science & Emerging Technologies*, 2013, 20: 230–237.
48. Singh, B., and A.K. Gupta, Mass transfer kinetics and determination of effective diffusivity during convective dehydration of pre-osmosed carrot cubes. *Journal of Food Engineering*, 2007, 79(2): 459–470.
49. Ramirez, C., E. Troncoso, J. Munoz, and J.M. Aguilera, Microstructure analysis on pre-treated apple slices and its effect on water release during air drying. *Journal of Food Engineering*, 2011, 106(3): 253–261.
50. Bassey, E.J., D.-W. Sun, O.J. Esua, and J.-H. Cheng, Effects of freeze-thaw pretreatments on the drying characteristics, physicochemical and phytochemical composition of red dragon fruit during mid- and near-infrared drying. *Drying Technology*, 2023, 41(4): 561–576.
51. Ciurzynska, A., H. Kowalska, K. Czajkowska, and A. Lenart, Osmotic dehydration in production of sustainable and healthy food. *Trends in Food Science & Technology*, 2016, 50: 186–192.
52. Rodrigues, S., and F.A.N. Fernandes, Dehydration of melons in a ternary system followed by air-drying. *Journal of Food Engineering*, 2007, 80(2): 678–687.
53. Ruiz-Lopez, I.I., I.R. Huerta-Mora, M.A. Vivar-Vera, C.E. Martinez-Sanchez, and E. Herman-Lara, Effect of osmotic dehydration on air-drying characteristics of chayote. *Drying Technology*, 2010, 28(10): 1201–1212.
54. de Oliveira, L.F., J.L. Gomes Correa, M.C. de Angelis Pereira, dL. Souza Ramos, and M.B. Vilela, Osmotic dehydration of yacon (Smallanthus sonchifolius): Optimization for fructan retention. *LWT-Food Science and Technology*, 2016, 71: 77–87.
55. Brochier, B., L.D. Ferreira Marczak, and C.P. Zapata Norena, Osmotic dehydration of yacon using glycerol and sorbitol as solutes: Water effective diffusivity evaluation. *Food and Bioprocess Technology*, 2015, 8(3): 623–636.
56. Teles, U., F. Fernandes, S. Rodrigues, A. Lima, G. Maia, and R. Figueiredo, Optimization of osmotic dehydration of melons followed by air-drying. *International Journal of Food Science & Technology*, 2006, 41(6): 674–680.
57. Nieto, A., M.A. Castro, and S.M. Alzamora, Kinetics of moisture transfer during air drying of blanched and/or osmotically dehydrated mango. *Journal of Food Engineering*, 2001, 50(3): 175–185.
58. An, K., H. Li, D. Zhao, S. Ding, H. Tao, and Z. Wang, Effect of osmotic dehydration with pulsed vacuum on hot-air drying kinetics and quality attributes of cherry tomatoes. *Drying Technology*, 2013, 31(6): 698–706.
59. Wang, L., B. Xu, B. Wei, and R. Zeng, Low frequency ultrasound pretreatment of carrot slices: Effect on the moisture migration and quality attributes by intermediate-wave infrared radiation drying. *Ultrasonics Sonochemistry*, 2018, 40: 619–628.
60. Serratosa, M.P., A. Lopez-Toledano, M. Medina, and J. Merida, Drying of Pedro Ximenez grapes in chamber at controlled temperature and with dipping pretreatments. Changes in the color fraction. *Journal of Agricultural and Food Chemistry*, 2008, 56(22): 10739–10746.
61. Esmaiili, M., R. Sotudeh-Gharebagh, K. Cronin, M.A.E. Mousavi, and G. Rezazadeh, Grape drying: A review. *Food Reviews International*, 2007, 23(3): 257–280.
62. Doymaz, I., Effect of dipping treatment on air drying of plums. *Journal of Food Engineering*, 2004, 64(4): 465–470.

63. Tarhan, S., Selection of chemical and thermal pretreatment combination for plum drying at low and moderate drying air temperatures. *Journal of Food Engineering*, 2007, 79(1): 255–260.
64. Sledz, M., A. Wiktor, M. Nowacka, and D. Witrowa-Rajchert, Drying kinetics, microstructure and antioxidant properties of basil treated by ultrasound. *Journal of Food Process Engineering*, 2017, 40(1): e12271.
65. Rojas, M.L., P.E.D. Augusto, and J.A. Carcel, Ethanol pre-treatment to ultrasound-assisted convective drying of apple. *Innovative Food Science & Emerging Technologies*, 2020, 61: 102328.
66. Carneiro da Cunha, R.M., S.C. Rupert Brandao, R.A. Batista de Medeiros, d. Silva Junior, J.H. Fernandes da Silva, and P.M. Azoubel, Effect of ethanol pretreatment on melon convective drying. *Food Chemistry*, 2020, 333: 127502.
67. de Arruda, G.M.P., S.C.R. Brandao, E.V. da Silva Junior, E.M. da Silva, Z.M.P. Barros, E.S. da Silva, et al., Influence of ultrasound and ethanol as a pretreatment on papaya infrared and convective drying characteristics and quality parameters. *Journal of Food Process Engineering*, 2023, 46(3): 1–10.
68. Gomes Correa, J.L., A.M. Pinheiro Braga, M. Hochheim, and M.A. Silva, The influence of ethanol on the convective drying of unripe, ripe, and overripe bananas. *Drying Technology*, 2012, 30(8): 817–826.
69. Ozdemir, Y., A. Ozturk, and T. Senem, Effect of two dipping pretreatment on drying kinetics of golden berry (Physalis peruviana L.). *African Journal of Agricultural Research*, 2016, 11: 40–47.
70. Corona, O., F. Torchio, S. Giacosa, S.R. Segade, D. Planeta, V. Gerbi, et al., Assessment of postharvest dehydration kinetics and skin mechanical properties of "Muscat of Alexandria" grapes by response surface methodology. *Food and Bioprocess Technology*, 2016, 9(6): 1060–1069.
71. Yanclo, L.A., G. Sigge, Z.A. Belay, F. October, and O.J. Caleb, Microstructural, biochemical and drying characteristics of dehydrated 'Sunectwentyone' nectarines as affected by sodium metabisulphite. *Food Science and Biotechnology*, 2022, 31(3): 311–322.
72. Zhu, Y., Z. Pan, and T.H. McHugh, Effect of dipping treatments on color stabilization and texture of apple cubes for infrared dry-blanching process. *Journal of Food Processing and Preservation*, 2007, 31(5): 632–648.
73. Rodriguez, O., V. Eim, C. Rossello, A. Femenia, J.A. Carcel, and S. Simal, Application of power ultrasound on the convective drying of fruits and vegetables: Effects on quality. *Journal of the Science of Food and Agriculture*, 2018, 98(5): 1660–1673.
74. Fan, K., M. Zhang, and A.S. Mujumdar, Application of airborne ultrasound in the convective drying of fruits and vegetables: A review. *Ultrasonics Sonochemistry*, 2017, 39: 47–57.
75. Nowacka, M., A. Wiktor, M. Sledz, N. Jurek, and D. Witrowa-Rajchert, Drying of ultrasound pretreated apple and its selected physical properties. *Journal of Food Engineering*, 2012, 113(3): 427–433.
76. Tao, Y., and D.-W. Sun, Enhancement of food processes by ultrasound: A review. *Critical Reviews in Food Science and Nutrition*, 2015, 55(4): 570–594.
77. Ade-Omowaye, B.I.O., A. Angersbach, K.A. Taiwo, and D. Knorr, Use of pulsed electric field pre-treatment to improve dehydration characteristics of plant based foods. *Trends in Food Science and Technology*, 2001, 12(8): 285–295.
78. Yu, Y., T.Z. Jin, X. Fan, and J. Wu, Biochemical degradation and physical migration of polyphenolic compounds in osmotic dehydrated blueberries with pulsed electric field and thermal pretreatments. *Food Chemistry*, 2018, 239: 1219–1225.
79. Parniakov, O., O. Bals, N. Lebovka, and E. Vorobiev, Pulsed electric field assisted vacuum freeze-drying of apple tissue. *Innovative Food Science & Emerging Technologies*, 2016, 35: 52–57.

80. Wiktor, A., M. Nowacka, M. Dadan, K. Rybak, W. Lojkowski, T. Chudoba, et al., The effect of pulsed electric field on drying kinetics, color, and microstructure of carrot. *Drying Technology*, 2016, 34(11): 1286–1296.
81. Hulle, N.R.S., and P.S. Rao, Effect of high pressure pretreatments on structural and dehydration characteristics of Aloe vera (Aloe barbadensis Miller) cubes. *Drying Technology*, 2016, 34(1): 105–118.
82. Yucel, U., H. Alpas, and A. Bayindirli, Evaluation of high pressure pretreatment for enhancing the drying rates of carrot, apple, and green bean. *Journal of Food Engineering*, 2010, 98(2): 266–272.
83. Zhang, L., Y. Qiao, C. Wang, L. Liao, D. Shi, K. An, et al., Influence of high hydrostatic pressure pretreatment on properties of vacuum-freeze dried strawberry slices. *Food Chemistry*, 2020, 331: 127203.
84. Verma, D., N. Kaushik, and P.S. Rao, Application of high hydrostatic pressure as a pretreatment for osmotic dehydration of Banana slices (Musa cavendishii) Finish-dried by dehumidified air drying. *Food and Bioprocess Technology*, 2014, 7(5): 1281–1297.
85. Santos, N.C., R.L. Jacinto Almeida, G.M. da Silva, S.S. Monteiro, V.H. de Alcantara Ribeiro, A.P. de Franca Silva, et al., Influence of high hydrostatic pressure (HHP) pre-treatment on plum (Prunus salicina) drying: Drying approach, physical, and morpho-structural properties of the powder and total phenolic compounds. *Journal of Food Processing and Preservation*, 2022, 46(11): e16968.
86. Zhang, S., Y. Zhao, X. Yao, Z. Zheng, C. Zheng, and Z. Jiang, Effect of high hydrostatic pressure pretreatment on flavour and physicochemical properties of freeze-dried carambola slices. *International Journal of Food Science and Technology*, 2022, 57(7): 4245–4253.
87. Ayala-Aponte, A., L. Serna-Cock, J. Libreros-Triana, C. Prieto, and K. Di Scala, Influence of osmotic pre-treatment on convective drying of yellow pitahaya X1 – Influencia de un pre-tratamiento osmótico sobre el secado convectivo de pitahaya amarilla. *DYNA*, 2014, 81(188): 145–151.
88. Chen, Q., J. Bi, X. Wu, J. Yi, L. Zhou, and Y. Zhou, Drying kinetics and quality attributes of jujube (Zizyphus jujuba Miller) slices dried by hot-air and short- and medium-wave infrared radiation. *LWT – Food Science and Technology*, 2015, 64(2): 759–766.
89. Kouhila, M., N. Kechaou, M. Otmani, M. Fliyou, and S. Lahsasni, Experimental study of sorption isotherms and drying kinetics of Moroccan Eucalyptus globulus. *Drying Technology*, 2002, 20(10): 2027–2039.
90. Lahsasni, S., M. Kouhila, M. Mahrouz, and J.T. Jaouhari, Drying kinetics of prickly pear fruit (Opuntia ficus indica). *Journal of Food Engineering*, 2004, 61(2): 173–179.
91. Duarte, Y., A. Chaux, N. Lopez, E. Largo, C. Ramirez, H. Nunez, et al., Effects of blanching and hot air drying conditions on the physicochemical and technological properties of yellow passion fruit (Passiflora edulis Var. Flavicarpa) by-products. *Journal of Food Process Engineering*, 2017, 40(3): e12425.
92. Dai, J.-W., H.-W. Xiao, L.-H. Zhang, M.-Y. Chu, W. Qin, Z.-J. Wu, et al., Drying characteristics and modeling of apple slices during microwave intermittent drying. *Journal of Food Process Engineering*, 2019, 42(6): e13212.
93. Kipcak, A.S., and I. Doymaz, Mathematical modeling and drying characteristics investigation of black mulberry dried by microwave method. *International Journal of Fruit Science*, 2020, 20: 1222–1233.
94. Puspasari, I., S.M. Tasirin, and Y. Gariépy, Quality changes of red pitaya (Hylocereus undatus) slices dried in hot air, microwave-hot air and microwave-vacuum dryers. *Iranica Journal of Energy & Environment*, 2014, 5(3): 313–322.
95. Liu, Y., Y. Sun, H. Yu, Y. Yin, X. Li, and X. Duan, Hot air drying of purple-fleshed sweet potato with contact ultrasound assistance. *Drying Technology*, 2017, 35(5): 564–576.

96. Aydin, D., S.E. Ezenwali, M.Y. Alibar, and X. Chen, Novel modular mixed-mode dryer for enhanced solar energy utilization in agricultural crop drying applications. *Energy Sources Part A-Recovery Utilization and Environmental Effects*, 2021, 43(16): 1958–1974.
97. Taghinezhad, E., M. Kaveh, and A. Szumny, Thermodynamic and quality performance studies for drying kiwi in hybrid hot air-infrared drying with ultrasound pretreatment. *Applied Sciences-Basel*, 2021, 11(3): 1297.
98. Hazervazifeh, A., P.A. Moghaddam, and A.M. Nikbakht, Microwave dehydration of apple fruit: Investigation of drying efficiency and energy costs. *Journal of Food Process Engineering* 2017, 40(3): e12463.
99. Raj, G.V.S.B., and K.K. Dash, Effect of intermittent microwave convective drying on physicochemical properties of dragon fruit. *Food Science and Biotechnology*, 2022, 31(5): 549–560.
100. Dash, K.K., and G.V.S. Bhagya Raj, Ultrasound assisted microwave vacuum drying of persimmon fruit: Modeling by artificial neural network and optimization by genetic algorithm. *Journal of Food Process Engineering*, 2023, 46(6): 1–11.

4 Quality of Dried Fruit Products

Maria Lucia Masson
Graduate Program of Food Engineering, Chemical Engineering Department, Federal University of Paraná
Curitiba, PR, Brazil

4.1 INTRODUCTION

Fruits were historically classified in accordance with these sensory attributes, such as appearance, color, and flavor; today, this is changing since consumers are looking for new properties; same happens to quality concepts for consumers and industry. The importance of fruit components is evolving more, and concepts related to health and wellness, such as looking for flavonoids, carotenoids, vitamins, and nutraceuticals, until they are called superfruits, are society's goals. Health issues go from people suffering from obesity and/or excess calories intake to a lack of nutrients and hunger, while wellness refers to desire for new aromas and flavors for consumers. Superfruits are fruits rich in nutritional qualities in an exceptional way, and we can cite as examples grape, guava, kiwi fruit, cranberry, jujube, and more. Those desirable characteristics need to be maintained after processing, and drying is one industrial process presenting high importance since it makes possible to extend shelf life by removing moisture, reducing microbial deterioration, and deteriorating reactions (Boateng, 2022).

According to FAO (2016) reports, within the next 10 years, annual production of the major exotic fruits is projected to increase by 3% per year, reaching approximately 116 million tons in 2024, and half of this production will be discarded as waste – peels and seeds underutilized. Tan et al. (2022) propose drying and bio-drying to valorize fruits and by-products. Drying can cause undesirable visual appearance, texture, flavor, odor, and color changes, which are incompatible with today's consumers' growing demand for premium quality dried products.

Dried fruit products (DFP) are a popular food product all around the world. Their quality assessment is a particularly key step for their secure commercialization. In this sense, many different quality analyses have been carried out in scientific reports over the last few years. Each food product needs to be dried in another way, using appropriate pre- and post-processing step(s) such as osmotic dehydration, blanching, salting, and soaking together with proper dryer type to add satisfactory value upon drying (Mujumdar, 2008; Mujumdar, 2015). Regarding fruits, those issues are critical, because most of their quality value is connected to sensory characteristics; nutritional analyses are considered important, but consumers look for sensory quality at first because they consider that sensory is an indication of nutritional and safety quality. And fruits are not produced at places where they are consumed; production

DOI: 10.1201/9781003402916-4

is far from the market area in general. This fact, together with the high perishability and moisture, leads to quality as dependent on several factors, including pre- and postharvest, transportation, packaging, commercialization, and preservation at home. Drying is an ancient process, one of those applied to foods that present more variation and innovation in techniques, and this is assumed to be possible to attend consumers' demands around the world with quality and safety.

4.2 PHYSICAL PROPERTIES AND THEIR EVOLUTION DURING DRYING

Properties of dried foods are classified into three categories: physical, chemical/nutritional, and microbiological. Physical properties are first to be perceived by human senses – perception of size, color, texture, and others. Those properties can be influenced by many factors, such as temperature and location (mass is a property dependent on Earth's gravitational acceleration). These variations need a reliable method to measure the properties of dried foods in general, DFP. Some organizations assume the responsibility to uniformalize and control methods of standards for all countries, and we present some in Table 4.1 (Figura & Teixeira, 2007).

4.2.1 Color

Color was defined by Francis and Clydesdale (1975) as the sensation experienced when energy, in the form of radiation within the visible spectrum, falls upon the retina of the eye, but this definition does not give the full picture. Color is considered an approximate quality because models cannot assure constancy in this area, reason because instrumental methods are used to measure color in standard conditions.

TABLE 4.1
Metrological Institutes around the World

ISO	International Organization for Standardization	ISO – International Organization for Standardization
NIST	National Institute of Standards and Technology	National Institute of Standards and Technology (nist.gov)
PTB	Physikalisch-Technische-Bundesanstalt	Physikalisch-Technische Bundesanstalt (ptb.de)
EURAMET	The European Association of National Metrology Institutes	Metrologie in Europa (PTB.de)
AIST	National Institute of Advanced Industrial Science and Technology (Japan)	AIST: 産業技術総合研究所
NML	National Metrology Laboratory (South Africa)	CSIR I
INMETRO	Instituto Nacional de Metrologia, Qualidade e Tecnologia	http://www.inmetro.gov.br/metlegal/metbrasil.asp

There are a wide variety of methods and instruments for measuring color that can be used by professionals in the food industry and in scientific research. The most commonly used techniques involve tristimulus colorimetry and reflectance spectrophotometry and are non-destructive. Destructive methods involve extraction of colorant pigments to evaluate spectrophotometrically or using high-performance liquid chromatography. Color is a physical property highly connected to chemical composition. A variety of natural dyes are present in composition of fruits and are responsible for color and nutritional importance as the source of nutrients, minerals, vitamins, and some other antioxidants such as flavonoids, carotenoids, chlorophyll, and a plethora of other good components for health (Jiang, Zhang, Mujumdar & Duan, 2010).

Krokida et al. (2001) stated that the color parameters L*, a*, and b* of dried banana and apple were significantly affected by convective, vacuum, and microwave drying techniques, meanwhile the same fruits preserved their original color after freeze and osmotic drying.

To see color, one requires to use a light source, an object, and an observer (Hunter Lab, 2015). Color is the first attribute perceived by consumers in food, and the color of DFP can be influenced by reactions like enzymatic browning and non-enzymatic browning (Reis et al., 2017). Color is composed of luminosity and chromaticity, and chromaticity is composed of hue and Chroma or saturation. Gloss, translucency, and transparency are properties related to color, but they are really properties of appearance (Caivano & Buera, 2012).

Studies dealing with color measurement of DFP show that the CIE L*a*b* color scale is commonly used for expressing their color (Al Maiman et al., 2021; Kumar et al., 2021; Li et al., 2019). CIE l*a*b* scale contains the L* (lightness) axis with a range between 0 and 100, in which 0 is black, 100 is white; parameter a* variates from red to green colors, in which range positive values are red, negative values are green, and the parameter b* (blue to yellow) variates as positive values are yellow, negative values are blue. It is usual to use the D65 illuminant, which represents noon daylight, and the 10° observer that best correlates with average visual assessments made with large fields of view, typical of most commercial applications. A parameter that is also very used is the total color difference, calculated as follows (Figura & Teixeira, 2007; Hunter Lab, 2015; Jiang, Zhang, Mujumdar & Duan, 2010; Reis et al., 2022):

$$\Delta E = \left[\left(\Delta L^*\right)^2 + \left(\Delta a^*\right)^2 + \left(\Delta b^*\right)^2\right]^{\frac{1}{2}} \quad \text{(Eq. 4.1)}$$

Where L* is lightness, a* is redness, and b* is yellowness. Other parameters can be analyzed as color intensity or Chroma C^* and hue angle h^* one qualitative attribute of color shades, and can be calculated by:

$$C^* = \left(a^{*2} + b^{*2}\right)^{\frac{1}{2}} \quad \text{(Eq. 4.2)}$$

$$h^* = \arctan\frac{b^*}{a^*} \quad \text{(Eq. 4.3)}$$

Reflectance spectrophotometry works in the entire visible wavelength range, measuring all the light reflected by the surface, and can graphically analyze the data obtained, resulting in what the perceived color should be. Reflectance is usually expressed as a fraction, with the ideal white tile having a value of 1 and black tile, which absorbs all incident light, having a reflectance of 0; colors fall between these two values.

Standard methodology officially accepted can be found at

- ISO/CIE 11664-1:2019 | EN ISO/CIE 11664-1:2019 Colorimetry — Part 1: CIE standard colorimetric observers,
- ISO/CIE 11664-2:2022 | EN ISO/CIE 11664-2:2022 Colorimetry — Part 2: CIE standard illuminants,
- ISO/CIE 11664-3:2019 | EN ISO/CIE 11664-3:2019 Colorimetry — Part 3: CIE tristimulus values,
- ISO/CIE 11664-4:2019 | EN ISO/CIE 11664-4:2019 Colorimetry — Part 4: CIE 1976 L*a*b* color space and ISO/CIE 11664-6:2022
- EN ISO/CIE 11664-6:2022 Colorimetry — Part 6: CIEDE2000 color-difference formula BL1>Color Analysis (ASTM E1347 with ASTM D2244),
- ASTM E313 WI Whiteness Index,
- ASTM D6866 or ISO 16620-2, ASTM E313-20 - Standard Practice for Calculating Yellowness and Whiteness Indices from Instrumentally Measured Color Coordinates.

Two sets of color-matching functions are specified: color-matching functions for the CIE 1931 standard colorimetric observer and color-matching functions for the CIE 1964 standard colorimetric observer."

One example is the Browning Indices (BI) employed to verify the effects of Maillard reactions, ascorbic acid oxidation, and/or enzymatic reactions. L* can be used to measure these effects, or the relation (Ruangchakpet & Sajjaanantakul, 2007):

$$BI = \frac{\left[100\left(\frac{a^* + 1,75L^*}{5,645L^* + a^* - 0,3012b^*} - 0,31\right)\right]}{0,17} \quad \text{(Eq. 4.4)}$$

Some case studies about drying fruits are presented using color measure for quality property analysis parameter. We can cite some of them here, having in focus that there is a straight relationship between the drying processes applied, temperature, and the final state of DFP (in pieces or powder).

Antoniolli et al. (2023) evaluated the stability of mango peel flakes obtained by drum drying at a temperature of 146°C/15 s with or without addition of 5% dry basis (d.b.) of regular corn starch and 1% d.b. of glycerol monostearate. They evaluated color measuring parameters a*.b* and L* and calculated C* and h^0. Products dried were lighter (L*) than fresh, and products without additives were darker than those with addition.

Li et al. (2023) evaluated quality of kiwifruit slices using treatment with infrared-assisted (IR-TTAID) tilted tray air impingement drying (TTAID). Color was measured, and ΔE was calculated for all treatments. Slices presented a thickness of 5 mm and a diameter of 42 mm. Kiwifruit slices were dried by IR-TTAID and TTAID alone at 50, 60, and 70°C. In the two drying methods, the a* values decreased and the ΔE values increased with the increase in drying temperatures. Authors indicated that IR-TTAID samples had higher a* values and lower ΔE values than TTAID samples. Due to elevated temperature and time for drying process, the color of kiwifruit slices was more prone to deterioration due to chlorophyll degradation and possible Maillard reaction.

Yanclo et al. (2022) described the effects of pretreatment for slices of "Sunectwentyone" nectarine (6.5 mm thick) using sodium metabisulfite before drying in a dehydrator tunnel. They measured the total difference of color, Chroma, and hue angle for slices with or without pretreatment. They concluded that browning was the main factor influencing values of parameters of color, strongly affecting shelf-life and consumers' purchase decisions. Other factors affecting color of nectarine slices were degradation of pigments and the browning and non-enzymatic browning.

Ashtiani et al. (2022) used cold plasma pretreatment followed by ultrasound-assisted convective drying to goldenberry and measured visual color values a*, b*, and L* to calculate ΔE between fresh and dried samples. For all samples, L* and b* decreased and a* increased; those results were attributed to browning reactions, like Maillard, and thermal degradation reactions of pigments. For samples treated with plasma, it minimizes darkening and color maintenance. ΔE values were higher than 5 for all drying conditions for the color difference between fresh and processed samples. The higher value for ΔE was obtained for sample with no pretreatment, only dried at 50°C and airspeed 1 m/s while the minor value was measured for sample treated with cold plasma and ultrasound before convective drying at 60°C.

Garcia et al. (2022) summited papaya slices to blanching and drying at two levels of temperature: 60°C and 70°C; they measured L*, a*, and b* to calculate C* and h^0. Blanching was responsible for resulting in significant decreases in the L*, a*, and b* values, thus reducing the intensity of C* but without affecting the tonality h^0. This suggests that blanching did not provide color stability for the samples. Fresh papaya was more stable than blanched samples during drying, which could be attributed to natural protection composed of latex secreted from the tissue after cutting, according to the authors.

Hot air drying of persimmons led to an increase in a*, which was attributed to Maillard reaction. Blanching of persimmons attenuated this effect (Oshima et al., 2021). Summarizing, the most common changes in fruit and vegetable color during drying are ascribed to browning reactions and natural pigment degradation, which can be attributed to the heat used for drying. New techniques such as blanching, osmotic dehydration, and innovative and mild drying methods are employed to improve quality of color for DFP pieces.

Rajoriya et al. (2020) dried apple slices using a far infrared-assisted refractance window at 50 and 60°C and hot air drying at 90°C. They measure L*, a*, and b* coordinates to calculate the whiteness index (WI) and browning index (BI).

Values for L* were lower for hot air drying than for refractance window; an increase of 14 and 16% in WI values was observed in the case of refractance window and far infrared-assisted refractance window, respectively, compared to hot air drying.

Apple slices dried by refractance window and hot air drying presented an increase in L*, a decrease in a*, and an increase in b*. In addition, the use of higher drying temperatures and osmotic dehydration coupled with a moderate electric field increased the value of ΔE (from fresh samples) (Hernández et al., 2020).

Orange slices presented lower total color difference (ΔE) from fresh samples when dried by vacuum microwave drying as compared to vacuum infrared drying and hot air drying (Bozkir, 2020).

Hot air drying, vacuum drying, and hot air drying followed by microwave vacuum drying produced dried kiwifruits with similar color (increased a* and b* and similar L* when compared to fresh samples), while freeze-drying yielded a product with a lighter color (Akar & Barutçu Mazı, 2019).

Jujube slices processed by Pulsed Air-Jet Impingement Drying presented a decrease in L*, a*, and b* with drying time, which was attributed to browning (Cao et al., 2018).

Mango slices dried by conductive multiflash drying (KMFD) with 14 heating vacuum pulse cycles and 2.55 kPa s^{-1} pressure drop rate and by KMFD with 4 cycles followed by vacuum drying (KMFD-VD) presented color closer to fresh mangoes, i.e., low values of total color difference (ΔE) (Link et al., 2018). These results were attributed to the degradation of carotenoids.

4.2.2 Apparent or Bulk Density

The density is a property of Materia expressed by the relation to mass of a solid per unit volume and is quite relevant for determining other properties such as bulk structure, particle size, shrinkage, and a correlation to porosity. It's an important property for DFP quality evaluation and engineering calculations.

Three types of density can be defined depending on how total volume is measured. The *true* particle density, the *apparent* particle density, and the *effective* (or aerodynamic) particle density are defined because particles can present cracks, flaws, hollows, and closed pores. And we cannot forget about overrun (a term used for products involving whipping or aeration) and density of foams, which are so important to the quality of products obtained through techniques like foam-mat drying, spray drying, and freeze-drying.

True particle density represents the mass of the particle divided by its volume, excluding open and closed pores, and is the density of the solid material of which the particle is made. For pure chemical substances, organic or inorganic, this is the density quoted in reference books with physical/chemical data.

However, in porous media like DFP, more than one type of density can be defined. Apparent or bulk density was the property of choice for expressing density of DFP pieces and fruits in powder. It can be defined as:

$$\rho_b = m/V_b \quad \text{(Eq. 4.5)}$$

where ρ_b is the apparent density (g.cm^3), m is the sample mass (g), and V_b is the bulk volume (cm^3). Density is expressed in SI (standard system of units) in kg/m^3, but g/cm^3 is a unit found in literature because 1000 cm^3 is equal to 1 liter, a popular measure. Specific volume is the reciprocal of density, and units are m^3/kg:

$$\vartheta = \frac{V}{m} = \frac{1}{\rho} \quad \text{(Eq. 4.6)}$$

The volume of fruits or other materials increases with temperature, so density decreases. We can calculate the changes in volume caused by temperature using the thermal expansion coefficient:

$$\gamma = \frac{1}{V}\frac{dV}{dT} \quad \text{(Eq. 4.7)}$$

where γ is thermal expansion coefficient in K^{-1} (Figura & Teixeira, 2007).

Control or measured for process control and quality control purposes, and recording of temperature is necessary when density is measured to describe this temperature dependency of density.

Fruits are considered as incompressible as one solid and, when subjected to high pressure, present an elastic behavior. If we consider high pressure process (100 bar), compressibility of water cannot be negligible, and energy of compression is converted to heat.

Specific density, or relative density (in the USA and Canada, specific gravity is used), is the ratio between absolute density and the density of a reference material, usually water at 4°C or water at 20°C (Figura & Teixeira, 2007).

Techniques that can be used to measure density are based on the weight of the mass of one volume known. Pycnometers are used to measure volume of liquids, and are constructed as bulbs with precisely known volume at a temperature of calibration for the materials used by pycnometers. To measure the absolute density, they must be used at a calibration temperature. To measure a relative density, the same pycnometer must be used to material and reference liquids. Fluctuations in atmospheric conditions make it necessary to weigh the true mass that is slightly higher than the mass that the balance displays, and it is used for true density. Without this correction, we have an apparent mass and an apparent density (Figura & Teixeira, 2007).

Hydrostatic balances follow the Archimedes principle; this technique can be applied using a Becker filled with water and submerged fully in a solid, and the weight will be displaced.

We can use it to calculate:

$$\rho_{solid} = \frac{m_{air}}{m_{air} - m_{fluid}} \rho_{fluid} \quad \text{(Eq. 4.8)}$$

Where ρ_{solid} is density of solid body in kg.m^{-3}, m_{air} is mass of body in air, m_{fluid} is mass of body submerged in fluid in kg, and ρ_{fluid} is density of fluid (Figura & Teixeira, 2007).

Powders and bulk products can present voids filled with air or any other gas. The density of this kind of material is the bulk density. The bulk material can be tapped before reading the volume because the solid particles will "settle" into the most stable situation they can reach – the void spaces will be smaller into a spatial situation, getting a maximum value for bulk density, or maximum bulk density that can be achieved without deformation of the particles. The time needed to reach this maximum depends on tapping speed and tapping amplitude – this is called tapped bulk density. Powders can be characterized for this property by the Hausner ratio, which is the quotient of tapped bulk density over untapped bulk density and bulk density and maximum tapped bulk density difference helps to evaluate the behavior of material under pressure (Figura & Teixeira, 2007).

Overrun is an important process parameter that is related to the density and foamability. Foaming causes an increase in volume of the foamed liquid far beyond the original volume of the liquid prior to whipping. This volume increase divided by the original volume, is called overrun. This can be expressed by:

$$A = \frac{V_{foam} - V_{liquid}}{V_{liquid}} \quad \text{(Eq. 4.9)}$$

Or:

$$A = \frac{\rho_{liquid}}{\rho_{foam}} - 1 \quad \text{(Eq. 4.10)}$$

Where A is overrun, V is volume in m^3, and ρ is density in $kg \cdot m^{-3}$ (Figura & Teixeira, 2007).

Some methodologies used to measure density are:

- **ISO 8130-3 –** Determination of density by liquid displacement pycnometer
- **ISO 7971-2:2019 –** Cereals Determination of bulk density, called mass per hectolitre – Part 2: Method of traceability for measuring instruments through reference to the international standard instrument
- **ISO 17034**-certified reference material: Density standards
- **ISO 8967 | IDF 134: 2005 –** Dried milk and dried milk products – Determination of bulk density
- **ASTM D792, ISO 1183 –** Density and Specific Gravity
- **ISO 4787:2021(en)** Laboratory glass and plastic ware – Volumetric instruments – Methods for testing of capacity and for use
- **ASTM D1481-17** Standard Test Method for Density and Relative Density (Specific Gravity) of Viscous Materials by Lipkin Bicapillary Pycnometer
- **ASTM D 1480-02e** Standard Test Method for Density and Relative Density (Specific Gravity) of Viscous Materials by Bingham Pycnometer
- **ASTM C128-04a** Standard Test Method for Density, Relative Density (Specific Gravity), and Absorption of Fine Aggregate
- **AOAC method 932.14 B** Solids and Syrups

- **AOAC 925.22-1925** Specific Gravity of milk. Pycnometer method
- **ISO 6883:2017** Animal and vegetable fats and oils – Determination of conventional mass per volume (liter weight in air)

The density of fresh and dried papayas was determined in four replicates by a volume dislocation technique using a Gehaka kit (Gehaka, BK-300 model, São Paulo, Brazil), which consists of an apparatus coupled with a semi-analytical balance. A 250 mL beaker was filled with the displacement fluid (methanol) and placed on the apparatus to measure the masses. Density was applied to measure shrinkage of fresh and dried or blanched dried papaya. (Garcia et al., 2022).

Chang et al. (2021) studied "kedondong" (*Spondias cytherea Sonnerat*) production in powder employing different drying methods. "Kedondong" is a tropical fruit also known as golden apple or ambarella. They dried the fruit by the following methods: conventional oven drying (48 h, 45°C), vacuum drying (24 h, 70°C, 100 mbar), spray drying (1 h, 170°C), drum drying (1 h, 130–140°C, rotate at 2 rpm), and freeze-drying (48 h, –55°C, vacuum 0.28 mbar). Bulk density and tapped density were measured and their results were respectively: 0,70±0.02 g/mL and 0,86±0.1 g/mL for conventional oven, 0.75±0.03 g/mL and 0.88 ±0.02 g/mL for vacuum drying, 0.32±0.01 g/mL and 0.38±0.03 g/mL for spray drying, 0.49±0.01 mg/L and 0.65±0.02 mg/L for drum drying and 0.55±0.02 mg/L and 0.70±0.02 mg/L for freeze-drying samples. Bulk densities of samples dried in oven and vacuum drying were higher than samples submitted to spray drying and freeze-drying.

Dried products presenting high bulk density are considered of low quality, while dried products presenting high porosity and low bulk density are considered of high quality (Rodríguez-Ramírez et al., 2012). High bulk density can occur because it fails in drying, causing a disruption of tissues and pores to be closed. Tapped density was highest for oven and vacuum drying while it was the lowest for spray drying samples. Authors justified the reasons for those results as elevated temperature that collapses pores for oven and vacuum drying, while elevated temperature used at spray drying did not alter structures of particles presenting intra-particle spaces. Another reason can be attributed to the high moisture content of the powder.

Shinde & Ramaswamy (2023) studied the drying of mango cubes using a combination of microwave-osmotic-vacuum processes. Bulk density variates from 214 to 261 kg.m^{-3}, and they found a correlation between bulk density and temperature, showing a positive effect during the drying process, and the power level for microwave vacuum-dried samples showed a negative effect.

When KMFD was combined with vacuum drying, the lowest values of bulk density were observed in mango slices when compared to KMFD alone (Link et al., 2018).

Pineapples pretreated by osmotic dehydration, ultrasound, and pectin-acid coating followed by hot air drying presented higher bulk density than control samples (without pretreatment) (Sakooei-Vayghan et al., 2020).

Jujube slices presented lower bulk density when dried by freeze-drying, followed by pressure-differential puffing drying (PPD) and hot air drying (Wang et al., 2021).

Bananas dried in a hybrid-solar-vacuum dryer presented the lowest values of apparent density when experiments were carried out without the use of a vacuum reservoir (Roratto et al., 2021).

In general, low apparent density is usually related to higher porosity and crispness, which are desirable aspects of DFP. Lower bulk density in snack foods is desirable since it suggests that the volume of the food may expand and produce a porous structure. Another physical quality criterion, volumetric expansion, assists in understanding the swelling degree of dried food products.

4.2.3 Porosity

Porosity (ε) is a property directly related to bulk density and to microstructure of DFP; it is defined as the reason between void volume (V_v) and total volume (V_t):

$$\varepsilon = 1 - \frac{V_v}{V_t}$$

In hygroscopic materials such as fruit tissues, voids can be partially or fully filled with gas, water, or oil. In fruits, these pores are full of gas and liquid. This property of fruits changes during processing by drying, and by understanding how this property evolves from its native state, we can work with characterization of the quality.

Porosity is susceptible to changes during pretreatment processes and is influenced by pressure and temperature employed at drying process, which impact the final quality of DFP. Blanching involves rapid heating with many goals, among those to preserve tissue structure and improve food processes by denaturing cells. Intensity of time-temperature treatment might destroy or preserve native structure of tissue. Sequences of low and high temperatures have been found as effective in improving porosity. According to Dadmohammadi and Datta (2022), many pretreatments have effect on final product, according to Table 4.2.

4.2.4 Microstructure

Food microstructure is comprised of microscale and nanoscale pores trapped in a cellular matrix, and it is comprised of biopolymers, solutes, and fluids in which pores are distributed at microscale (cell wall), mesoscale (cell cytoplasm), and macroscale (tissue) (Dadmohammadi & Datta, 2022). Microstructure suffers the influence of drying process conditions, and all those changes are measured by porosity alterations, texture alterations, shrinkage, and visual measurements. Alterations can be considered a lack of quality when changes show collapse of tissues, but those possible defects can be controlled by application of knowledge about new processes.

Knowledge of the microstructure of DFP can provide important data for understanding their quality on macroscale (Reis et al., 2022). Microstructure and pores can also be induced in tissue employing puff-technologies or foam elaboration. Studies on this theme have been using techniques like scanning electron microscopy (Farhaninejad et al., 2017; Hernández et al., 2020; Li et al., 2021; Roratto et al., 2021; Vallespir et al., 2019; Wang et al., 2021; Yang et al., 2019), transmission electron microscope (Zhang et al., 2018), light microscope (Zhou et al., 2021), and X-ray

TABLE 4.2
Effects of Pretreatments and Processes on Final Product

Pretreatment/Process	Fruit	Main Alteration	Evolution of Effective Porosity	Reference
Ultrasonic/hot air dry	Banana, Genipap, Jambo, Melon, Papaya, Pineapple	Extracellular breakdown ↑ Cell collapse and rupture ↑	Increase	Walde et al., 2006
Sucrose solution + ultrasound/ultrasound-assisted osmotic dehydration	Strawberry	Cell swelling and rounding up ↑ Extracellular breakdown ↑	Increase	Hayman et al., 1998
Coating (pectin and starch solution)/air + microwave drying	Apple (Golden Delicious)	Cell swelling and rounding up ↑ Starch gelatinization ↑ Cell wall Firmness ↑ Extracellular breakdown ↑	Increase	Askhari et al., 2006

Adapted from Dadmohammadi & Datta (2022).

microtomography (μCT) scanning (Li et al., 2020; Peng et al., 2018a, 2018b) for obtaining the microstructural images.

The microstructure analysis of banana slices subjected to osmotic dehydration associated with ultrasound and mechanical agitation showed that control samples presented less pores and lower tissue damage, while mechanically agitated samples presented increased pores and tissue damage, and ultrasound-treated samples presented the highest porosity and tissue damage (Farhaninejad et al., 2017). Related results were found for apples (Vallespir et al., 2018).

Apple slices pretreated by osmotic dehydration coupled with moderate electric field and dried by hot air drying (control) and refractance window presented a less collapsed structure due to pretreatment, in the case of hot air drying, and an open structure with the presence of more pores for both pretreated and unpretreated samples, in the case of refractance window (Hernández et al., 2020).

The use of hot air drying followed by moisture equilibrium process (MEP – store the apple cubes at 0.4 g g^{-1} of relative humidity for 48 h before DIC) and DIC provided apple cubes with a homogeneous pore size that resulted in a better expansion behavior (Li et al., 2020).

The microstructure of bananas dried in a hybrid-solar-vacuum dryer was better preserved when only one vacuum pulse was used, resulting in a higher number of pores (Roratto et al., 2021).

PPD was the best method for preserving the original microstructure of jujubes when compared to freeze-drying and hot air drying. When dried by PPD, jujubes presented a honeycomb-like microstructure (Wang et al., 2021).

4.2.5 Rehydration

Rehydration is a property used as factor of quality for dehydrated fruits and is fully correlated to the method of drying and fully depends on porosity, high rehydration ratio means higher reabsorption and less damage to the tissue structure. This property consists of moistening dry material, and according to Ratti (2001), different methods of drying applied can result in many changes in the tissue structure of fresh fruit, which impact the maintenance of volume, mass, and water content in DFP. Processing at low temperature (freeze-drying, convection-microwave-vacuum drying technique with "puffing" effect) can protect primary structure. Treatment and pretreatments like osmotic treatment can increase mechanical resistance and positively influence rehydration. Usually, it is done by addition of amount of water to sample at a temperature of 20°C and time till 5 h (Kowalska et al., 2018). In general, the higher the temperature, the higher the increase in the rate of rehydration in the early stage of the process (Rahman, 2015). The calculation of rehydration capacity can be made as follows (Ostermeier et al., 2020):

$$RC = \frac{M_f - M_d}{M_r - M_d} 100 \qquad \text{(Eq. 4.11)}$$

where M_r is the mass of the rehydrated sample (g), M_d is the mass of the dried samples, and M_f is the mass of the fresh samples (g). Rehydration has also been used as a quality parameter for dried pieces of fruits and vegetables. For example, one of these reports showed that freeze-drying was the best method for rehydration of kiwifruit pieces, followed by hot air–microwave-assisted drying, hot air drying, and vacuum drying. A preserved porous structure, such as the one achieved in freeze-dried products, is favorable for rehydration (Akar & Barutçu Mazı, 2019).

The use of rapid freezing pretreatment and pulsed electric fields (PEF) enhanced rehydration capacity of strawberries, which was attributed to the increased number of pores in the membrane produced by PEF (Fauster et al., 2020).

The use of osmotic dehydration, ultrasound, and coating with pectin and ascorbic acid was the best strategy for increasing the rehydration capacity of hot air-dried pineapples. The result was attributed to the formation of cavities, fractures, and microchannels due to ultrasound application (Sakooei-Vayghan et al., 2020).

Ohmic heating pretreatment of hot air-dried pineapple cubes was effective for increasing the mass gain after rehydration, i.e., ohmic heating showed a positive effect on rehydration capacity (Kumar et al., 2021).

For apple slices dried using different methods, they showed significant differences in values of rehydration ratio. Comparing freeze-drying, hot air and non-thermal ultrasound contact drying processes, freeze-fried samples show highest values for rehydration, while minor of the values were measured for hot air dried samples. The higher the drying temperature, the lower the rehydration capacity (Kahraman et al., 2021).

Results can be extended to other raw materials since high drying temperatures destroy the original microstructure, which is unfavorable to rehydration. For example, jujube slices that were hot air dried at high temperature presented collapse of the capillaries and reduced water retention capacity. On the other hand, PPD and especially freeze-drying reduced the collapsing area of the capillaries, which resulted in better rehydration (Wang et al., 2021).

4.2.6 Shrinkage

Shrinkage is a change in product shape, and sometimes cracking is caused by an unbalance between the internal and external pressures of food while moisture is removed. Plant foods shrink during drying proportionally to the elimination of moisture from the tissue (Jangam et al., 2010). When drying rates are low, the product shrinks and becomes denser. When drying rates are high, case hardening takes place, which limits shrinkage to some extent (Ordoñez, 2005). Shrinkage of fruit and vegetable pieces after drying has been assessed during the last few years. In this sense, the solid displacement method and the toluene displacement method can be used to calculate percentual shrinkage by the following equation:

$$S = \frac{V_0 - V_f}{V_0} 100 \tag{Eq. 4.12}$$

Where S is shrinkage, V_0 is the sample volume before drying, and V_f is the sample volume after drying.

Porosity as function of shrinkage and density follows this equation (Martynenko, 2008):

$$\varepsilon = \left(1 - \frac{\rho}{\rho_l}\right) + \left(\frac{\alpha(\beta - 1)}{1 + X_0}\right)\left(\frac{1}{\xi}\right) \tag{Eq. 4.13}$$

Where ε is porosity, α and β are density ratio coefficients, X is moisture content in kg kg^{-1} in dry basis, ρ is bulk density (kg/m^3), and ξ is a bulk shrinkage. Subscripts 0 are for initial and l for liquid.

The combination of image analysis and a Vernier caliper was used for measuring shrinkage of strawberries (Fauster et al., 2020). Results showed that the use of PEF prior to freeze-drying reduced shrinkage, which was attributed to pore formation.

Shrinkage is fully connected to porosity and suffers from the influence of various factors. The intrinsic factors are chemical composition, inclusion of volatile components (such as alcohol and carbon dioxide), and initial structure before processing (Rahman, 2001, 2004). Genskow (1990) and Achanta and Okos (1995) mentioned several mechanisms that affect the degree of collapse or shrinkage and formation of pores.

Glass transition is important in drying; fruit samples have high moisture content and are rubbery at the start of drying, while shrinkage would be ideal. If profiles of moisture are developed inside the fruit during drying, it can occur that the superficial portion becomes less mobile while the core is still in a rubbery state, causing tensional stresses through the hardened crust and the core being compressed, leading to formation of fractures. If the crust's integrity is held, volume reduction is restricted, and shrinkage is not ideal – case known as casehardening (Katekawa & Silva, 2007).

Osmotic dehydration and ultrasound-assisted osmotic dehydration (UOD) reduced shrinkage in pectin-acid coated, hot air-dried pineapples, which was attributed to the uptake of solids from the osmotic solution by the fruits and to the microcavities created by ultrasound, respectively (Sakooei-Vayghan et al., 2020).

Summarizing, usually, the use of pretreatments like coatings, PEF, osmotic dehydration, and ultrasound leads to decrease in shrinkage of DFP, independent of the drying method used, which results in a better product quality.

4.2.7 Caking

The caking of an amorphous powder is a time-dependent phenomenon and follows the same mechanism of agglomeration. The rate of caking attributable to the viscous flow property of the material is a function of ($T - Tg$) (Aguilera et al., 1995).

Caking event is time dependent, whereas sticking is a relatively instantaneous phenomenon. Longer contact times increase the tendency toward sticking and caking, when all other things are equal (Bhandari, Datta, & Howes, 1997). The dried product needs to be cooled immediately to an appropriate temperature before packaging. The interesting consequence of this is that the viscosity (and thus temperature) requirements for storage are much more stringent than those for processing; however, it is still possible to estimate critical values for both. Many powders can be a mixture of amorphous and crystalline materials (Fitzpatrick et al., 2007).

4.2.8 Texture

Food texture reflects human sensory perception when tactile on food, including the bite ability, mainly in the form of biting, yawing, grinding, etc., destroying or changing its structure and overall shape, making it suitable to be transferred to the stomach (Lu & Cen, 2013). The most important factor affecting texture of foods during drying is moisture content. Structural collapse in food due to moisture removal from the food product results in significant changes in texture. This causes shrinkage and a change in porosity of the dried product. Texture attributes, such as hardness, fracturability, springiness, chewiness, gumminess, cohesiveness, and resilience, can be measured by instrumental methods. Those measurements are made by means of a texture analyzer, according to Reis et al. (2022). The most common parameter used for expressing DFP texture is hardness (Aamir & Boonsupthip, 2017; Chottanom et al., 2020; Fauster et al., 2020; Hernández et al., 2020; Kowalska, Marzec, Domian, Masiarz, et al., 2020; Peng et al., 2018a, 2018b; Ullah et al., 2021; Wang et al., 2021).

Firmness of strawberries decreased with the application of PEF, a behavior that was attributed to the creation of pores due to PEF (Fauster et al., 2020).

Hot air drying and refractory window drying methods reduced the firmness of apple slices. Furthermore, osmotic pretreatment coupled with moderate electric field was not indicated as a pretreatment for refractory window drying of apples because it does not affect firmness at all (Hernández et al., 2020).

The use of osmotic pretreatment combined with convection-microwave-vacuum drying led to higher breaking force and higher work in Japanese quince-dried slices when compared to convective drying and freeze-drying (Kowalska, Marzec, Domian, Masiarz, et al., 2020).

Bananas dried in a hybrid-solar-vacuum dryer presented a crispy texture when one vacuum pulse was used, which was confirmed by force-deformation curves with more significant irregularities (Roratto et al., 2021).

Results suggest that DFP crispy textures are well accepted by consumers. Pretreatments that improve crispness, which is related to higher porosity, include freezing, PEF, and osmotic dehydration. Innovative drying methods that improve crispness include convection-microwave-vacuum drying and hybrid-solar-vacuum drying.

4.2.9 Water Activity

The water activity (a_w) of a food is the ratio between the vapor pressure of the food itself, when in a completely undisturbed balance with the surrounding air media, and the vapor pressure of distilled water under identical conditions (FDA, 2014). In drying studies, it is very interesting to measure water activity as it provides information on dried product shelf-life.

Link et al. (2018) found that mango slices dried by KMFD with four cycles followed by vacuum drying (KMFD-VD) achieved very low values of water activity (0.226).

Osmotic pretreatment with sucrose solution coupled with moderate electric field led to a decrease in drying time needed to achieve an a_w of 0.4, which was attributed to electroporation (Hernández et al., 2020). Combined UOD pretreatment at 35 kHz caused a higher decrease in water activity of dried apricots when compared to osmotic dehydration, which was attributed to an increase in the mass fraction of soluble solids (Sakooei-Vayghan et al., 2020).

4.3 CHEMICAL ANALYSES

4.3.1 Antioxidant Activity

One of the greatest challenges facing food science and medical science is to effectively combat oxidation in food and human cells. Oxidative stress is a pathological condition in which reactive oxygen species/nitrogen species (ROS/RNS) modify biological macromolecules, causing tissue damage, accelerating cell death in the body (Apak et al., 2016), and nutritional loss in food matrices. Unfortunately, not only one kind of analysis is enough to describe those processes of oxidation, but they also occur in various ways, and each assay describes one important way. So usually, two or more methods are combined in each study. The most common methods used in food studies are 2,2-Diphenyl-1-picrylhydrazyl - DPPH, [2, 2′ -Azinobis-(3-Ethylbenzothiazolin-6-Sulfonic Acid)] - ABTS (Mixed-Mode methods; electron transfer ET + hydrogen atom transfer HAT), Fluorescence recovery after photobleaching - FRAP, Cupric Ion Reducing Antioxidant Capacity - CUPRAC and Oxygen radical absorbance capacity - ORAC (ET methods). The meaning of antioxidant essays in drying studies is to verify the effect of temperature, drying methods, and treatments in the reduction of antioxidant activity. Lopez et al. (2017) stated that time was more significant than temperature when studying the vacuum drying of murta berries, as the temperature of 50°C got lower values of antioxidant activity than 90°C, as the process was longer. In general, DPPH and ORAC methods presented around 30% loss of antioxidant activity at the optimal temperature. Drying is one of the most harmful procedures for bioactive compounds, due to the degradation of biologically active compounds at high temperatures from chemical, enzymatic, or thermal decomposition (López et al., 2017).

Nehra and Deen (2017) concluded that Kankoda fruits lost 10–20% of antioxidant activity after 4 days of oven-drying at 55°C. Thus, even the low temperature, drying promotes losses of bioactive compounds.

Filiz and Seydim (2018) produced hot air dried apple chips and assessed several quality parameters, including ORAC and Trolox equivalent antioxidant capacity – TEAC. They found that antioxidant activity in apple chips increased when compared to raw apples, especially at higher hot air drying temperatures (75°C). The authors hypothesized that bound phenolics may have been released from the breakdown of cell constituents, which occurred by the effect of drying heat, thus increasing antioxidant activity. Comparison of methods is one of the most common drying studies, alongside the evaluation of time and temperature parameters' optimization.

In Sehrawat et al. (2019) investigation, Indian gooseberry and papaya showed that vacuum drying gets higher antioxidant activity (by DPPH) than hot air drying.

Vallespir et al. (2019) stated that freezing treatments, before convective drying, may help with microstructure, although they significantly decrease antioxidant activity (FRAP, CUPRAC, and ABTS) for the apple slices. Hence, usually, antioxidant essays demonstrate how negatively the drying process affects food matrices.

Another technique used for increasing antioxidant activity of DFP pieces is osmotic dehydration, which presents novel features such as the use of fruit juices as osmotic solutions. In this sense, Kowalska, Marzec, Domian, Masiarz, et al. (2020) used chokeberry juice as osmotic solution for treating apples prior to drying by freeze-drying and hot air/microwave vacuum drying (hybrid method). They discovered that no difference between drying methods was found regarding DPPH antioxidant activity, while the use of chokeberry juice improved significantly when compared to sucrose. Such difference was attributed to the presence of phenolic compounds in chokeberry juice.

4.3.2 Total Phenolics

Usually, research papers relate total phenolic content with antioxidant activity, as phenolic compounds act as natural antioxidants, and results are often expressed in mg of GAE (gallic acid equivalent, used as standard) per g or 100 g (d.m. – dry matter). López et al. (2017) corroborated antioxidant activity, stating that 50°C drying temperature and its longer associated time are more harmful to phenolics than 60 and 70°C. As well as Nistor et al. (2017), in which total phenolics found in the combination of methods, namely microwave and forced convection drying, resulted in higher results by 2–3 times.

Filiz and Seydim (2018) observed that drying increased phenolic content of apple slices, such effect was more intense at higher drying temperatures. It was attributed to the release of bound phenolics from the tissue due to hot air heat.

Pretreatments are used to increase the quality of DFP pieces, but the objective is not completely achieved; the correct selection of processes of pretreatment and drying must be careful. Vallespir et al. (2019) used freezing before hot air drying of apples, finding that such pretreatment increased the losses of phenolic compounds caused by drying.

In Sehrawat et al. (2019), results for papaya showed a decay in phenolic content with drying up to 50%, while the Indian gooseberry samples, which have naturally less water in their composition, got up to 30% phenolic loss.

Sehrawat et al. (2019) and Bozkir (2020) tested equipment for vacuum drying and compared it to hot air drying and stated the superiority of the vacuum options, this time for orange slices.

Bozkir (2020) did not use antioxidant activity as confirmation as Kowalska, Marzec, Domian, Galus, et al. (2020) study in apple slices. Phenolic content confirmed the significant improvement no matter the drying method (puffing or freeze-drying) by applying a type of osmotic solution so-called "sucrose with concentrated chokeberry juice". This highlights the importance of not focusing only on the drying method but also on pretreatments to protect bioactive compounds (Reis et al., 2022).

Bozkir (2020) and Kowalska, Marzec, Domian, Galus, et al. (2020) used the vitamin C analysis to confirm phenolic results. Innovative osmotic solutions can also be used for increasing the phenolic content of DFP pieces.

Low drying time, vacuum methods, and the use of pretreatments are fundamental to preserve more phenolic compounds, which is possibly related to the protection against oxidative reactions possibly occurring in such conditions and techniques.

Regarding the evaluation of methods, vitamin C content may also help with optimization. Wang et al. (2021) found that, for jujube crisp slices, hot air drying (80°C–15 h) is highly more harmful to vitamin C than freezing and puffing drying, which confirms the negative effect of high temperatures. Summarizing, given the high susceptibility of vitamin C to oxidation, it was expected that vacuum methods would preserve this compound better than traditional methods. Surprisingly, solar drying showed high potential for preservation of vitamin C as well. Long drying times and high drying temperatures are also known for destroying vitamin C, and for DFP, this was also observed, along with disadvantages for the processing of DFP.

4.4 CHANGES IN MICROBIOLOGICAL ATTRIBUTES

Microbiological contamination of fruits depends on how safely they are manipulated from the field to table. Enzymes, microorganisms, molds, and viruses can be listed among all agents of microbiological contamination. Drying is a technique that helps to preserve fruits in contamination aspects, so moisture and water activity are main parameters to control growing and survival of microorganisms, helping to guarantee safety for consumers.

Enzymatic reactions can occur in low moisture foods if the enzymes are not inactivated by heating. Hence, reduction of water activity in the final product is a very important way to assure the stability and safety of the dried foods. The final product with sufficiently low water activity is safe from enzymatic spoilage in general because active water is not available for microbial growth (Chieh, 2006).

The study of inactivation of *Listeria monocytogenes* during drying and storage of peach slices treated with acidic or sodium metabisulfite solutions showed that after 6 h of dehydration, populations on control or water immersed slices were reduced by 3.2–3.4 log cfu/g, whereas populations on slices treated with sodium metabisulfite or acidic solutions were reduced by 4.3–5.1 log cfu/g and 5.3–6.2 log cfu/g respectively (DiPersio et al., 2004).

The effect of hot air treatment at 38°C for 36 h in combination with *Pichia guilliermondii* on postharvest anthracnose rot of loquat fruit revealed that the combined treatment significantly reduced natural decay, disease incidence, and lesion diameter in artificially inoculated fruit (Liu et al., 2010).

Fruit given the combined treatment maintained low activities of catalase during the early storage period, but catalase and superoxide dismutase activities both rose later in storage. According to Machado et al. (2010), using moderate electric fields below 25°C showed that electric fields above 220 V cm^{-1} have greater electric fields, while presumably overcoming the thermal degradation caused by conventional high-temperature treatments.

In general, a_w in most DFP is typically less than 0.6, which is the recommended level for safe storage (Ong and Law, 2010). Low a_w inhibits the growth of various microorganisms and prevents oxidation and enzymatic reactions in fruits (Rahman, 2005).

Results presented in Table 4.3 show that DFP summited to technologies employing new pretreatments like cold plasma, ultrasound, and drying was conducted at mild or low temperature, innovative methods in general, presents higher quality,

TABLE 4.3
Studies Conducted over the Past Years on Quality Assessment of DFP Obtained by Innovative Pretreatment and Drying Methods

Materials	Pretreatment/ Drying Methods	Main Results	References
White grapes	Cold plasma/hot air	Increased antioxidant (DPPH = 435 µmol/L Trolox and FRAP = 415 µmol/L Trolox). The TOTAL PHENOLIC CONTENT and TOTAL FLAVONOIDS CONTENT were not evaluated.	(Huang et al., 2019)
Jujube slices	Cold plasma/hot air	Increased TOTAL FLAVONOIDS CONTENT (33.89%), TOTAL PHENOLIC CONTENT (13.85%), antioxidant activity (36.85%), and procyanidins (53.81%).	(Bao et al., 2021)
Whole jujube	Cold plasma/hot air	Cold plasma-assisted drying increased TOTAL PHENOLIC CONTENT by 12%, enhancing antioxidant activity by 11%. In addition, it reduced the generation of 5-hydroxymethylfurfural by 15%.	(Bao et al., 2022)

(Continued)

TABLE 4.3 *(Continued)*
Studies Conducted over the Past Years on Quality Assessment of DFP Obtained by Innovative Pretreatment and Drying Methods

Materials	Pretreatment/ Drying Methods	Main Results	References
Wolfberry	Cold plasma/hot air	The cold plasma significantly decreased TOTAL PHENOLIC CONTENT and TOTAL FLAVONOIDS CONTENT ($p < 0.05$).	(Zhou et al., 2020)
Grape	Cold plasma/hot air	Improved TOTAL PHENOLIC CONTENT (3.06–30.53%), antioxidant activity (7.31–62.29%), and ascorbic acid (17.87–168.73%) compared to the untreated.	(Miraei Ashtiani et al., 2020)
Banana	Cold plasma/hot air	More than a 50% increment in antioxidant activity. Higher Vitamin B6 (0.8 µg/100 g).	(Pour et al., 2022)
Tucumã	Cold plasma/hot air	Improved total phenolic content (45.3 mg GAE/g) and antioxidant activity (799.8 µM Trolox equivalent).	(Loureiro et al., 2021)
Bitter mellon	Ultrasound/hot air	Higher ascorbic acid (57.83) than the control (48.56 mg/100g).	(Jin et al., 2018)
Honeyberry	Ultrasound/hot air + vacuum drying	The highest TOTAL PHENOLIC CONTENT (445 mg GAE/100 g), TOTAL FLAVONOIDS CONTENT (0.94 mg GAE/100 g), ascorbic acid (1.187 mg/100 g), and anthocyanin (2.33 mg/kg) were obtained after ultrasound + vacuum drying.	(Šic Žlabur et al., 2019)
Apple	Pulsed-electric field/ freeze-drying	The TOTAL PHENOLIC CONTENT increased significantly ($p < 0.05$). However, as compared to untreated dried apple tissue, it can be seen that PEF-treated dried material's antioxidant activity dropped by up to 52.7% (DPPH assay) – 60.0% (ABTS assay).	(Lammerskitten et al., 2019b)
Cranberry	Pulsed-electric field/ hot air	The PEF pretreatment significantly increased the TOTAL PHENOLIC CONTENT (926 mg GAE/100 g), anthocyanins (667 mg C3G/100 g), and TOTAL FLAVONOIDS CONTENT (10.26 mg/g).	(Nowacka et al., 2019)
Blueberries	Pulsed-electric field/ hot air + vacuum dryer	The PEF significantly minimized the loss of anthocyanin (4.11 mg C3G/g), TOTAL PHENOLIC CONTENT (10.01 mg GAE/g), vitamin C (10.85 mg/100 g), and antioxidant activity (3.41 mg/g) in dried blueberries.	(Yu et al., 2017)
Apricots	Pulsed-electric field/ hot air	The PEF-treated apricots had higher antioxidant activity than the control in all three antioxidant assays (ORAC, DPPH, and FRAP), with a significant difference. PEF treatment retained more β-carotene (87.5%).	(Huang et al., 2019)

(Continued)

TABLE 4.3 *(Continued)*
Studies Conducted over the Past Years on Quality Assessment of DFP Obtained by Innovative Pretreatment and Drying Methods

Materials	Pretreatment/ Drying Methods	Main Results	References
Apple	Edible coating/hot air	The edible-coated apple had a higher TOTAL PHENOLIC CONTENT (160 mg GAE/100 g), antioxidants (~90%), and vitamin C (~8 mg/100g).	(Santagata et al., 2018)
Mellon cantaloupe	Edible coating/hot air	The edible-coated cantalope had a higher polyphenol (~165 mg GAE/100 g), antioxidants (~70%), and vitamin C (~10 mg/100g).	(Santagata et al., 2018)
Mango	Edible coating/hot air	The edible-coated mango had a higher polyphenol (~100 mg GAE/100 g), antioxidants (~30%), and vitamin C (~14 mg/100g).	(Santagata et al., 2018)
Pineapple	Edible coating/hot air	The edible-coated pineapple shows a higher polyphenol (~250 mg GAE/100 g), antioxidants (~80%), vitamin C (~10mg/100g), and improvements in volatile compounds.	(Santagata et al., 2018)
Red guava	Edible coating/hot air	The edible coating significantly increases the retention of ascorbic acid (6.03–13.48 g/kg), carotenoids (466.72–839.57 mg/kg), and phenolic (17.76–8.17 mg GAE/g) of the dried red guava.	(Todisco et al., 2018)
Andean blackberry	Ultrasound/ convective	Application of ultrasound improves the migration of antioxidant compounds to the liquid medium, which reduces the retention of these components in the fruit, and hence contributes to the loss of its functional quality.	(Romero and Yépez 2015)
Blackberry	Ultrasound/hot air	Depletion of antioxidant activity with increase in treatment time.	(Romero and Yépez 2015)
Nectarine slices	1% sodium Metabisulfite/hot air	DE for untreated and treated nectarines were 32.4 and 38.9. Changes observed in redness are associated with the decrease in anthocyanin and carotenoids concentrations. Changes observed in the nectarine yellowness could be linked to its tissue breakdown and other mechanisms such as the degradation of pigments and the occurrence of browning and non-enzymatic browning, which occurred during the drying process. Microstructural damages at dried samples compared to fresh-cut samples – fewer pores, no intracellular spaces.	Yanclo et al., 2022

(Continued)

TABLE 4.3 *(Continued)*
Studies Conducted over the Past Years on Quality Assessment of DFP Obtained by Innovative Pretreatment and Drying Methods

Materials	Pretreatment/ Drying Methods	Main Results	References
White mulberry Black mulberry Himalayan mulberry	Hot air, open sun, microwave	Hardness (g) and stiffness (g.mm^{-1}) were affected by drying methods – hardness variates as Microwave>hot air>open sun>fresh fruit. Antioxidants retention – hot air>microwave>open sun.	Khattak et al, 2019
Acerola pomace extract	Maltodextrin or cashew tree gum added/Spray drier	Degree of ascorbic acid and anthocyanins was impaired by increasing the inlet temperature and favored by increasing drying aid-to-acerola ratio. Higher inlet temperature favored the desired physical properties of powder. Drying aids decreased the powder hygroscopicity.	Moreira et al. (2010) Moreira et al. (2009)
Acerola fruits	Freeze-drying	Variation of water activity along the drying and sorption isotherms obtained for different freezing techniques. Cryogenic freezing using liquid N2 was the most recommended technique for freezing.	Marques et al. (2007)
Strawberry	vacuum-microwave drying (240, 360, and 480 W), freeze-drying (–60°C), vacuum-drying (50°C and 100 Pa), and convective drying (70°C and air velocity of 1 ms^{-1})	Drying techniques and cultivars have a degrading effect on ellagic acid, flavanols, anthocyanins, and ascorbic acid content, with a significant decrease in antioxidant activity.	Wojdyło & Oszmiańsk (2009)

considering attributes, and properties studied. Additional results confirm that DFP with high antioxidant activity, total phenolics, and vitamin C are considered as having a high chemical/bioactive quality. DFP must present low microbial counts and high sensory acceptability. In conclusion, the quality of DFP is directly affected by the drying method chosen.

4.5 CONCLUDING REMARKS

Food manufacturers have been, from the beginning, concerned about assuring food products with multiple quality features simultaneously. For example, good sensory acceptability, stability for long-term storage, and safety for consumption are typical quality features assessed. Recently, health-promotion capacity of the so-called functional foods has been extensively investigated. In addition, environmental-friendly food manufacturing is on the spot as well.

This chapter showed that much attention has been paid to DFP quality over the last few years. Such trend suggests that customers are expecting DFP with multiple quality features. To fulfill the needs of modern customers, a holistic approach to food quality must be put in place, dealing with its various aspects like physical, chemical/nutritional/bioactive, sensory, and microbiological.

REFERENCES

Aamir, M., & Boonsupthip, W. (2017). Effect of microwave drying on quality kinetics of okra. Journal of Food Science & Technology, 54(5), 1239–1247. https://doi.org/ 10.1007/s13197-017-2546-3

Achanta, S., Okos, M. R., Cushman, J. H., and Kessler, D. P. (1995) Drying of shrinking biological materials. Proceedings of ISOPOW Practicum - II, Food Preservation by Moisture Controls (Mexico), eds G.V. Barbosa-Canovas and J. Welti-Chanes, Technomic Publishing, Lancaster, PA.

Aguilera J. M., del Valle J. M., and Karel M. (1995). Caking phenomena in amorphous food powders. Trends in Food Science and Technology, 6, 149–154.

Akar G, and Barutçu Mazı I. (2019) Color change, ascorbic acid degradation kinetics, and rehydration behavior of kiwifruit as affected by different drying methods. J Food Process Eng. 42, e13011. https://doi.org/10.1111/jfpe.13011

Al Maiman, S. A., Albadr, N. A., Almusallam, I. A., Al-Sa′ad, M. J., Alsuliam, S., Osman, M. A., and Hassan, A. B. (2021) The potential of exploiting economical solar dryer in food preservation: Storability, physicochemical properties, and antioxidant capacity of solar-dried tomato (*Solanum lycopersicum*) fruits. Foods, 10 (4), 734. https://doi.org/10.3390/foods10040734.

Antoniolli, P. R., Ferrari, C. C., Borges, D. F., and Germer, S. P. M. (2023) Sorption isotherms, glass transition and storage stability of drum-dried mango peels obtained with and without process additives. Drying Technology, 41:3, 378–389. DOI:10.1080/07373937.2022.2093897.

Apak, R., ¨Ozyürek, M., Güçlü, K., & Çapanoılu, E. (2016). Antioxidant activity/capacity measurement. 1. Classification, physicochemical principles, mechanisms, and electron transfer (ET)-based assays. Journal of Agricultural and Food Chemistry, 64(5), 997–1027. https://doi.org/10.1021/acs.jafc.5b04739

Ashtiani, S.-H. M., Rafiee, M., Morad, M. M., and Martynenko, A. (2022) Cold plasma pretreatment improves the quality and nutritional value of ultrasound-assisted convective drying: The case of goldenberry. Drying Technology, 40:8, 1639–1657. DOI:10.1080/07373937.2022.2050255.

Askari, G., Emam-Djomeh, Z., and Mousavi, S. (2006) Effects of combined coating and microwave assisted hot-air drying on the texture, microstructure and rehydration characteristics of apple slices. Revista de Agaroquimica y Tecnologia de Alimentos 12(1), 39–46.

ASTM (American Society for Testing and Materials). (2005) E 313–05 Standard Practice for Calculating Yellowness and Whiteness Indices from Instrumentally Measured Color Coordinates. ASTM International, West Conshohocken, PA. https://www.astm.org/e0313-20.html, accessed on 06/24/2023.

Bao, T., Hao, X., Shishir, M. R. I., Karim, N., and Chen, W. (2021) Cold plasma: An emerging pretreatment technology for the drying of jujube slices. Food Chemistry, 337, 127783. DOI:10.1016/j.foodchem.2020.127783.

Bao, T., Hao, X., Shishir, M. R. I., Karim, N., and Chen, W. (2022) Green alternative methods for pretreatment of whole jujube before the drying process. Journal of the Science of Food and Agriculture, 102(3), 1030–1039. DOI:10.1002/jsfa.11438.

Bhandari, B. R., Datta, N., and Howes, T. (1997). Problems associated with spray drying of sugar-rich foods. Drying Technology 15(2) 671–684. https://doi.org/10.1080/07373939708917253

Boateng, I. D. (2022) Recent processing of fruits and vegetables using emerging thermal and non-thermal technologies. A critical review of their potentialities and limitations on bioactives, structure, and drying performance. Critical Reviews in Food Science and Nutrition, 1–35. DOI:10.1080/10408398.2022.2140121.

Bozkir H. (2020) Effects of hot air, vacuum infrared, and vacuum microwave dryers on the drying kinetics and quality characteristics of orange slices. J Food Process Eng. 43, e13485. https://doi.org/10.1111/jfpe.13485

Caivano, J. L., and Buera, M. P. (2012). Color in Food. Technological and Psychophysical Aspects. CRC Press, Taylor & Francis Group. CRC Press, Taylor & Francis Group, Boca Raton, FL.

Cao Y., Wang G., Huang Y., Zhu R., Wang Q., Wei S., and Yao X. (2018). Colour change and kinetics of winter jujube slices during pulsed air-jet impingement drying. Czech J. Food Sci. 36(4), 307–313. DOI: 10.17221/259/2017-CJFS.

Chang, L. S., Lau, K. Q., Tan, C. P., Yusof, Y. A., Nyam, K. L., and Pui, P. L. (2021) Production of 'Kedondong' (*Spondias Cytherea Sonnerat*) powder as affected by different drying methods. Acta Scientiarum Polonorum Technologia Alimentaria, 20:4, 417–427. http://dx.doi.org/10.17306/J.AFS.2021.0903.

Chen, X. D. and Mujumdar, A. S. (2008). *Drying Technologies in Food Processing*. John Wiley & Sons Ltd., United Kingdom, 322 p.

Chieh, C. (2006). Water Chemistry and Biochemistry. In Food Biochemistry and Food Processing. Hui, Y. H., Ed., Blackwell Publishing, USA. DOI: 10.1002/9780470277577.

Chottanom, P., Amornsin, A., Yodthava, N., & Wunnapong, S. (2020). Effect of edible coating on antioxidants and certain properties of dried Jerusalem artichoke. Pakistan Journal of Biological Sciences, 23(3), 271–277. https://doi.org/10.3923/pjbs.2020.271.277

Dadmohammadi, Y., and Datta, A. K. (2022) Food as porous media: A review of the dynamics of porous properties during processing. Food Reviews International, 38:5, 953–985. DOI:10.1080/87559129.2020.1761376.

DiPersio, P. A., Kendalla, P. A. and Sofos, J. N. (2004). Inactivation of Listeria monocytogenes during drying and storage of peach slices treated with acidic or sodium metabisulfite solutions. Food. Microbiol. 21, 641–648. https://doi-org.ez22.periodicos.capes.gov.br/10.1016/j.fm.2004.03.011

FAO. Medium-term Prospects for Raw Materials, Horticulture and Tropical Products. 2016. http://www.fao.org/3/i5579e/i5579e.pdf, accessed on 06/18/2023.

Farhaninejad, Zeynab et al. (2017) Osmotic dehydration of banana slices using direct and indirect sonication: Optimization and microstructure analysis. Journal of Food Process Engineering, 40(1), e12336. https://doi.org/10.1111/jfpe.12336

Fauster, T., Giancaterino, M., Pittia, P., & Jaeger, H. (2020). Effect of pulsed electric field pretreatment on shrinkage, rehydration capacity and texture of freeze-dried plant materials. Lebensmittel-Wissenschaft & Technologie, 121, Article 108937. https://doi.org/10.1016/j.lwt.2019.108937. August 2019.

FDA (2014). Water activity in foods. Available at: https://www.fda.gov/inspections-compliance-enforcement-and-criminal-investigations/inspection-technical-guides/water-activity-aw-foods

Figura, L. O., and Teixeira, A. A. (2007) Food Physics: Physical Properties – Measurement and Applications. Springer-Verlag, Berlin Heidelberg. https://link.springer.com/content/pdf/10.1007/978-3-540-34194-9.pdf.

Filiz, B., & Seydim, A. C. (2018). Kinetic changes of antioxidant parameters, ascorbic acid loss, and hydroxymethyl furfural formation during apple chips production. Journal of Food Biochemistry, 42(6), 1–10. https://doi.org/10.1111/jfbc.12676

Fitzpatrick, J. J., Hodnett, M., Twomey, M., Cerqueira, P. S. M., O'Flynn, J., and Roos, Y. H. (2007) Glass transition and the flowability and caking of powders containing amorphous lactose. Powder Technol., 178, 119–128. https://doi.org/10.1016/j.powtec.2007.04.017

Francis, F. J., and Clydesdale, F. M. (1975) Food Colorimetry: Theory and Applications. AVI Publishing Co, Westport.

Garcia, C. C., Caetano, L. C., Canizares, D., Silva, K. S., and Mauro, M. A. (2022) Blanching effects on convective air-drying and sorption properties of papaya. Drying Technology, 40:16, 3509–3519. DOI:10.1080/07373937.2022.2062378.

Genskow, L. R. (1990) Considerations in drying consumer products. In *Drying'89*, eds, A. S. Mujumdar and M. Roques, Hemisphere, New York.

Hayman, D. A., Hoseney, R. C., and Faubion, J. M. Bread Crumb grain development during baking. Cereal Chem. 1998, 75(5), 577–580. DOI: 10.1094/CCHEM.1998.75.5.577.

Hernández Y, Ramírez C, Moreno J, et al. (2020) Effect of refractance window on dehydration of osmotically pretreated apple slices: Color and texture evaluation. J Food Process Eng. 43, e13304. https://doi.org/10.1111/jfpe.13304

Huang, W., Feng, Z., Aila, R., Hou, Y., Carne, A., and Bekhit, A. E. D. A. (2019). Effect of pulsed electric fields (PEF) on physico-chemical properties, β-carotene and antioxidant activity of air-dried apricots. Food Chemistry, 291, 253–262. DOI:10.1016/j.foodchem.2019.04.021.

Hunterlab. (2015). Basics of color theory. Available at: https://www.hunterlab.com/blog/basics-color-theory/. (Accessed August 2021).

ISO, Colorimetry – Part 1: CIE standard colorimetric observers. In: **ISO/CIE 11664-1:2019 (en),** https://www.iso.org/obp/ui/en/#iso:std:iso-cie:11664:-1:ed-1:v1:en, accessed on 06/12/2023.

ISO, Colorimetry – Part 2: CIE standard illuminants. In: **ISO/CIE 11664-2:2022 (en),** https://www.iso.org/obp/ui/en/#iso:std:77215:en, accessed on 06/12/2023.

ISO, Colorimetry – Part 4: CIE 1976 L*a*b* colour space. In: **ISO/CIE 11664-4:2019 (en),** https://www.iso.org/obp/ui/en/#iso:std:iso-cie:11664:-4:ed-1:v1:en, accessed on 06/12/2023.

ISO, Colorimetry – Part 6: CIEDE2000 colour-difference formula. In: **ISO/CIE 11664-6:2022 (en),** https://www.iso.org/obp/ui/en/#iso:std:iso-cie:11664:-6:ed-2:v1:en, accessed on 06/12/2023.

Jangam, S. V. and Mujumdar, A. S. (2010) Basic concepts and definitions, in Drying of Foods, Vegetables and Fruits - Volume 1, Ed. Jangam, S. V., Law, C. L. and Mujumdar, A. S., Published.

Jiang, H., Zhang, M., Mujumdar, A.S., Duan, X. Microwave vacuum freeze drying of fruit and vegetables. In Drying of Foods, Vegetables and Fruits - Volume 3, Ed. Jangam, S.V., Law, C.L. and Mujumdar, A.S., 2011, ISBN - 978-981-08-9426-9, Published in Singapore, pp. 1–16.

Jin, W., Zhang, M., and Shi, W. (2018) Evaluation of ultrasound pretreatment and drying methods on selected quality attributes of bitter melon (*Momordica charantia* L.). Drying Technology, 37:3, 1–10. DOI:10.1080/07373937.2018.1458735.

Kahraman, O., Malvandi, A., Vargas, L., and Feng, H. (2021) Drying characteristics an quality attributes of apple slices dried by a non-thermal ultrasonic contact drying method. Ultrasonics Sonochemistry, 73, 105510. DOI:https://doi.org/10.1016/j.ultsonch.2021.105510.

Katekawa, M. E. and Silva, M. A. (2007). On the influence of glass transition on shrinkage in convective drying of fruits: A case study of banana. Drying Technology, 25, 1659–1666.

Khattak, M. K., Hanif, M., Khan, M., Ramzan, M., and Abdurab (2019) Comparison of drying process and preservatives on drying kinetics, texture and antioxidants retention in mulberry fruits. The Journal of Animal & Plant Sciences, 29:3, 803–808.

Kowalska, H., Belka, M., and Lenart, A. (2018) Rehydration properties of hybrid method dried fruit products enriched by natural components. International Agrophysics, 32, 175–182. DOI:10.1515/intag-2016-0100.

Kowalska, H., Marzec, A., Domian, E., Masiarz, E., Ciurzy´nska, A., Galus, S., Małkiewicz, A., Lenart, A., & Kowalska, J. (2020). Physical and sensory properties of Japanese quince chips obtained by osmotic dehydration in fruit juice concentrates and hybrid drying. Molecules, 25(23), 1–20. https://doi.org/10.3390/molecules25235504

Krokida, M. K., Maroulis, Z. B., and Saravacos, G. D. (2001) The effect of the method of drying on the colour of dehydrated products. International Journal of Food Science & Technology, 36, 53–59. DOI:10.1046/j.1365-2621.2001.00426.x.

Kumar, A., Begum, A., Hoque, M., Hussain, S., and Srivastava, B. (2021) Textural degradation, drying and rehydration behaviour of ohmically treated pineapple cubes. Lebensmittel-Wissenschaft & Technologie, 142(January), Article 110988. https://doi.org/10.1016/j.lwt.2021.110988.

Lammerskitten, A., Wiktor, A., Siemer, C., Toepfl, S., Mykhailyk, V., Gondek, E., Rybak, K., Witrowa-Rajchert, D., and Parniakov, O. (2019b) The effects of pulsed electric fields on the quality parameters of freeze-dried apples. Journal of Food Engineering, 252, 36–43. DOI:10.1016/j.jfoodeng.2019.02.006.

Li, L., Chen, J., Zhou, S., Ren, G., and Duan, X. (2021). Quality evaluation of probiotics enriched Chinese yam snacks produced using infrared-assisted spouted bed drying. Journal of Food Processing and Preservation, 45(4), 1–14. https://doi.org/10.1111/jfpp.15358

Li, M., Ai, Z., Xiao, H., Mowafy, S., Pei, Y., and Liu, Y. (2023) Improvement of drying efficiency and physicochemical quality of kiwifruit slices using infrared-assisted tilted tray air impingement drying. Drying Technology, 41:7, 1159–1170. DOI:10.1080/07373937.2022.2125526.

Li, X., Bi, J., Jin, X., Wu, X., Lyu, J., Li, X., and Hu, L. (2020). Effect of the moisture equilibrium process on the expansion behavior of instant controlled pressure drop (DIC) drying of dried apple cubes. Journal of the Science of Food and Agriculture, 100 (4), 1635–1642. https://doi.org/10.1002/jsfa.10175

Li, Y. H., Wu, Z. F., Wang, X. C., Yu, F., and Yang, M. (2019). Effects of combined infrared and steam blanching on enzyme inactivation and product quality of *Chrysanthemum indicum* L. Flower. Journal of Food Processing and Preservation, 43:10, 1–8. https://doi.org/10.1111/jfpp.14119.

Link J. V., Tribuzi G., and Laurindo J. B. (2018). Conductive multi-flash drying of mango slices: Vacuum pulse conditions on drying rate and product properties. J Food Process Preserv. 42, e13440. https://doi.org/10.1111/jfpp.13440

Liu, F., Tua, K., Shaob, X., Zhaoa, Y., Tuc, S., Sua, J., Houa, Y. and Zoua, X. (2010). Effect of hot air treatment in combination with Pichia guilliermondii on postharvest anthracnose rot of loquat fruit. Postharvest. Biol. Tech. 58, 65–71. https://doi.org/10.1016/j.postharvbio.2010.05.009

López, J., Vega-G´alvez, A., Bilbao-Sainz, C., Chiou, B., Sen Uribe, E., & Quispe-Fuentes, I. (2017). Influence of vacuum drying temperature on: Physico-chemical composition and antioxidant properties of murta berries. Journal of Food Process Engineering, 40 (6), 1–9. https://doi.org/10.1111/jfpe.12569

Loureiro, A. C., Souza, F. C. A., Sanches, E. A., Bezerra, J. A., Lamarao, C. V., Rodrigues, S., Fernandes, F. A. N., and Campelo, P. H. (2021) Cold plasma technique as a pre-treatment for drying fruits: Evaluation of the excitation frequency on drying process and bioactive compounds. Food Research International (Ottawa, ON), 147, 110462. DOI:10.1016/j.foodres.2021.110462.

Lu, R., & Cen, H. (2013). Non-destructive methods for food texture assessment. In D. Kilcast (Ed.), *SawstonInstrumental Assessment of Food Sensory Quality*. Woodhead Publishing.

Machado, L. F., Pereira, R. N., Martins, R. C., Teixeira, J. A. and Vicente, A. A. (2010). Moderate electric fields can inactivate Escherichia coli at room temperature. J. Food. Eng. 96, 520–527. https://doi-org.ez22.periodicos.capes.gov.br/10.1016/j.jfoodeng.2009.08.035

Marques, L. G., Ferreira, M. C., and Freire, J. T. (2007) Freeze-drying of acerola (*Malpighia glabra* L.). Chemical Engineering Process, 46:5, 451–457. DOI:10.1016/j.cep.2006.04.011.

Martynenko, A. (2008) The system of correlations between moisture, shrinkage, density and porosity. Drying Technology, 26, 1497–1500.44.

Miraei Ashtiani, S. H., Rafiee, M., Mohebi Morad, M., Khojastehpour, M., Khani, M. R., Rohani, A., Shokri, B., and Martynenko, A. (2020) Impact of gliding arc plasma pretreatment on drying efficiency and physicochemical properties of grape. Innovative Food Science & Emerging Technologies, 63, 102381. DOI:10.1016/j.ifset.2020.102381.

Moreira, G. E. G., Azeredo, H. M. C. D., Medeiros, M. D. F. D. D., Brito, E. S. D., and Souza, A. C. R. D. (2010) Ascorbic acid and anthocyanin retention during spray drying of acerola pomace extract. Journal of Food Processing Preservation, 34, 915–925. DOI:10.1111/j.1745-4549.2009.00409.x.

Moreira, G. E. G., Costa, M. G. M., Souza, A. C. R. D., Brito, E. S. D., Medeiros, M. D. F. D. D., and Azeredo, H. M. C. D. (2009) Physical properties of spray dried Acerola pomace extract as affected by temperature and drying aids. LWT – Food Science Technology, 42:2, 641–645. DOI:10.1016/j.lwt.2008.07.008.

Mujumdar, A. S. (2008) Guide to Industrial Drying: Principles, Equipments and New Developments. CRC Press, Boca Raton, FL.

Mujumdar, A. S. (2015) *Handbook of Industrial Drying* (4th ed.). CRC Press, Boca Raton.

Nehra, S., and Deen, M. K. (2017). Activity of fruits of momordica dioica (Kankoda). Asian Journal of Chemistry, 29(3), 480–484. https://doi.org/10.14233/ajchem.2017.20185

Nowacka, M., Wiktor, A., Anuszewska, A., Dadan, M., Rybak, K., and Witrowa-Rajchert, D. (2019). The application of unconventional technologies as pulsed electric field, ultrasound, and microwave-vacuum drying in the production of dried cranberry snacks. Ultrasonics Sonochemistry, 56, 1–13. DOI:10.1016/j.ultsonch.2019.03.023.

Nistor, O. V., Seremet Ceclu, L., Andronoiu, D. G., Rudi, L., & Botez, E. (2017). Influence of different drying methods on the physicochemical properties of red beetroot (Beta vulgaris L. var. Cylindra). Food Chemistry, 236, 59–67. https://doi.org/10.1016/j. foodchem.2017.04.129

Ong, S. P. and Law, C. L. (2010). Hygrothermal properties of various foods, vegetables and fruits. In *Drying of Foods, Vegetables and Fruits*, pp. 31, Jangam, S. V., Law, C. L. and Mujumdar, A. S., Eds., Singapore.

Ordoñez, J. A. (2005). Tecnologia de Alimentos: Componentes dos Alimentos e Processos. Porto Alegre. Artmed.

Oshima, T., Kato, K., and Imaizumi, T. (2021). Effects of blanching on drying characteristics, quality, and pectin nanostructures of dried cut-persimmons. Lebensmittel-Wissenschaft & Technologie, 143(January), Article 111094. https://doi. org/10.1016/j.lwt.2021.111094.

Ostermeier, R., Parniakov, O., T¨opfl, S., & J¨ager, H. (2020). Applicability of pulsed electric field (PEF) pre-treatment for a convective two-step drying process. Foods, 9(4), 9–12. https://doi.org/10.3390/foods9040512

Peng, J., Bi, J., Yi, J., Wu, X., Zhou, M., Lyu, J., and Liu, J. (2018a). Engineering texture properties of instant controlled pressure drop (DIC) dried carrot chips via modulating osmotic conditions. Food and Bioprocess Technology, 11(9), 1674–1685. https://doi.org/10.1007/s11947-018-2133-2

Peng, J., Yi, J., Bi, J., Chen, Q., Wu, X., Zhou, M., and Liu, J. (2018b). Freezing as pretreatment in instant controlled pressure drop (DIC) texturing of dried carrot chips: Impact of freezing temperature. LWT –Food Science and Technology, 89(2), 365–373. https://doi.org/10.1016/j.lwt.2017.11.009

Pour, A. K., Khorram, S., Ehsani, A., Ostadrahimi, A., and Ghasempour, Z. (2022). Atmospheric cold plasma effect on quality attributes of banana slices: Its potential use in blanching process. Innovative Food Science & Emerging Technologies, 76, 102945. DOI:10.1016/j.ifset.2022.102945.

Rahman, M. S. (2001). Toward Prediction of Porosity in Foods During Drying: A Brief Review. Drying Technology 19(1), 1–13. https://doi.org/10.1081/DRT-100001349

Rahman, M. S., Guizani, N., Al-Ruzeiki, M. H. (2004). D- and Z-values of microflora in tuna mince during moist- and dry-heating, LWT. Food Sci Technol 37(1), 93–98.

Rahman, M. S. (2015). Drying of fish and seafood. In A. S. Mujumdar (Ed.), Handbook of industrial drying (4th ed.). CRC Press, Boca Raton.

Rajoriya, D., Shewale, S.R. & Hebbar, H.U. Refractance window drying of apple slices: Mass transfer phenomena and quality parameters. Food Bioprocess Technol 12, 1646–1658 (2019). https://doi.org/10.1007/s11947-019-02334-7

Ratti, C. (Ed.) (2001). *Advances in Food Dehydration* p. cm. – (Contemporary food engineering; 5)

Reis, F. R., Marques, C., Moraes, A. C. S., and Masson, M. L. (2022) Trends in quality assessment and drying methods used for fruits and vegetables. Food Control, 142, 109254. DOI:10.1016/j.foodcont.2022.109254.

Richter Reis, F., Ivahashi, M. M., & Gu´eniat Rosa, A. H. (2017). Effect of vacuum drying temperature on drying kinetics, effective moisture diffusivity and quality of peeled Litchi (Litchi chinensis sonn.). *Journal of Food Process Engineering*, *40*(2). https://doi.org/10.1111/jfpe.12419

Rodríguez-Ramírez, J., Méndez-Lagunas, L., López-Ortiz, A. and Torres, S.S. (2012), True density and apparent density during the drying process for vegetables and fruits: A review. Journal of Food Science, 77, R146–R154. https://doi.org/10.1111/j.1750-3841.2012.02990.x

Romero J., Carlos A. and Yépez V., Byron D. (2015). Ultrasound as pretreatment to convective drying of Andean blackberry (Rubus glaucus Benth). Ultrasonics Sonochemistry, 22, 205–210, https://doi.org/10.1016/j.ultsonch.2014.06.011

Roratto, T. B., Monteiro, R. L., Carciofi, B. A. M., & Laurindo, J. B. (2021). An innovative hybrid-solar-vacuum dryer to produce high-quality dried fruits and vegetables. Lebensmittel-Wissenschaft & Technologie, 140. https://doi.org/10.1016/j. lwt.2020.110777. June 2020.

Ruangchakpet, A., and Sajjaanantakul, T. (2007) Effect of browning on total phenolic, flavonoid content and antioxidant activity in Indian gooseberry (*Phyllanthus emblica* Linn.). Kasetsart Journal (Natural Science), 41, 331–337. 10471519.

Sakooei-Vayghan, R., Peighambardoust, S. H., Hesari, J., Soltanzadeh, M., and Peressini, D. (2020). Properties of dried apricots pretreated by ultrasound-assisted osmotic dehydration and application of active coatings. Food Technology and Biotechnology, 58(3), 249–259. https://doi.org/10.17113/ftb.58.03.20.6471

Santagata, G., Mallardo, S., Fasulo, G., Lavermicocca, P., Valerio, F., Di Biase, M., Di Stasio, M., Malinconico, M., and Volpe, M. G. (2018) Pectin-honey coating as novel dehydrating bioactive agent for cut fruit: Enhancement of the functional properties of coated dried fruit products. Food Chemistry, 258, 104–110. DOI:10.1016/j.foodchem.2018.03.064.

Sehrawat, R., Chandra, A., Nema, P. K., & Arora, V. K. (2019). Drying of fruits and vegetables in a developed multimode drying unit and comparison with commercially available systems. Journal of The Institution of Engineers (India): Series A, 100(3), 381–386. https://doi.org/10.1007/s40030-019-00371-1

Shinde, B., and Ramaswamy, H. S. (2023) Microwave–osmotic–vacuum combination drying of mango cubes with maltodextrin moderated sucrose solution. Drying Technology, 41:6, 1046–1059. DOI:10.1080/07373937.2023.2201304.

Šic Žlabur, J., Colnar, D., Voća, S., Lorenzo, J. M., Munekata, P., Barba, E. S., Dobričević, F. J., Galić, N., Dujmić, A., Pliestić, F., et al. (2019) Effect of ultrasound pre-treatment and drying method on specialized metabolites of honeyberry fruits (Lonicera caerulea var. kamtschatica). Ultrasonics Sonochemistry, 56, 372–377. DOI:10.1016/j.ultsonch.2019.04.034.

Tan, C. H., Hii, C. L., Borompichaichartkul, C., Phumsombat, P., Kong, I., and Pui, L. P. (2022) Valorization of fruits, vegetables, and their by-products: Drying and bio-drying. Drying Technology, 40:8, 1514–1538. DOI:10.1080/07373937.2022.2068570.

Todisco, K. M., Janzantti, N. S., Santos, A. B., Galli, F. S., and Mauro, M. A. (2018) Effects of temperature and pectin edible coatings with guava by-products on the drying kinetics and quality of dried red guava. Journal of Food Science and Technology, 55:12, 4735–4746. DOI:10.1007/s13197-018-3369-6.

Ullah, F., Hasrat, K., Iqbal, S., Hussain, I., Hussain, A., & Mumtaz, Y. (2021). An approach to evaluate dehydration of apples (malus domestica L) with the effect of temperature and time interval under the response surface method. International Journal of Fruit Science, 21(1), 657–669. https://doi.org/10.1080/15538362.2021.1920553

Vallespir, F., Rodrígue, ´O., Eim, V. S., Rossell´o, C., and Simal, S. (2018). Freezing pretreatments on the intensification of the drying process of vegetables with different structures. Journal of Food Engineering, 239(January), 83–91. https://doi.org/ 10.1016/j.jfoodeng.2018.07.008

Vallespir, F., Rodríguez, ´O., Eim, V. S., Rossell´o, C., and Simal, S. (2019). Effects of freezing treatments before convective drying on quality parameters: Vegetables with different microstructures. Journal of Food Engineering, 249(January), 15–24. https://doi.org/10.1016/j.jfoodeng.2019.01.006

Walde, S. G., Velu, V., Jyothirmayi, T., and Math, R. G. Effects of pretreatments and drying methods on dehydration of mushroom. J. Food Eng. 2006, 74(1), 108–115. https://doi.org/10.1016/j.jfoodeng.2005.02.008

Wang, X., Gao, Y., Zhao, Y., Li, X., Fan, J., and Wang, L. (2021). Effect of different drying methods on the quality and microstructure of fresh jujube crisp slices. *Journal of Food Processing and Preservation*, 45(2), 1–9. https://doi.org/10.1111/jfpp.15162

Wojdyło, A., Figiel, A., and Oszmiański, J. (2009) Effect of drying methods with the application of vacuum microwaves on the bioactive compounds, color, and antioxidant activity of strawberry fruits. Journal of Agricultural and Food Chemistry, 57:4, 1337–1343. DOI: https://doi.org/10.1021/jf802507j.

Yanclo, L. A., Sigge, G., Belay, Z. A., October, F., and Caleb, O. J. (2022) Microstructural, biochemical, and drying characteristics of dehydrated 'Sunectwentyone' nectarines as affected by sodium metabisulphite. Food Science and Biotechnology, 31:3, 311–322. DOI:10.1007/s10068-022-01039-6.

Yang, Z., Li, H., Xu, Y., Liu, Y., Kan, H., & Fan, F. (2019). Vacuum impregnation and drying of iron-fortified water chestnuts. Journal of Food Processing and Preservation, 43(12), 1–11. https://doi.org/10.1111/jfpp.14260

Yu, Y., Jin, T. Z., and Xiao, G. (2017) Effects of pulsed electric fields pretreatment and drying method on drying characteristics and nutritive quality of blueberries. Journal of Food Processing and Preservation, 41:6, e13303. DOI:10.1111/jfpp.13303.

Zhang, Z., Wei, Q., Nie, M., Jiang, N., Liu, C., Liu, C., Li, D., & Xu, L. (2018). Microstructure and bioaccessibility of different carotenoid species as affected by hot air drying: Study on carrot, sweet potato, yellow bell pepper and broccoli. Lebensmittel-Wissenschaft & Technologie, 96(February), 357–363. https://doi.org/10.1016/j.lwt.2018.05.061

Zhou, C., Wang, Z., Wang, X., Yagoub, A. E. A., Ma, H., Sun, Y., & Yu, X. (2021). Effects of tri-frequency ultrasound-ethanol pretreatment combined with infrared convection drying on the quality properties and drying characteristics of scallion stalk. Journal of the Science of Food and Agriculture, 101(7), 2809–2817. https://doi.org/10.1002/jsfa.10910

Zhou, Y.-H., Vidyarthi, S. K., Zhong, C.-S., Zheng, Z.-A., An, Y., Wang, J., Wei, Q., and Xiao, H.-W. (2020) Cold plasma enhances drying and color, rehydration ratio and polyphenols of wolfberry via microstructure and ultrastructure alteration. LWT, 134, 110173–110177. DOI:10.1016/j. lwt.2020.110173.

5 Aroma Retention in Dried Fruit Products

Maria Terezinha Santos Leite Neta,
Rafael Donizete Dutra Sandes,
Hannah Caroline Santos Araujo,
Mônica Silva de Jesus, and Narendra Narain
Laboratory of Flavor and Chromatographic Analysis/PROCTA,
Federal University of Sergipe
São Cristóvão-SE, Brazil

5.1 INTRODUCTION

Aroma retention is one of the most important challenges in food processing due to its sensitivity to processing. The aroma of food has several functions, viz., providing variety and interest to the consumer, and is possible with the help of aroma handling tools that could possibly control the bad quality characteristics in foods such as generating compounds, which are formed by rancid reactions and consequently become unsafe food (Parker, 2015).

In fruits, the aroma compounds are very important since they provide the consumer with the best organoleptic characteristics that can influence the decision to buy a fruit. Fruits are known to possess a very diverse aroma, which is formed by several chemical compounds pertaining to different classes. Although different fruits often share many volatile compounds, each fruit has a peculiar aroma that depends upon the combination of volatiles, their concentrations, and the odor perception threshold of individual volatile compounds (El Hadi et al., 2013). Thus, the identification of key volatile flavor metabolites that impart the unique character of any natural fruit is essential, as it provides the principal sensory identity and characteristic flavor of the fruit (Cheong et al., 2010).

The big challenge in the food chain is to transform fruits, which are very perishable, into stable products that can conserve their nutritional and organoleptic characteristics of fruit. One of the promising alternatives is the application of drying methods to obtain products that can be stored for long periods, whereby these can be available in times of shortage and also at faraway places. Thus, the use of dehydration in foods has become increasingly a frequent operation in the food industry. This growth is due to other advantages that dehydrated foods have over fresh foods, such as reduced packaging, product mass, ease of transport, longer shelf life, and storage at room temperature, among others. Due to the growing change in eating habits, consumers are increasingly seeking to consume food free of artificial additives such as colors and flavors.

During the process of dehydration, aroma plays an important role in the quality of dried fruit products, since the aroma compounds retention is one of the most difficult

DOI: 10.1201/9781003402916-5

tasks in food dehydration process due to sensitivity of these compounds to high temperatures. Considering these aspects, this chapter will review initially the aroma compounds that are usually present in fruits, the different methods of dehydration, and their impact on the aroma compounds in tropical fruits.

5.2 FORMATION OF VOLATILE COMPOUNDS IN FRUITS

5.2.1 Bioformation of Volatile Compounds

The aroma of any food is characterized by the presence of a large number of volatile compounds of different chemical classes. Some foods can naturally contain more than a thousand substances that, together, result in their characteristic aroma. One example of this is the volatile profile of fruits that can contain more than 200 different compounds. However, not all these compounds are important, and from the aroma point of view, only a few of them are responsible for the aroma characteristics of the fruit.

The fruit aroma is generally a complex mixture of a large number of volatile compounds generated from several pathways derived from amino acid compounds, lipid-derived phenolic derivatives, and mono- and sesquiterpenes, whose composition depends on the combination of volatiles, their concentrations, and the perception threshold of each individual volatile compound, which is specific to the variety of fruit (Schwab et al., 2008; Cheong et al., 2010; El Hadi et al., 2013).

These aroma compounds could be produced by anabolic or catabolic pathways. These pathways include different mechanisms that can use substances like carbohydrates, fatty acids, and amino acids as substrates for the biochemical pathways. A brief summary of these pathways for aroma formation in fruits is detailed in the following section.

5.2.1.1 Fatty Acids Pathway

Fatty acids play an important role in the formation of volatile compounds in fruits. They are responsible for the generation of aldehydes and alcohols that give fresh and green notes to fruits. There are two main oxidative pathways to transform fatty acids in these compounds: β-oxidation and Lipoxygenase (LOX) (Feussner & Wasternak, 2002; Baysal & Deminrdoven, 2007; Siegmund, 2015).

The β-oxidation mechanism is considered to be the main pathway for the degradation of straight chain fatty acids in the intact fruits. This mechanism involves the participation of acetylcoenzime A (acetyl-CoA) and the fatty acids metabolized are acetyl-CoA derivatives. In each cycle of this mechanism, the loss of two carbons occurs until the complete breakdown of the compound occurs. Figure 5.1 describes the β-oxidation metabolism.

The other pathway for the formation of volatile compounds involving fatty acids is due to the enzyme LOX. This pathway consists of the liberation of free fatty acids by the action of lipases. The main precursors in the LOX pathway are the polyunsaturated fatty acids that, with a (Z,Z)-1,4-pentadiene moiety, such as linolenic acids, are oxidized into 9-, 10-, or 13-chain hydroperoxides, which are catalyzed by LOX. The hydroperoxides formed can be converted into a range of volatile products

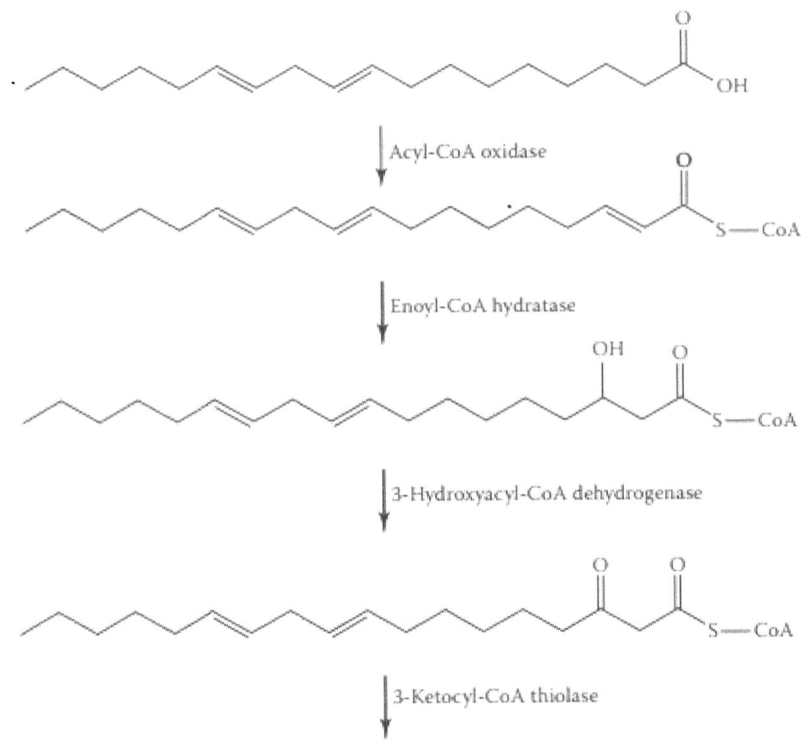

FIGURE 5.1 Cycle of β-oxidation of linolenic acid. (Adapted from Christensen et al., 2007.)

by the action of hydroperoxide lyases, such as unsaturated C6- and C9-aldehydes. One example of the aldehydes that could be formed is (*E*)-2-hexenal that plays a role in the flavor of many fruits contributing to intense fresh green and green apple-like odor notes. Figure 5.2 shows the mechanism of LOX pathway in the degradation of linolenic acid.

5.2.1.2 Amino Acids Pathway

A variety of amino acids, such as alanine, leucine, isoleucine, phenylalanine, and aspartic acid, are important in the formation of primary flavor compounds in the intact fruit, known as nonvolatile precursors. The metabolism of amino acids is responsible for the formation of some volatile compounds like aliphatic alcohols, carbonyls, acids, and esters. These compounds are formed by three enzymatic activities: aminotransferase, decarboxylase, and alcohol dehydrogenase (Figure 5.3).

Enzymes in the pathway remove both amine and carboxyl groups from the amino acids and produce aldehydes. The aldehydes are then reduced to produce esters. All compounds formed by amino acids pathway have a great influence on the aroma of several fruits, and these are highly dependent on the availability of free amino acids (Dalal et al., 1968; Reineccius, 2006; Reeve et al., 2008; Oms-Oliub et al., 2011; Prinsi et al., 2011).

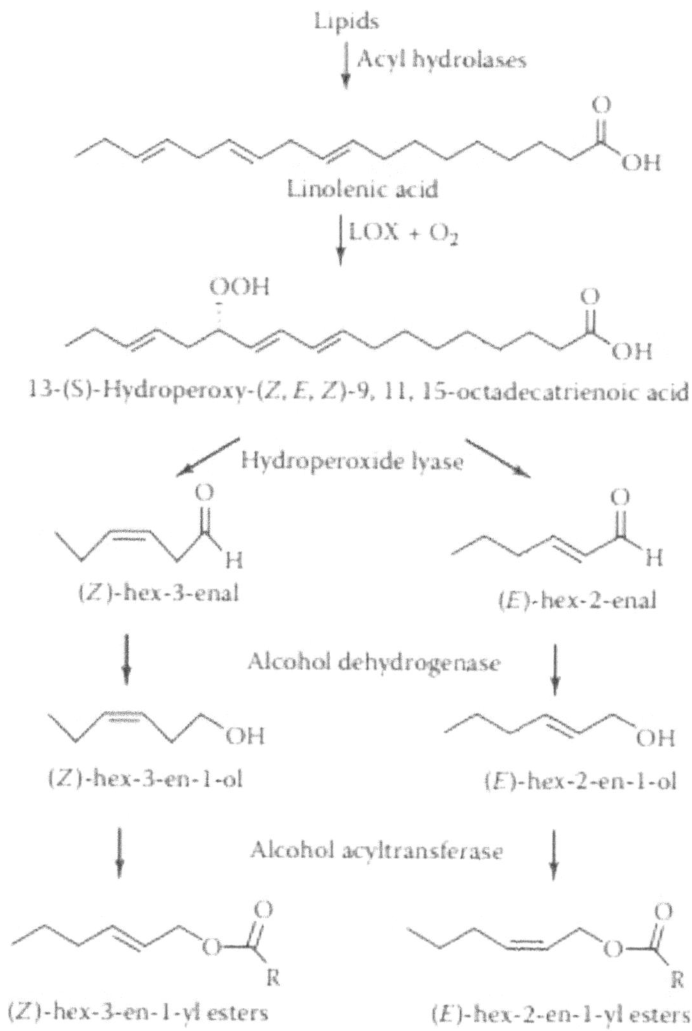

FIGURE 5.2 LOX pathway in the degradation of linolenic acid.

5.2.1.3 Carbohydrates Pathway

The major constituents in fruits and vegetables are carbohydrates. However, only two groups of volatile compounds are derived from carbohydrates which are terpenes and furanones. In the formation of terpenes, there are two routes of biosynthesis: the mevalonate pathway (MVA), in cytosol, and the mevalonate-independent methylerythritol phosphate pathway (MEV) (Mahmoud & Croteau, 2002). Figure 5.4 describes both mechanisms for the formation of terpenes.

The formation of terpenes starts with the production of isoprene units (isopentenyl diphosphate and dimethylallyl diphosphate) that are condensed by prenyltransferase, producing geranyl diphosphate (GPP) as the basis for the monoterpene

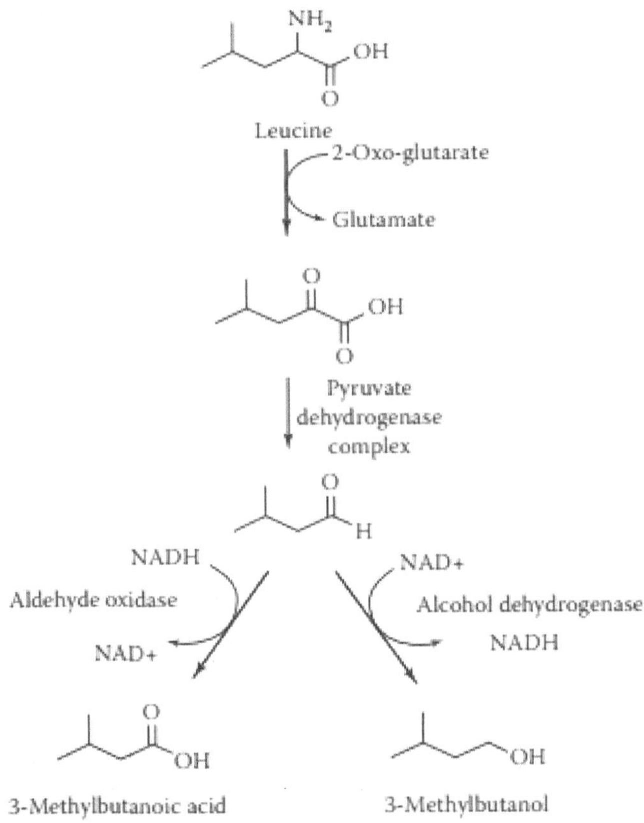

FIGURE 5.3 Scheme of formation of volatile compounds from amino acids. (Adapted from Fisher et al., 1997.)

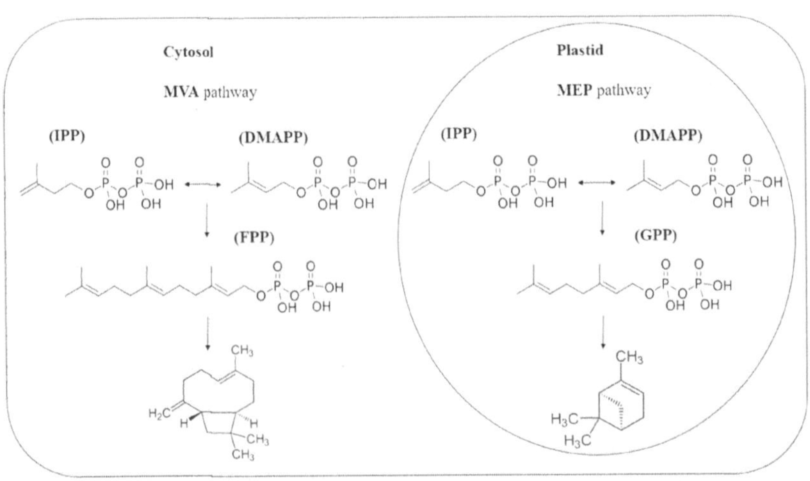

FIGURE 5.4 Mechanisms of biosynthesis of terpenoids.

biosynthesis as well as farnesyl diphosphate (FPP) as the basis for sesquiterpene biosynthesis.

After the formation of GPP and FPP, a sequence of reactions including hydrolysis, cyclization, ionizations, isomerizations, oxidoreductions, or hydride shifts occur which finally lead to the formation of many members of the large terpene compounds that are important for the flavor of many fruits (Tholl, 2006; Degenhardt et al., 2009).

The differences in MVA and MEV pathways are mainly due to the localization where the reaction occurs. For the MVA pathway, it occurs in the cytosol of the cell and is generally considered to produce the precursors for sesqui- and triterpenes, whereas for the MEP pathway, it is active in the plastid and delivers the precursor compounds for the production of mono-, di-, and triterpenes (Rohmer, 1999; Tholl, 2006; Perez & Sanz, 2008; Hemmerlin et al., 2012).

The second group of aroma compounds that can be formed from carbohydrates are furanones. Furanones are compounds that can be found in a range of highly cooked foodstuffs, where they are important flavor compounds with aroma threshold values as low as 20 micrograms kg^{-1} water (approximately 0.14 mμmol L^{-1}). These compounds can be formed during Maillard reaction, which occurs between sugars and amino acids, during the course of several fermentation processes, as well as being biosynthesized by several plants, such as strawberries or pineapples, and during the fermentation stages in the production of soy sauce and beer (Schwab, 1998; Slaughter, 1999; Zabetakis et al., 1999).

Some furanones, such as 2,5-dimethyl-4-hydroxy-3(2H)-furanone (DMHF, known as furaneol) and 2,5-dimethyl-4-methoxy-3(2H)-furanone (known as mesifuran), are the most prominent examples. DMHF is considered to be a key aroma compound for strawberries and pineapples (Schwab, 1998; Siegmund, 2015).

5.3 CLASSES OF VOLATILE COMPOUNDS

5.3.1 Terpenes

Terpenes belong to the most important group of natural products, such as essential oils, and are formed by the fusion of five-carbon isoprene units (C_5H_{10}). These are compounds that occur in all plants which comprise a class of secondary metabolites with a great structural variety possessing flavoring properties. These can be synthesized as secondary metabolites by a wide variety of higher plants and microorganisms, mainly Ascomycetes and Basidiomycetes, capable of producing terpenes such as citronellol, geraniol, linalool, nerol, and α-terpineol (Janssens et al., 1992; Raven et al., 2001).

Until the present time, approximately 25,000 different compounds have been cataloged within the category of terpenics. One of the most famous terpenes is limonene, which is very common in citrus fruit peel, but this compound has a weak odorant power and is more commonly used as solvent in the industry. Other terpenes present in fruits are α-thujene (woody note) and sabinene (citrus note) that are present in fruits and spices, and the sesquiterpenes such as α-valencene imparting classic citrus notes, and both farnesene and humulene resulting in a woody spicy note.

Pinene, myrcene, and ocimene are major components of basil aroma (Bicas et al., 2010; Chen et al., 2012; Parker, 2015).

5.3.2 Esters

Esters are among the most important compounds for the flavor industry. They are important both for natural foods, such as fruits, and in fermented foods, being present in low concentrations (1–100 ppm) (Kempler, 1983). In fruits, esters are fundamental to aroma of most fruits, for example, melons, apples, pineapple, and strawberries (Chen et al., 2012; El Hadi et al., 2013).

The most abundant ester is ethyl acetate, which is present in most ripe or ripening fruits. In ripening fruits, the formation of esters is boosted by the production of ethanol during the ripening (Gonzalez et al., 2009). Esters were the first compounds produced synthetically, but it is known that these can also be produced by microbiological or enzymatic routes (Janssens et al., 1992; Jong & Birgmingham, 1993).

Some esters are responsible for the aroma identity of some fruits, such as ethyl butanoate, which resembles strawberry aroma, while ethyl hexanoate is characteristic of fresh pineapple and some other tropical fruits. Other examples of esters can be quite characteristic of specific fruits: 3-methylbutyl acetate is characteristic of pear or pear drops; allyl hexanoate is typically pineapple; cis-3-hexenyl butanoate imparts the green leafy aroma of the parent alcohol; and the C9 esters are important for melon aroma (Chen et al., 2010; El Hadi et al., 2013; Parker, 2015).

5.3.3 Aldehydes

Aldehydes are very volatile compounds with characteristic penetrating and nauseating aromas that are extremely common in most foods, and these have a low-odor threshold. The straight-chain unbranched aldehydes are ubiquitous. Acetaldehyde is a crucial component of many fruit flavorings, imparting fruity ether notes, whereas the C3–C5 aldehydes (propanal, butanal, and pentanal) tend to have a rather chemical/malty/green odor to them that is hard to define, depending on their chain length and its concentration present. C5 aldehydes, namely, 2- and 3-methylbutanal (1), have low odor thresholds and are found in most cooked foods as well as in fresh fruits and vegetables.

In general, aldehydes up to eight carbon atoms, such as acetaldehyde (C_2H_4O), formaldehyde (CH_2O), and furfural ($C_5H_4O_2$), have undesirable aromas, while those with longer chains, having more than ten carbons, have pleasant aromas and are generally formed by the reduction of fatty acids, occurring during alcoholic fermentation (Nykänen, 1986; Alcarde et al., 2011). Aldehydes are of great commercial importance, with emphasis on acetaldehyde since they contribute to the freshness and pungency note in foods and drinks.

5.3.4 Alcohols

Alcohols are important constituents of aroma and are found abundant in foods, but their contribution to aroma tends to be less than that of the aldehydes. Alcohols could be formed from the reduction of aldehydes from the amino acids valine and leucine by Strecker degradation, like 2-methylpropanol and 3-methylbutanol, or by reducing methyl ketones to their corresponding alcohols; most frequently obtained from this pathway are 2-heptanol and 2-nonanol. 2- and 3-methylbutanol characterize fruity aromas in concentrations of 50 mg/L for 2-methylbutanol and 1 mg/L for 3-methylbutanol (Smogrovicová et al., 1999).

In fruits, there is a greater presence of straight-chain alcohols that have their quantity increased during the fruit maturation, such as 2-methyl-1-propanol, which can be associated with brown apples and bruised fruit. Other examples of important alcohols are geosmin and 2-methylisoborneol. Both bicyclic alcohols that have aromatic notes are described as imparting earthy, musty notes, although they have very low odor thresholds (Murray & Whitfield, 1975; Parker, 2015).

5.3.5 Lactones

Lactones are cyclic esters of hydroxy acids and are potent aroma compounds found in a wide variety of foods. They are associated with fruit aromas such as coconut (δ-dodecalactone) and peach (4-decalactone), butter (jasmine), sweet (Δ-dodecalactone), and walnuts (γ-decalactone) (Kempler, 1983; Janssens et al., 1992). However, they can also be responsible for undesirable odors such as rancidity. Lactones have high commercial value due to their low concentrations of odor thresholds (Jong & Birgmingham, 1993). The odor thresholds of lactones decrease significantly as the number of constituent carbons increases (Parker, 2015).

These compounds can be formed from amino acids (Heath & Reineccius, 1986) or fatty acids (Feron et al., 1996). Most lactones are produced chemically, but the use of microorganisms has the advantage of producing optically active lactones (Gatfield, 1996). Some fungi of the genus Tricoderma produce 6-pentyl-α-pyrone, which has a coconut aroma (Parker, 2015). Some lactones, like 5-hydroxydec-2-enoic acid lactone, were found in the cocoa that had been used in the production of the chocolate. Meanwhile, the jasmine lactone provides a floral petal-like aroma to green tea (Katsuno et al., 2014).

5.3.6 Ketones

Ketones are characterized by the presence of the carbonyl group and can be classified as aliphatic, aromatic, or phenol derived. Aliphatic ketones, such as acetone and diacetyl, are important in the aroma of mold-ripened cheeses. The ketones produced by biotechnology are mostly aliphatic, such as the 2-alkanones (methyl ketones), produced by fungi.

Other examples of aromatic ketones that are widely used in aroma industries include butanone, raspberry ketone (raspberry flavoring), phenylbutan-2-one (flower odor), and zingerone (ginger aroma) (Lomascolo et al., 1999). Diacetyl and pentan-2,3-dione (vicinal diketones) are ketones that have aromas and flavor characteristics described as buttery, honey, or coffee, being perceived at a concentration of 0.1–0.14 mg/L, while in concentrations above 1 mg/L, the perceived aroma is of cheese.

5.4 VOLATILE MARKERS

Odor can be defined as a complex mixture of volatile chemical molecules, and for these molecules to be odorous; they must react with the nasal mucosa causing the reception of a stimulus by the olfactory sensory system. The types of answers humans provide to these stimuli depend on the sensory properties of the substance evaluated, including on its odor intensity, detection concentration, and the type of sensation, pleasant or unpleasant it provides. In this way, the sensation of odor is associated with a variety of compounds that, when present sufficiently high in the air, trigger responses in exposed persons (Le Cloirec et al., 1991; Cudmore & Dons, 2000).

Gas chromatography is a powerful tool that could be helpful to unveil the volatile composition of a food. According to Grosch (2001), until 1997, about 8,000 volatiles had been identified and reported by the scientific literature using gas chromatography. However, among the identified compounds, several did not possess any odor and only a part of them effectively contributed to the aroma and flavor of food (Stephan et al., 2000).

To facilitate the identification of compounds that contribute to the aroma of food, Rothe and Thomas (1963) and Guadagni et al. (1966) proposed the concept of odor unit – Uo or OAV (Odor Activity Value), which expresses the ratio between the concentration of the volatile compound in the aromatic extract and its detection threshold (threshold) in aqueous or oily solution depending on the nature of the matrix of the food (Guadagni et al., 1972). It should be noted that the higher the OAV value, the greater is the odorant importance of the compound for the product's aroma and taste.

5.5 VOLATILE COMPOUNDS EXTRACTION METHODS

Volatile and semi-volatile compounds play a central role in the quality and sensory properties of food and represent one of the main tools used to monitor the stages of the manufacturing process while studying the changes involved in food products (Franchina et al., 2020 Diez-Simon et al., 2020).

When studying aroma compounds, sample preparation is a critical step as a large amount of analyte can be lost, which consequently leads to significant errors. Thus, the choice of an adequate extraction procedure is of fundamental importance (Diez-Simon et al., 2020; Marín-San Román et al., 2020). Several techniques have been used to extract volatile compounds from fruit samples. Some examples of these methods are the techniques as Purge and Trap (P&T), Solid-Phase Micro-Extraction (SPME), and Stir Bar Sorptive Extraction (SBSE) (Marín-San Román et al., 2020; Lasekan et al., 2022).

5.5.1 Purge & Trap

P&T extraction is one of the most common techniques for the preparation of aqueous samples and consists of a dynamic headspace technique commonly applied for the preparation of these samples for the determination of low concentration organic compounds present in the volatile fraction of food products (Nogueira et al., 2018). In the P&T technique, an inert gas is bubbled continuously through the sample, and the volatile compounds released are trapped by the adsorbent material and subsequently desorbed, either by thermal or solvent desorption (Karimkhan et al., 2022; Eker et al., 2023).

The P&T method is simple, fast, solvent-free, and has as its main advantages the high recovery of very volatile compounds and the low dispersion associated with the use of a fully automated system (Marín-San Román et al., 2020; Xinyu et al., 2020). However, the extraction efficiency is affected by several factors (sample temperature, inert gas flow rate, material used in the trap, and extraction time), indicating the need for experimental optimization of these parameters to obtain adequate amounts of representative compounds analyzed by mostly GC-MS systems (Eker et al., 2023).

5.5.2 Solid Phase Micro-Extraction (SPME)

It is an organic solvent-free extraction technique based on the principle of adsorption, in which analytes from a gas or liquid sample are adsorbed onto a fused silica fiber coated with an absorbent polymeric material (stationary phase) (Marín-San Román et al., 2020; Karimkhan et al., 2022). Among the SPME application modes, headspace extraction (HS-SPME), where the analytes are extracted from the gaseous phase in equilibrium with the sample, has been considered the most suitable for high volatility compounds and has advantages such as low cost, high sensitivity, easy operation, and above all is solvent free (Petronilho et al., 2014; Marín-San Román et al., 2020; Alahmad et al., 2023; Hou et al., 2023).

The extraction efficiency, selectivity, and reproducibility of the SPME technique depend on several experimental parameters, among the most important of which can be mentioned: the SPME fiber coating (phases, polarities, and thicknesses), sampling mode, temperature and extraction time, amount of sample, and addition of salt (salting-out effect) (Petronilho et al., 2014; San et al., 2014; Nogueira et al., 2018; Diez-Simon et al., 2020).

5.5.3 Stir Bar Sorptive Extraction (SBSE)

SBSE is based on sorption extraction, where volatile compounds are extracted onto a coated polymer on a glass-coated magnetic stir bar (Diez-Simon et al., 2020; Liang et al., 2022). The SBSE consists of two steps: extraction and desorption, and the analyte extraction process can be performed both in immersion mode (SBSE) and in headspace mode (HSSE). Both methods have great capacity for extraction at trace levels of volatile to semi-volatile compounds in food matrices and can be used on different types of samples as shown in Table 5.1 (Diez-Simon et al., 2020; Alahmad et al., 2023).

The principles of SBSE are similar to those of SPME, but in this method, higher sensitivities can be achieved due to the larger volume of extraction phase present in the Twister (extractor bar) (Hou et al., 2023; Liang et al., 2022). However, the main disadvantages of the SBSE method are the limited number of adsorbents available on the market and the need for a specific thermal desorption unit (Marín-San Román et al., 2020).

TABLE 5.1
Extraction Techniques for Dehydrated Fruits

Fruit	Sample	Extraction Technique	Adsorbent Material	Reference
Apple	Fresh peel and pulp	HS-SPME	PDMS	Wandjou et al., 2019
Jujube	Fresh and dehydrated fruit	HS-SPME	DVB/CAR/PDMS	Liu et al., 2022
Mandarin	Juice and juice powder	HS-SPME	DVB/CAR/PDMS	Chen et al., 2021
Mango	Fresh and freeze-dried fruit	SPME	DVB/CAR/PDMS	Fratianni et al., 2020
Melon	Juice	P&T	C18	Dagli et al., 2022
Melon	Juice	P&T	ENVI-Carb	Fredes et al., 2016
Mulberry	Dried fruit	P&T	Tenax GR	Hwang et al., 2020
Passion fruit	Juice	HS-SPME and SBSE	CAR/PDMS and PDMS	Xiao et al., 2022
Pear	Pulp powder	SPME	DVB/CAR/PDMS	Wang et al., 2019
Persimmon	Pulp powder	SBSE	PDMS	de Jesus et al., 2023
Sapota	Pulp powder	SBSE	PDMS	Araujo et al., 2023
Strawberry	Fresh and pulp powder	HS-SPME	CAR/PDMS	Frabetti et al., 2021
Tamarillo	Fresh and freeze-dried fruit	SPME	DVB/CAR/PDMS	Diep et al., 2021
Tomato	Fresh and dehydrated fruit	HS-SPME	DVB/CAR/PDMS	Rajkumar et al., 2021
Watermelon	Juice	P&T	ENVI-Carb	Fredes et al., 2016

5.6 DRYING METHODS

5.6.1 Hot Air Drying

Conventional drying methods are predominantly used in industries, as they are low-investment cost methods; among these methods is the hot air drying (HAD), which is one of the most important preservation techniques for fruits and vegetables (Qiu et al., 2019). This process reduces water activity, preventing the food from being spoiled by microorganisms and enzymatic activity, which are the main causes of changes in the organoleptic properties of foods. In addition to these benefits, drying also reduces transport and storage costs due to the low weight of these dehydrated products (Srikanth et al., 2019; Chen et al., 2021; Ojediran et al., 2020).

Although HAD is an extremely advantageous process with regard to food preservation, due to its low cost, this process has some disadvantages. One of the main disadvantages of hot air drying is precisely linked to its process, where, due to different mechanisms of exchange of heat, mass, water, and energy transport, this drying process becomes a complex problem. This method requires high-energy demand and prolonged drying time, which can cause sample shrinkage, reduced density, and rehydration capacity due to high temperature (Onwude et al., 2016). In addition, the prolonged period in which the food is exposed to high temperatures alters its organoleptic characteristics, worsening the taste, color, and nutritional content of the product, and limiting its applications (Lin et al., 2020).

All the characteristics, such as texture, flavor, color, and the aroma of food, are also affected during the HAD process. The heat used in this process causes losses of some volatile compounds. These losses are directly linked to some parameters, such as temperature, concentration of solids in food, solubility of these compounds in water vapor, and vapor pressure of these substances. The reduction of aroma compounds in dehydrated foods can be avoided by controlling the drying conditions, which may minimize the loss of flavor in these products (Celestino, 2010).

5.6.2 Freeze Drying

Being a process that consists of removing water by sublimation of ice, freeze drying, also known as lyophilization, is a technique already known since the end of the 19th century, but it was only used in industrial processes after the Second World War. This drying process was initially used in the pharmaceutical industry. In the food industry, this technique produces foods with superior quality when compared to foods dehydrated through other techniques, as this process is carried out at low temperatures, whereby it preserves the flavor, color, and appearance of the food, without any adverse effects caused by heat, such as sample shrinkage (Berk, 2013).

Freeze drying is based on the passage of the water present in the samples from the solid to the gaseous state without going through the liquid phase (Capozzi et al., 2019). During this process, some changes occur due to freezing, such as changes in pH, dehydration, and formation of interfaces between water and ice, among others (Tang & Pikal, 2004). Although the samples undergo some changes during lyophilization, this type of drying has the advantage of causing minimal changes in structure, texture, and appearance; preserving aroma, flavor, biological activity, and nutrients, in addition to facilitating their transport and storage (Varzakas & Tzia, 2014).

The aroma is a very important sensorial characteristic to certify the quality of dehydrated fruits; however, fruits that go through the lyophilization process may have loss of aroma. The changes in the aroma of freeze-dried fruits are a complicated and complex physical and chemical process, which is related to the process conditions. Some causes of the loss of aroma compounds are related to their vapor pressure, where volatile compounds that have low vapor pressure can be combined with the vapor from the sublimated water, which is trapped in the cold trap of the lyophilizer. Another cause of aroma loss is linked to the vacuum pump, where these compounds are condensed and discharged (Yan et al., 2019; Gou et al., 2023).

5.6.3 Spray Drying

Drying by atomization consists of removing water or organic solvents from the droplets formed in the process, dried by spraying, in a short period of time, producing particles with sizes ranging from nanometers to micrometers. In this type of drying, spray dryer is used (Jubaer et al., 2020; Wang et al., 2021). This drying technique aims to create an external membrane or coating of one material over another, applied for the protection and/or preservation of sensitive substances that are biochemically and thermally degraded, such as bioactive and volatile compounds (Shishir & Chen, 2017).

Widely used in the food, chemical, pharmaceutical, and cosmetic industries to transform liquids into powders, spray drying extends shelf life, concentrates, and preserves components such as aromatic compounds. This drying process has been applied in various types of food production: egg products, drinks, plant proteins, carbohydrates, fruit and vegetable extracts, yogurt, etc. One of the food industries that shows great potential is by using this type of drying for fruit pulp, consequently enabling the emergence of new products, in addition to providing greater conservation and even facilitating transportation (Shishir & Chen, 2017).

What's more, for preserving and concentrating food components, atomization is an easy, continuous process that features aseptic drying conditions. Despite the advantages, this drying process has disadvantages such as having a high cost, low thermal efficiency, and the possibility of degradation of the product by heat. If the drying temperature is high, this ends up affecting compounds in the food, such as volatile and bioactive components, which are sensitive to high temperatures (Varzakas & Tzia, 2014).

5.7 AROMA ASPECTS OF DRIED FRUITS

Dehydration is defined by the application of heating temperature under controlled conditions to remove a large amount of water present in food by evaporation. The dehydration technique has advantages such as reducing the weight and volume of the food, reducing transport and storage costs, and making these products easy to handle many times. However, it can also cause both sensory and nutritional deterioration of the food, so it is necessary to study processing conditions that lead to the minimization of these changes, mostly in the aroma compounds, which are very sensitive to high temperatures.

Considering the aforementioned information, it is necessary to study deeper the parameters of all dehydration processes, like temperature, dehydration time, air velocity, and the use of encapsulating agents, which do not impact the aroma compounds of dried fruit products, especially the volatile compounds that are responsible for the specific aroma characteristics of the fruit. This way it will be possible to obtain products that will last for longer time than the fresh fruits but will also retain the main aroma compounds of the fresh fruit.

5.8 DRYING OF TROPICAL FRUITS

5.8.1 Soursop

Soursop is a fruit that is one of the most important fruit species of the *Annonaceae* family and it is highly appreciated for its succulence, sweet and sub-acid flavor, and having a pleasant and distinctive aroma. A great variety of volatile compounds of different chemical classes are present in soursop pulp; among these, mostly aliphatic esters represent quantitatively the largest chemical class of volatile compounds present in soursop pulp (Cheong et al., 2010; Cheong et al., 2014; Santana et al., 2017). These compounds are very susceptible to high temperatures used in drying process and hence require more studies about the impact of drying process on aroma aspects of soursop.

Leite Neta et al. (2019) studied the impact of atomization process on the soursop pulp aroma using SBSE technique, which uses neither solvent nor higher temperatures in the extraction process. The authors found that in the dehydrated soursop pulp, the total amount of volatiles in the rehydrated dried powder was reduced to 74.31%, resulting in a significant loss of these compounds. The authors also concentrated their analysis of soursop pulp on OAV. The use of OAV data can be an effective method to analyze the real contribution of the volatile compounds which have a greater impact on the aroma (Zhu et al., 2016). Following an established procedure, the OAV is calculated by dividing the concentration of the individual compounds by their threshold concentrations. Compounds with OAV > 1 were considered promising. Table 5.2 presents the data obtained for the aroma analyses of soursop fresh pulp and rehydrated dried powder, organized by chemical classes with OAV, which were calculated based on odor thresholds of the compounds reported in the literature.

Among the compounds that had OAVs > 1 but could not be found in the soursop rehydrated dried powder were 2-phenylethanal, ethyl octanoate, α-pinene, methyl

TABLE 5.2
Volatile Compounds of Fresh Soursop Pulp and Its Dehydrated Powder

Compounds	LRI^E	LRI^L	OT	Fresh Pulp	OAV	Dried Powder	OAV	Odor Descriptors
Methyl (*E*)-2-butenoate	755	756		13.59 ± 2.35		5.62 ± 0.97		green
Methyl 2-butenoate	760	755		4.57 ± 0.79		2.73 ± 0.47		
Methyl butanoate	762	723	0.059	2.73 ± 0.47	46	1.38 ± 0.23	23	fruity, sweet
3-(*Z*)-hexenal	802	795	0.00025	0.31 ± 0.05	1240	0.17 ± 0.02	680	green, fatty, grassy
(*E*)-2-hexenal	858	844	0.017	2.44 ± 0.42	144	2.85 ± 0.49	168	green, banana, aldehydic
(*Z*)-3-hexen-1-ol	865	859	0.07	7.34 ± 1.27	105	5.04 ± 0.87	72	green
(*E*)-2-hexen-1-ol	876	862	6.7	1.54 ± 0.26	< 1	0.43 ± 0.07	< 1	fresh, green, leafy
1-hexanol	878	862	2.5	2.21 ± 0.38	< 1	1.20 ± 0.20	< 1	herbal
Heptanal	903	896	0.003	0.08 ± 0.01	27	0.07 ± 0.00	23	green, fresh, aldehydic
Methyl 5-hexenoate	916	894		0.13 ± 0.01		0.06 ± 0.00		
γ-Butyrolactone	925	914	20	0.06 ± 0.00	< 1	0.04 ± 0.00	< 1	creamy, oily, fatty, caramel
Methyl hexanoate	932	936	0.07	33.34 ± 5.78	476	5.25 ± 0.90	75	fruity
α-pinene	934	931	0.062	0.35 ± 0.05	6	Nd		herbal, fresh, camphor
Methyl (*Z*)-3-hexenoate	938	917		4.53 ± 0.78		0.38 ± 0.06		earthy, green, fruity
Ethyl 3-hydroxybutyrate	955	949		0.06 ± 0.00		0.04 ± 0.00		fruity, green
Methyl (*E*)-2-hexenoate	976	953		91.01 ± 15.80		12.88 ± 2.23		fatty, fruity, green
β-pinene	977	970	0.14	0.45 ± 0.07	3	0.36 ± 0.05	3	herbal, dry woody

(Continued)

TABLE 5.2 ***(Continued)***
Volatile Compounds of Fresh Soursop Pulp and Its Dehydrated Powder

Compounds	LRI[E]	LRI[L]	OT	Fresh Pulp	OAV	Dried Powder	OAV	Odor Descriptors
Methyl-2-hydroxy-4-methylpentanoate	991	983		2.58 ± 0.44		1.58 ± 0.27		sweet, fruity, musty
Methyl 2-hydroxy-3-methylpentanoate	995	989	0.0024	1.03 ± 0.17	429	0.69 ± 0.11	288	fruity, estery, caramellic
Hexanoic acid	999	999	0.42	0.64 ± 0.10	2	0.48 ± 0.04	1	fatty, sour, sweat
Ethyl hexanoate	1001	980	0.001	0.71 ± 0.12	710	0.34 ± 0.05	340	sweet, fruity, waxy
Methyl 2,4-hexadienoate	1018	991		2.64 ± 0.45		0.35 ± 0.05		
4-methyl-5-vinylthiazole	1028	1011		2.27 ± 0.39		Tr		musty, nutty
Limonene	1031	1030	0.2	0.64 ± 0.10	3	0.53 ± 0.09	3	citrus
2-ethyl-1-hexanol	1032	1020		0.42 ± 0.07		0.28 ± 0.04		citrus
Phenylethanal	1045	1048	0.004	0.10 ± 0.01	25	Nd		green, sweet, floral
Ethyl 2-hexenoate	1047	1020		0.12 ± 0.01		Nd		rum, fruity, green
α-ocimene	1051	1050		0.11 ± 0.01		0.21 ± 0.03		fruity
Butyl isovalerate	1055	1044		0.26 ± 0.04		0.51 ± 0.08		fruity, fruit, sweet
Acetophenone	1068	1066	0.065	0.14 ± 0.02	2	0.11 ± 0.01	2	floral
cis-linalool oxide	1075	1070	0.32	0.37 ± 0.06	1	Nd		citrus, green
trans-Linalool oxide	1090	1089	3	0.17 ± 0.02		Nd		floral
Methyl benzoate	1096	1097	0.52	0.77 ± 0.13	1	0.11 ± 0.01		phenolic, wintergreen,
Linalool	1104	1100	0.025	12.84 ± 2.22	514	0.86 ± 0.14	34	citrus, floral, sweet
Nonanal	1106	1104	0.008	0.46 ± 0.07	58	0.24 ± 0.03	30	aldehydic, waxy, rose
Methyl octanoate	1126	1123	0.2	0.28 ± 0.02	1	0.29 ± 0.03	1	waxy, green, sweet
Methyl nicotinate	1140	1137		1.96 ± 0.33		1.43 ± 0.24		herbal, warm
Methyl (*E*)-2-octenoate	1172	1163		0.46 ± 0.07		0.61 ± 0.10		hairy, costus, sweet
Octanoic acid	1180	1181	3	0.19 ± 0.03	< 1	0.20 ± 0.03	< 1	fatty, waxy, rancid
α-terpineol	1193	1195	0.25	0.18 ± 0.02	< 1	0.10 ± 0.01	< 1	terpene, citrus
Ethyl octanoate	1198	1183	0.005	0.08 ± 0.01	16	Nd		fruity, wine, waxy

(Continued)

TABLE 5.2 *(Continued)*
Volatile Compounds of Fresh Soursop Pulp and Its Dehydrated Powder

Compounds	LRI[E]	LRI[L]	OT	Fresh Pulp	OAV	Dried Powder	OAV	Odor Descriptors
Decanal	1199	1209	0.006	0.30 ± 0.05	50	0.25 ± 0.04	42	sweet, aldehydic, waxy
Benzothiazole	1222	1218	0.05	0.07 ± 0.00	1	0.09 ± 0.01	2	sulfury, rubbery
Dimethyl hexanedioate	1241	1237		0.10 ± 0.01		Nd		mild, nutty
Geraniol	1253	1258	0.04	0.96 ± 0.16	24	0.10 ± 0.01	3	sweet, floral, fruity
γ-octalactone	1258	1261	0.007	0.70 ± 0.11	100	0.26 ± 0.04	37	sweet, coconut, waxy
Dibutylnitrosoamine	1266	1269		0.14 ± 0.02		Nd		
Methyl 3-phenylpropionate	1277	1278		0.67 ± 0.11		0.08 ± 0.01		honey, fruity, wine
Dibutylformamide	1306	1319		2.03 ± 0.34		0.21 ± 0.03		
N,N-dibutylacetamide	1360	1332		0.33 ± 0.05		Nd		vanilla
γ-nonalactone	1366	1341	0.0097	0.47 ± 0.07	48	0.17 ± 0.02	18	creamy, sweet
Farnesane	1374	1382		0.08 ± 0.01		Nd		
Methyl (*E*)-cinnamate	1387	1386	0.191	9.42 ± 1.63	49	0.89 ± 0.15	5	sweet, balsam
β-caryophyllene	1430	1424	0.064	5.99 ± 1.03	94	4.02 ± 0.69	63	spicy
Geranyl acetone	1455	1454	0.06	0.19 ± 0.03	3	0.11 ± 0.01	2	fresh, floral
α-humulene	1463	1467	0.16	0.46 ± 0.07	3	0.31 ± 0.05	2	woody
γ-decalactone	1472	1472	0.0011	0.21 ± 0.03	191	0.09 ± 0.01	82	fruity, fresh, oily
Geranyl isovalerate	1475			0.04 ± 0.00		0.04 ± 0.00		green, fruity
β-copaene	1490			0.37 ± 0.06		Nd		
Methyl vanillate	1522			0.06 ± 0.00		Nd		floral
β-sesquiphellandrene	1530	1524		0.08 ± 0.01		0.11 ± 0.01		herbal, fruity
Dihydroactinolide	1540	1532		0.18 ± 0.02		0.13 ± 0.02		fruity, ripe, apricot
Dodecanoic acid	1564	1559		0.15 ± 0.02		Nd		mild, fatty, coconut
Caryophyllene oxide	1595	1606		0.08 ± 0.01		0.09 ± 0.01		woody
Benzophenone	1636	1625		0.09 ± 0.01		0.22 ± 0.03		balsam, rose
Methyl jasmonate	1653	1628	5.7	0.28 ± 0.04	< 1	Nd		floral, jasmin
Methyl hydrojasmonate	1658	1657		0.28 ± 0.04		0.18 ± 0.02		floral
(*Z,E*)-α-bergamotol	1706	1700		0.11 ± 0.01		Nd		woody
Tetradecanoic acid	1763	1763	10	0.39 ± 0.06	< 1	0.15 ± 0.02		waxy, fatty, soapy

(Continued)

TABLE 5.2 *(Continued)*
Volatile Compounds of Fresh Soursop Pulp and Its Dehydrated Powder

Compounds	LRI^E	LRI^L	OT	Fresh Pulp	OAV	Dried Powder	OAV	Odor Descriptors
(*Z*)-9-hexadecenal	1814	1800		0.06 ± 0.00		Nd		
Pentadecanoic acid	1861	1860	10	0.14 ± 0.02	< 1	Nd		waxy
Methyl hexadecanoate	1926	1912	1	0.06 ± 0.00	< 1	Nd		oily, waxy, fatty
Hexadecanoic acid	1964	1965	10	0.50 ± 0.08	< 1	0.44 ± 0.07		slightly, waxy, fatty
Octadecanal	1994	1999		0.08 ± 0.01		Tr		oily
Methyl 10-octadecenoate	2102			0.14 ± 0.02		0.14 ± 0.02		
Ethyl linoleate	2164	2159		0.13 ± 0.01		Nd		
Ethyl *Z*-9-octadecenoate	2169	2175	130	0.30 ± 0.05	< 1	Nd		fatty, oily, dairy

LRI^E, Linear retention index obtained experimentally; LRI^L, Linear retention index from literature; OT, Odor threshold values; OAV, Odor activity value

benzoate, and cis-linalool oxide. Thus, these compounds were considered to be of low aroma intensity for soursop. The classes of volatile compounds which had higher reduction (40.19% to 79.10%) in their concentrations in soursop rehydrated dried powder were the esters, terpenes, acids, and alcohol.

The use of spray drying process in soursop pulp resulted in higher reductions in the concentrations of the main volatile compounds, which varied from 32.89% to 93.3%. Because of this decrease in compound concentrations, the decreases in OAVs of these compounds are also natural. However, the odoriferous compounds, which decreased in concentrations expressed in OAV between soursop pulp and rehydrated dried powder, continued to be responsible for the pleasant aroma of soursop powder, and these compounds were: 3-(Z)-hexenal (1240.00 to 680, OAV), ethyl hexanoate (710 to 340, OAV), methyl hexanoate (476.29 to 75.00, OAV), methyl 2-hydroxy-3-methyl pentanoate (429.17 to 287.50, OAV), γ-decalactone (190.91 to 81.82, OAV), (E)-2-hexenal (143.53 to 167.65, OAV), (Z)-3-hexen-1-ol (104.86 to 72.00, OAV), and β-caryophyllene (93.59 to 62.81, OAV), with the exception of linalool (513.60 to 34.40, OAV) and γ-octalactone (100.00 to 37.14, OAV).

All compounds with OAVs > 1 in the fresh pulp also had OAVs > 1 in the soursop dehydrated powder. This way, the rehydrated dried powder maintained the presence of all the key aroma compounds, preserving the pleasant soursop aroma. Thus, the soursop powder can be considered as a viable product as it retains the volatile and bioactive compounds present in the fresh pulp. The significant contribution of these compounds in spray dried soursop powder can be attributed to the ability of the effective microencapsulation process, which was facilitated using maltodextrin (Forero et al., 2015).

5.8.2 Tomato

Tomato (*Lycopersicon esculentum* Mill.) is an important vegetable crop cherished all over the world for its consumption as a raw vegetable or as a processed product. Tomatoes are subjected to various drying processes so as to increase product stability and decrease postharvest loss. However, drying process causes loss of volatile compounds, and hence there is a need for more studies to perform so as to deepen our knowledge on the volatile profile of tomatoes subjected to hot air, freeze, and spray drying processes, and also to understand the variations in aroma profile for attaining overall acceptability of consumers for dehydrated tomato products.

Some publications report that, in general, aldehydes, alcohols, terpenes, and esters are considered as the main products that contribute to the aroma of any foodstuff. Martina et al. (2021) reported that (Z)-3-hexanal, (Z)-3-hexanol, hexanal, 1-penten-3-one, 3-methyl butanal, (E)-2-hexenal, 6-methyl-5-hepten-2-one, methyl salicylate, 2-isobutylthiazole, and β-ionone are the characteristic compounds contributing to the aroma of tomatoes. The interaction of cis-hex-3-enal, 2-methylhept-2-en-6-one, eugenol, and β-ionone resulting in desirable effects in the aroma of tomatoes has been reported by Jeyaprakash et al., (2020).

Rajkumar et al. (2021) studied the influence of drying methods on the aroma profile of tomatoes. Table 5.3 shows the volatile compounds and OAV of the compounds found in fresh and dehydrated tomatoes. In this work, the contents of aldehydes, alcohols, and esters underwent variations, which were expected to contribute to aroma changes among the dehydrated products studied. The compounds such as aldehydes, (E)-hex-2-enal, (E,E)-2,4-heptadienal, 2,3- dimethyl pentenal, 2-butanol, (Z)-2-penten-1-ol, and undecalactone contributing to fruity, and sweet and partly green, and fat odors increased in all tomato products subjected to drying. Some compounds were found only in the dehydrated tomato products, such as 2-methyl butanal, 3-methyl butanal, hexanol, and terpinyl acetate. Another interesting fact about the drying process of tomatoes was the formation of new compounds such as 2, 3-pentadione, 3-nonenone, β-ionone, and 2-ethyl furan.

The increase in volatile compounds is one of the consequences of the drying process since a large quantity of water present in tomatoes gets removed. Tandon et al. (2003) reported that the increased levels of compounds contributing to floral (β-ionone) and fruity, green ((Z)-3-hexenal, and geranyl acetone) are beneficial to tomato flavor.

5.9 MANGO

Mango is one of the most consumed and appreciated fruits in the world, mainly due to its peculiar and characteristic flavor, its succulence, and its sweet and exotic taste (Jahurul et al., 2015; Matheyambath et al., 2016). Dehydration process is one of the methods of conservation and consists in reducing the water activity of the product; several publications relate to these techniques, viz., convection drying and lyophilization are among the methods used to increase the shelf life of fruits (Fongin et al., 2017; Link et al., 2017; Pu & Sun, 2017; Rajkumar et al., 2017).

Leite Neta et al. (2018) analyzed the effects of drying methods: convection drying and lyophilization in the aroma compounds of mango. Table 5.4 shows the

TABLE 5.3
Volatile Compounds of Fresh Tomato Pulp and Its Dehydrated Powder

Compounds	LRI[E]	OT	Fresh	OAV	Hot Air Drying	OAV	Freeze Drying	OAV	Spray Drying	OAV	Odor Descriptors
Acetaldehyde	402	0.01	1.01 ± 0.0	101	1.49 ± 0.3	149	2.53 ± 0.57	253	2.02 ± 0.01	202	Fruity
Dimethyl sulfide	506	0.0011	2.31 ± 0.62	2100	3.48 ± 0.95	3164	4.07 ± 0.43	3700	3.97 ± 0.14	3609	Sulfur
Propanol	529		0.88 ± 0.25		0.35 ± 0.01				0.52 ± 0.12		Alcohol, pungent
2-Butanol	592		1.32 ± 0.01		0.56 ± 0.12		1.79 ± 0.94		1.47 ± 0.63		Alcohol
2-Methylfuran	635		0.23 ± 0.10		2.04 ± 0.14				0.92 ± 0.04		Sweet
2-Methyl butanal	637	0.0125							0.3 ± 0.01	24	
3-Methylfuran	638				0.07 ± 0.01				0.02 ± 0.00		
4-Penten-2-ol	642		0.46 ± 0.07								
3-Methyl butanal	645	0.009					0.14 ± 0.02	16	0.02 ± 0.00	2	Malt
1-Penten-3-one	652	0.001							0.03 ± 0.00	30	Strong
1-Butanol	661	0.5	1.23 ± 0.02	2	2.59 ± 0.21	5	2.96 ± 0.18	6	2.92 ± 0.52	6	Fruit, sweet
2,3-Pentadione	670		0.17 ± 0.01		0.19 ± 0.01		0.34 ± 0.02		0.21 ± 0.15		Cream, butter
Diethyl sulfide	697								0.82 ± 0.06		
Thiophene	699		0.04 ± 0.00								Pungent
3-Methyl-1-butanol	703	0.25	1.17 ± 0.16	5	0.41 ± 0.01	2			0.69 ± 0.24	3	Wine
2-Ethylfuran	731		0.06 ± 0.01		0.13 ± 0.01		0.59 ± 0.21		0.41 ± 0.05		
4-Methyl-3-penten-2-one	780								0.06 ± 0.00		
(Z)-3-Hexanal	813		3.52 ± 0.3		2.09 ± 0.88		3.01 ± 0.56		3.22 ± 0.14		Grass, tallow
(E)-Hex-2-enal	822	0.017	0.84 ± 0.21	49	2.56 ± 0.98	151	3.44 ± 0.45	202	2.11 ± 0.63	124	Sweet, fruit, green
5-Methyl-2-hexanone	828		2.15 ± 0.11		0.07 ± 0.01						Fruity
Hexanol	858	2.5			3.63 ± 0.98	1	4.21 ± 1.01	2	4.04 ± 0.12	2	Flower, green
(E)-4-Hexen-1-ol	881		0.13 ± 0.02		0.06 ± 0.01						Green

(Continued)

TABLE 5.3 *(Continued)*
Volatile Compounds of Fresh Tomato Pulp and Its Dehydrated Powder

Compounds	LRI^E	OT	Fresh	OAV	Hot Air Drying	OAV	Freeze Drying	OAV	Spray Drying	OAV	Odor Descriptors
Heptanal	901	0.003	1.57 ± 0.02	523	1.20 ± 0.11	400	1.64 ± 0.84	547	1.21 ± 0.06	403	Citrus, fat
α-Pinene	938	0.006	0.61 ± 0.04	102			0.21 ± 0.02	35	0.19 ± 0.01	32	Turpentine
6-Methyl-5-hepten-2-one	942	0.05			0.27 ± 0.00	5			0.01 ± 0.00	0.2	Pungent, fruity
(E,E)-2,4-Heptadienal	1010	0.056	0.15 ± 0.04	3	0.92 ± 0.54	16	1.12 ± 0.23	20	0.06 ± 0.01	1	Nut, fat
3-δ-Carene	1014	0.77	0.34 ± 0.07	0.4	1.02 ± 0.54	1	1.15 ± 0.75	1	0.66 ± 0.21	1	Lemon, resin
(E)-2-Octenal	1029	0.003	0.32 ± 0.04	107					0.42 ± 0.10	140	Green
β-Phellandrene	1031		0.53 ± 0.15		1.07 ± 0.52		1.42 ± 0.35		0.88 ± 0.06		Mint, turpentine
(E)-Linalool oxide	1064	3					0.31 ± 0.06	0.1			Flower
Linalool	1082	0.025	1.11 ± 0.13	44	2.42 ± 0.45	97	2.74 ± 0.11	110	2.40 ± 0.90	96	Flower
(E)-2-Nonenal	1134	0.00008	0.03 ± 0.01	375			0.21 ± 0.01	2625	0.16 ± 0.02	2000	Cucumber
Eugenol	1135	0.006	1.12 ± .021	187	1.73 ± 0.41	288	2.06 ± 0.32	343	1.95 ± 0.16	325	Sweet
γ-Terpineol	1163		0.03 ± 0.01		0.56 ± 0.04		0.94 ± 0.41		0.62 ± 0.06		Pine
Hydroxymethylfurfural	1164				0.09 ± 0.01		0.02 ± 0.00		0.16 ± 0.01		Caramel
α-Terpineol	1174	0.33	3.18 ± 0.91	10	4.22 ± 1.01	13	5.62 ± 1.24	17	5.44 ± 0.32	16	Floral, sweet
Methyl salicylate	1276	0.04	0.83 ± 0.01	21	0.11 ± 0.00	3	0.25 ± 0.01	6.25	0.14 ± 0.00	4	Smoky
Terpinyl acetate	1334				1.08 ± 0.01		1.59 ± 0.25		1.36 ± 0.38		Wax
Geranyl acetone	1418	0.06							0.16 ± 0.11	3	Green
α-Ionone	1423	0.0026	0.03 ± 0.01	12			0.05 ± 0.01	19			Wood
β-Damascenone	1437		1.08 ± 0.13								Apple
β-Ionone	1457	0.000007	0.09 ± 0.02	12857	0.14 ± 0.01	20000	0.33 ± 0.13	47143	0.25 ± 0.01	35714	Flower
γ-Undecalactone	1601	0.007	1.58 ± 0.25	226	2.13 ± 0.04	304	2.89 ± 0.82		2.40 ± 0.68	343	Fruity
(E,E)-Farnesyl acetate	1830		0.05 ± 0.01				0.11 ± 0.01		–		Oil, wax

LRI^E, Linear retention index obtained experimentally; OT; Odor threshold values reported: Pino and Mesa (2006), De Jesus et al. (2023), Neta et al. (2019); OAV, Odor activity value compounds and their respective OAVs of the fresh pulp and of rehydrated dried powder. The volatile profile of the fresh mango was characterized predominantly by terpenes (18 compounds), as already reported in other mango cultivars (Pino et al., 2005; Pandit et al., 2009; Bonneau et al., 2016), esters (5 compounds), alcohols (6 compounds), ketones (3 compounds), aldehydes (3 compounds), lactones (5 compounds), and acid (1 compound). However, in the dehydrated products of mango, there was a higher variation when compared to the fresh samples. The total amount of volatiles present in fresh pulp decreased between 30 and 48% in all dehydrated products. Terpene compounds (81.48%) were again the main volatile compounds in fresh mango, followed by esters (17.26%), which together accounted for almost 98% of the volatile composition of this fruit.

Other publications show that the main volatile markers for the characteristic aroma of mango are the δ-3-carene, α-pinene, limonene, terpinolene, β-caryophyllene, β-ionone, the ethyl butanoate ester, and (*E*)-2-nonenal and (2*E*, 6*Z*)-nonadienal aldehydes (Andrade et al., 2000; Pino & Mesa, 2006; Canuto et al., 2009).

In the drying process, considerable losses occurred in relation to the mango aroma compounds (Table 5.4). δ-3-carene was the predominant compound in fresh mango, followed by ethyl-butanoate, which are the main responsible compounds for the characteristic aroma of mango. The authors observed that lyophilization showed higher retention of both compounds compared to that in the convection drying. Other compounds retained better were terpinolene and (*E*)-2-nonenal. The losses could be explained by the physicochemical properties of the terpenoids, which are hydrocarbons prone to evaporative losses (Choia et al., 2018). These changes in the volatile compounds may still be the result of formation and degradation reactions occurring in tissue associated with cell stress induced by drying processes, such as enzymatic and Maillard reactions (Fennema et al., 2010). The number of terpenes decreased in the dehydrated samples, while some esters were formed in the samples submitted to convection drying, possibly due to degradation of alcohols and aldehydes.

The volatile compounds responsible for the aroma of the mango were higher than in the dehydrated mango powder, indicating that they resist to drying process, while the main active odorant contributors remained were (2*E*, 6*Z*)-nonadienal, ethyl butanoate, (*E*)-β-ionone, (*E*)-2-nonenal, and δ-3-carene.

5.9.1 Sapota

Sapota (*Manilkara zapota*) is a fruit belonging to the Sapotaceae family and was originated in Mexico. It is now the sixth most important fruit in India and is largely grown throughout the subcontinent of the country (Ghani, 2003; National Horticulture Board, 2021). Since it has a sweet and soft pulp with a granular texture, sapota is a very popular fruit (Siddiqui et al., 2014). Another important sensory characteristic of this fruit is its aroma, which is sweet and slightly astringent (Oliveira et al., 2011).

There are some publications relating to dehydration processes of sapota fruit and fruits belonging to the same family (Ganjyal et al., 2003; Jangam et al., 2008; Chong & Wong, 2015; Bala et al., 2017; Kumar et al., 2018; Lasekan & Yap, 2018). For the production of a dehydrated product, a spray dryer is widely used, which is an

TABLE 5.4
Volatile Compounds of Fresh Mango Pulp and Its Dehydrated Powder

						Convection Drying						Freeze Drying		
Compounds	LRI[E]	LRI[L]	OT	Fresh	OAV	40°C	OAV	50°C	OAV	60°C	OAV	−55°C	OAV	Odor Descriptors
α-Pinene	1021	1024	0.006	2.53 ± 0.30	422	2.99 ± 0.35	498	0.19 ± 0.02	32	2.25 ± 0.26	375	2.82 ± 0.03	469	herbal, fresh, camphor
Ethyl butanoate	1040	1041	0.001	7.04 ± 0.82	7040	3.92 ± 0.46	3919	2.99 ± 0.35	2987	2.34 ± 0.27	2342	5.15 ± 0.60	5149	sweet, fragrant, fruity
3-Carene	1145	1145	0.77	49.13 ± 5.73	64	20.62 ± 2.41	27	28.14 ± 3.28	37	25.19 ± 2.94	33	30.97 ± 3.61	40	lemon, resin
Ethyl (E)-crotonate	1169	1166		6.00 ± 0.70		4.72 ± 0.55		1.40 ± 0.16		3.52 ± 0.41		7.96 ± 0.93		fermented
Limonene	1199	1198	0.01	2.66 ± 0.31	266	0.27 ± 0.03	27	0.07 ± 0.01	7	1.92 ± 0.22	192	0.39 ± 0.05	39	citrus
3-Methyl-1-butanol	1207	1207	0.25	0.09 ± 0.01	< 1	0.10 ± 0.01	< 1	0.23 ± 0.03	1			0.03 ± 0.00	< 1	Wine
Ethyl hexanoate	1233	1234	0.005			2.67 ± 0.31	533	3.16 ± 0.37	631	0.40 ± 0.05	80			sweet, fruity, pineapple
(E)-β-Ocimene	1235	1235	0.034	0.15 ± 0.02	4									sweet, herb
(Z)-β-Ocimene	1246	1242	0.055	0.19 ± 0.02	4					0.75 ± 0.09	14	0.06 ± 0.01	1	citrus, herb, flower
p-Cymene	1263	1265	0.15	0.14 ± 0.02	1	0.01 ± 0.00	< 1	0.01 ± 0.00	< 1	0.20 ± 0.02	1	0.01 ± 0.00	< 1	citrus
γ-Terpinene	1274	1275	1	0.32 ± 0.04	< 1	0.02 ± 0.00	< 1	0.01 ± 0.00	< 1	0.16 ± 0.02	< 1	0.01 ± 0.00	< 1	turpentine
Terpinolene	1277	1280	0.2	4.79 ± 0.56	24	0.52 ± 0.06	3	0.09 ± 0.01	< 1	4.32 ± 0.50	22	0.65 ± 0.08	3	floral, sweet
3-(Z)-Hexenol	1389	1389	0.07	0.04 ± 0.00	1	0.06 ± 0.01	1	0.03 ± 0.00	< 1			0.04 ± 0.00	1	green
p-Cymenene	1431	1438		0.12 ± 0.01						0.05 ± 0.01		0.04 ± 0.00		
1-Heptanol	1460	1462	0.425	0.02 ± 0.00	< 1	0.06 ± 0.01	< 1	0.05 ± 0.01	< 1	0.01 ± 0.00	< 1	0.02 ± 0.00	< 1	herbal, green, sweet

(Continued)

TABLE 5.4 *(Continued)*
Volatile Compounds of Fresh Mango Pulp and Its Dehydrated Powder

						Convection Drying						Freeze Drying		Odor
Compounds	LRI^E	LRI^L	OT	Fresh	OAV	40°C	OAV	50°C	OAV	60°C	OAV	−55°C	OAV	Descriptors
α-Copaene	1485	1488		0.26 ± 0.03		0.09 ± 0.01		0.06 ± 0.01		2.02 ± 0.24		0.10 ± 0.01		wood, spice
Pinocamphone	1536	1539		0.03 ± 0.00		0.02 ± 0.00		0.02 ± 0.00		0.01 ± 0.00		0.01 ± 0.00		
(E)-2-Nonenal	1537	1536	0.00008	0.04 ± 0.00	492	0.03 ± 0.00	384	0.04 ± 0.00	457	0.03 ± 0.00	403	0.04 ± 0.04	470	cucumber, fat, green
Ethyl 2-hydroxy-4-methylpentanoate	1547	1548				0.01 ± 0.00		0.02 ± 0.00		0.01 ± 0.00		0.01 ± 0.00		
Linalol	1554	1553	0.006	0.07 ± 0.01	12	0.02 ± 0.00	3	0.01 ± 0.00	2	0.01 ± 0.00	2			citrus, floral, sweet
1-Octanol	1563	1563	0.19	0.02 ± 0.00	< 1	0.01 ± 0.00	< 1	0.01 ± 0.00	< 1	0.01 ± 0.00	< 1	0.01 ± 0.00	< 1	green, citrus, floral
5-Methylfurfural	1578	1578		0.03 ± 0.00		0.01 ± 0.00		0.03 ± 0.00		0.14 ± 0.02		0.01 ± 0.00		almond, caramel
(2E,6Z)-Nonadienal	1582	1587	0.00002	0.39 ± 0.05	19420	0.04 ± 0.00	1945	0.23 ± 0.03	11674		0	0.05 ± 0.01	2742	cucumber, wax, green
(3Z)-3-Hexenyl (2E)-2-butenoate	1585	1543				0.04 ± 0.00		0.01 ± 0.00		0.01 ± 0.00				
β-Caryophyllene	1590	1591	0.064	1.15 ± 0.13	18	0.77 ± 0.09	12	1.10 ± 0.13	17	3.66 ± 0.43	57	0.43 ± 0.05	7	spice
Butyrolactone	1632	1630	0.035	0.01 ± 0.00	< 1	0.10 ± 0.01	3	0.04 ± 0.00	1	0.10 ± 0.01	3	0.01 ± 0.00	< 1	creamy, oily, caramel

(Continued)

TABLE 5.4 *(Continued)*
Volatile Compounds of Fresh Mango Pulp and Its Dehydrated Powder

						Convection Drying						Freeze Drying		
Compounds	LRI^E	LRI^L	OT	Fresh	OAV	40°C	OAV	50°C	OAV	60°C	OAV	−55°C	OAV	Odor Descriptors
Butyric acid	1636	1638	4.8	0.04 ± 0.00	< 1	0.36 ± 0.04	< 1	0.40 ± 0.05	< 1	0.35 ± 0.04	< 1	0.06 ± 0.01	< 1	rancid, cheese, sweat
Ethyl decanoate	1643	1642	6.3			0.23 ± 0.03	< 1	0.11 ± 0.01	< 1	0.04 ± 0.00	< 1			soapy
α-Caryophyllene	1663	1663	0.16	0.60 ± 0.07	4	0.33 ± 0.04	2	0.28 ± 0.03	2	2.21 ± 0.26	14	0.19 ± 0.02	1	woody
β-Selinene	1714	1712		0.24 ± 0.03		0.12 ± 0.01		0.14 ± 0.02		0.40 ± 0.05		0.06 ± 0.01		herb
β-Chamigrene	1724	1718		0.02 ± 0.00		0.01 ± 0.00		0.02 ± 0.00		0.08 ± 0.01		0.01 ± 0.00		
Verbenone	1735	1746		0.12 ± 0.01		0.03 ± 0.00		0.03 ± 0.00		0.03 ± 0.00		0.02 ± 0.00		camphoreous
p-Mentha-1,5-diene-8-ol	1738	1721		0.07 ± 0.01		0.05 ± 0.01		0.02 ± 0.00		0.03 ± 0.00		0.02 ± 0.00		
δ-Cadinene	1755	1753		0.05 ± 0.01		0.03 ± 0.00		0.03 ± 0.00		0.09 ± 0.01		0.01 ± 0.00		thyme, wood
Ethyl phenylacetate	1785	1787				0.02 ± 0.00		0.01 ± 0.00		0.01 ± 0.00				fruit, sweet
δ-Hexalactone	1792	1790		0.01 ± 0.00		0.01 ± 0.00		0.01 ± 0.00		0.01 ± 0.00		0.01 ± 0.00		
Ethyl dodecanoate	1848	1846	5.9	0.02 ± 0.00	< 1	0.04 ± 0.00	< 1	0.02 ± 0.00	< 1	0.02 ± 0.00	< 1	0.02 ± 0.00	< 1	leaf
Hexanoic acid	1855	1853	3			0.20 ± 0.02	< 1	0.24 ± 0.03	< 1	0.18 ± 0.02	< 1	0.01 ± 0.00	< 1	fatty, sour, sweat
p-Cymen-8-ol	1856	1851		0.03 ± 0.00		0.01 ± 0.00		0.01 ± 0.00		0.01 ± 0.00		0.01 ± 0.00		
β-Phenylethanol	1913	1920	1	0.01 ± 0.00	< 1	0.10 ± 0.01	< 1	0.12 ± 0.01	< 1	0.15 ± 0.02	< 1	0.01 ± 0.00	< 1	sweet, floral, fresh
γ-octalactone	1915	1914	0.007	0.01 ± 0.00	2	0.01 ± 0.00	2	0.02 ± 0.00	3	0.03 ± 0.00	4	0.01 ± 0.00	1	fruity

(Continued)

TABLE 5.4 *(Continued)*
Volatile Compounds of Fresh Mango Pulp and Its Dehydrated Powder

						Convection Drying						Freeze Drying		
Compounds	LRI^E	LRI^L	OT	Fresh	OAV	40°C	OAV	50°C	OAV	60°C	OAV	−55°C	OAV	Odor Descriptors
(E)-β-Ionone	1933	1935	0.00001	0.01 ± 0.00	805	0.02 ± 0.00	2469	0.02 ± 0.00	2559	0.01 ± 0.00	1753	0.01 ± 0.00	1083	flower
δ-Octalactone	1966	1964	0.4	0.01 ± 0.00	< 1	0.01 ± 0.00	< 1	0.01 ± 0.00	< 1	0.01 ± 0.00	< 1	0.01 ± 0.00	< 1	
γ-Nonanolactone	2025	2026	0.007			0.01 ± 0.00	1	0.01 ± 0.00	1	0.01 ± 0.00	1	0.01 ± 0.00	1	creamy, sweet
Ethyl tetradecanoate	2053	2052	4			0.03 ± 0.00	< 1	0.03 ± 0.00	< 1	0.01 ± 0.00	< 1	0.01 ± 0.00	< 1	ether
Octanoic acid	2069	2069	3			0.03 ± 0.00	< 1	0.03 ± 0.00	< 1	0.02 ± 0.00	< 1			sweat, cheese
Ethyl (E)-cinnamate	2130	2128	0.165			0.01 ± 0.00	< 1	0.01 ± 0.00	< 1	0.01 ± 0.00	< 1			flower, honey
γ-Decalactone	2141	2142	0.011	0.03 ± 0.00	3	0.02 ± 0.00	2	0.02 ± 0.00	2	0.01 ± 0.00	1	0.01 ± 0.00	1	fruity, fresh, oily
δ-Decalactone	2186	2189	0.41	0.03 ± 0.00	< 1	0.02 ± 0.00	< 1	0.02 ± 0.00	< 1			0.01 ± 0.00	< 1	coconut
Ethyl hexadecanoate	2259	2258	2	0.01 ± 0.00	< 1	0.01 ± 0.00	< 1	0.01 ± 0.00	< 1					waxy, fruity

LRI^E, Linear retention index obtained experimentally; LRI^L, Linear retention index from literature; OT, Odor threshold values; OAV, Odor activity value economical process since it is possible to transform liquid food directly into powder (Schweiggert et al., 2008; Ahmed et al., 2010; Ray et al., 2016; Shishir & Chen, 2017).

Araujo et al. (2023) studied the volatile compounds found in the sapota juice and in their rehydrated powders and found that the main chemical classes in the sapota fruit were of alcohols (22.4% of the identified compounds), followed by aldehydes (20.6%), esters (25.8%), ketones (10.3%), terpenes (17.2%) and lactones (1.7%). The alcohol was the main class, which had a large number (58) of volatile compounds identified in the fresh and dried samples. The data on identified compounds, their respective retention índices, concentration, and OAV present in each sample are shown in Table 5.5. It was observed that most of the aroma compounds identified in sapota juice were also found in samples of its rehydrated powders. However, only 14 compounds were identified in sapota juice, and some of these were lost during the drying process. It is possible to observe in Table 5.5 that for the class of alcohols, aldehydes, esters, and ketones, there was a reduction in concentrations of these when sapota juice was compared to the volatile profile of its rehydrated powders. This reduction can be explained, since during the atomization process, due to the hot air flow that is used in the process, a small quantity of solids is also retained in the drying chamber (Forero et al., 2015).

There are already some publications (MacLeod & Troconis, 1983; Pino et al., 2003; Laohakunjit et al., 2007; Uekane et al., 2017; Lasekan & Yap, 2018; Monteiro et al., 2018) which report the aroma compounds present in the sapota pulp, dehydrated fruit, and other fruits of the same family. Among the volatile compounds identified in these publications, those with the highest concentrations were 1-hexanol, 3-methyl-1-butyl acetate, and ethyl acetate, which were observed in both, the sapota juice sample and in dehydrated sapota powder.

5.10 FUTURE PERSPECTIVES

Dried fruits are a good source of important nutrients such as proteins, fibers, minerals, and vitamins. With the increase in the population that seeks to be a healthier consumer of foods, it is expected that the consumption of dried fruits will grow. The expectation of the dried fruit market profits is 11.62 billion until 2027 (Dried Fruits Global Market Report, 2023). In view of this perspective, the development of new dried fruit products is necessary to keep up with market needs. For that, it is essential that more research needs to be produced about the impact of drying parameters process in the characteristics of these products. One of these characteristics that deserves more attention is the aroma compounds that are very sensitive to the conditions of the drying process. Maintaining the similarity of the aroma between fresh fruits and dried fruit products is the big challenge in development of new dried products. In this way, by exploring new fruits, new methods, and improving drying conditions, it is a field with great prospects for research and development.

5.11 CONCLUDING REMARKS

This chapter reviews recent works about the impact of the drying process on the aroma of dried fruits. There are a few studies focused on aroma composition of dried products, in general, the data focused on bioactive compounds, antioxidant capacity,

TABLE 5.5
Volatile Compounds of the Fresh Sapota Pulp and Its Dehydrated Powder

Compounds	LRI[E]	LRI[L]	OT	Juice	OAV	SM	OAV	SG	OAV	SMG	OAV	Odor Descriptors
Ethyl acetate	584	584	5	6.060 ± 0.002	1	0.060 ± 0.007	<1	1.237 ± 0.002	<1	0.087 ± 0.001	<1	fruity, sweet, grape
Hexanal	781	799	0.0045	0.094 ± 0.000	21	0.066 ± 0.000	15	0.036 ± 0.000	8	0.038 ± 0.000	8	green, leafy, fruity, woody
Ethyl butanoate	782	799	0.001	0.329 ± 0.000	329			0.013 ± 0.001	13			fruity, juicy fruit, pineapple
2,3-butanediol	783	782				0.096 ± 0.001		0.022 ± 0.001				fruity, creamy, buttery
1,3-butanediol	784	785								0.002 ± 0.000		-
Butyl acetate	786	789	0.069	0.298 ± 0.001	4			0.008 ± 0.000	<1			fruity, banana
2-ethyl-1-butanol	794	834		0.038 ± 0.000								alcoholic
Ethyl 2-methyl-butanoate	794	830	0.000006	0.023 ± 0.001	3820							sweet, green, apple, fruity
(*E*)-2-hexenal	797	821	0.017	0.050 ± 0.001	3	0.057 ± 0.000	3	0.017 ± 0.002	1	0.024 ± 0.001	1	green leafy
3-hexen-1-ol	799	828						0.043 ± 0.000		0.044 ± 0.000		green leafy
1-hexanol	819	827	2.5	7.235 ± 0.001	3	0.066 ± 0.001	<1	0.588 ± .001	<1	0.062 ± 0.001	<1	pungent, fruity, sweet
3-methyl-1-butyl acetate	839	853		3.983 ± 0.001		0.042 ± 0.001		0.688 ± 0.002				banana, sweet, fruity

(Continued)

TABLE 5.5 *(Continued)*
Volatile Compounds of the Fresh Sapota Pulp and Its Dehydrated Powder

Compounds	LRI[E]	LRI[L]	OT	Juice	OAV	SM	OAV	SG	OAV	SMG	OAV	Odor Descriptors
3-methyl-3-butenyl acetate	848	861		0.022 ± 0.001								fruity
Heptyl pentanoate	865	899		0.034 ± 0.001								fruity, apple, green apple
2-heptanone	870	870	0.14					0.012 ± 0.000	<1			fruity, spicy, herbal, coconut
Heptanal	892	896	0.003			0.006 ± 0.001	2	0.026 ± 0.000	9	0.013 ± 0.001	4	fresh, fatty, green
(*E,E*)-2,4-hexadienal	908	908				0.025 ± 0.000		0.021 ± 0.001		0.009 ± 0.000		green, fruity, citrus, waxy
Butyrolactone	919	917	20			0.015 ± 0.001	<1					creamy, oily, fatty, caramel
Methyl hexanoate	924	924	0.07	0.035 ± 0.000	1							fruity, pineapple
(Z)-2-heptenal	956	958				0.005 ± 0.000						green, fatty, oily
Benzaldehyde	960	960	0.35	0.429 ± 0.001	1	0.095 ± 0.001	<1	0.110 ± 0.001	<1	0.096 ± 9.001	<1	sweet, almond, cherry

(Continued)

TABLE 5.5 (Continued)
Volatile Compounds of the Fresh Sapota Pulp and Its Dehydrated Powder

Compounds	LRI[E]	LRI[L]	OT	Juice	OAV	SM	OAV	SG	OAV	SMG	OAV	Odor Descriptors
1-heptanol	972	972	0.425	0.029 ± 0.000	<1							musty, leafy, herbal, sweet,
6-methyl-5-hepten-2-one	989	988	0.05	0.042 ± 0.000	1	0.005 ± 0.001	<1			0.005 ± 0.001	<1	citrus green, apple
(*E,E*)-2,4-heptadienal	999	997	0.056	0.019 ± 0.001	<1							fatty, green, oily, aldehydic
Ethyl hexanoate	1002	1002	0.001	0.089 ± 0.001	89							sweet, fruity, pineapple
Octanal	1005	1005	0.0007			0.004 ± 0.001	6	0.020 ± 0.000	29	0.006 ± 0.001	9	waxy, citrus, orange peel,
Hexyl acetate	1017	1017	0.002	0.320 ± 0.001	160							fruity, apple, sweet
Benzyl alcohol	1041	1041	20	0.256 ± 0.001	<1	0.096 ± 0.000	<1	0.160 ± 0.002	<1	0.079 ± 0.001	<1	floral, rose, phenolic
Benzeneacetaldehyde	1048	1047	0.004			0.026 ± 0.001	6			0.034 ± 0.001	8	green, sweet, floral, honey
Acetophenone	1071	1071	0.065	0.202 ± 0.000	3	0.027 ± 0.000	0	0.014 ± .000	<1	0.011 ± 0.001	<1	sweet, pungent, acacia

(Continued)

TABLE 5.5 (Continued)
Volatile Compounds of the Fresh Sapota Pulp and Its Dehydrated Powder

Compounds	LRI^E	LRI^L	OT	Juice	OAV	SM	OAV	SG	OAV	SMG	OAV	Odor Descriptors
1-nonanol	1075	1074	0.05					0.032 ± 0.001	1			fresh, clean, floral, orange
(*E*,*E*)-3,5-octadien-2-one	1075	1068								0.082 ± 0.001		fruity, green, grassy
1-octanol	1076	1054	0.19	0.413 ± 0.000	2							waxy, green, citrus, floral
Myrcenol	1076	1056				0.027 ± 0.001						fresh, citrus, floral, clean
Linalool oxide	1078	1074						0.017 ± 0.000				floral, herbal, earthy, green
Methyl benzoate	1100	1084	0.52	0.105 ± 0.000	<1							phenolic, almond, floral
Nonanal	1108	1107	0.001			0.035 ± 0.000	35	0.023 ± 0.001	23	0.031 ± 0.001	31	waxy, rose, orange peel
Phenylethyl alcohol	1119	1119	1.1	0.936 ± 0.002	1	0.123 ± 0.001	<1	0.387 ± 0.001	<1	0.173 ± 0.000	<1	sweet, floral, fresh,
Camphor	1152	1151	0.46	0.015 ± 0.001	<1	0.006 ± 0.001				0.007 ± 0.001	<1	camphor, minty, herbal

(Continued)

TABLE 5.5 *(Continued)*
Volatile Compounds of the Fresh Sapota Pulp and Its Dehydrated Powder

Compounds	LRI[E]	LRI[L]	OT	Juice	OAV	SM	OAV	SG	OAV	SMG	OAV	Odor Descriptors
Benzyl acetate	1171	1169	0.364	0.061 ± 0.000	<1							sweet, floral, fruity
Isoborneol	1174	1172		0.286 ± 0.001		0.034 ± 0.001		0.030 ± 0.001		0.027 ± 0.000		camphoreous, herbal
Ethyl benzoate	1178	1177	0.06	0.207 ± 0.000	3							fruity, sweet
Levomenthol	1180	1172				0.012 ± 0.000						peppermint, minty
Terpinen-4-ol	1185	1184	0.13							0.013 ± 0.001	<1	pepper, woody, earth, musty
Diethyl butanedioate	1186	1188		0.054 ± 0.001				0.011 ± 0.000				fruity, apple
Decanal	1205	1205	0.0001	0.015 ± 0.001	150	0.014 ± 0.001	143	0.012 ± 0.001	122	0.015 ± 0.002	151	sweet, waxy, orange peel
3-phenylpropanol	1235	1232		0.564 ± 0.001		0.253 ± 0.001		0.214 ± 0.000		0.088 ± 0.001		spicy, fruity, floral
Undecanal	1276	1300	0.0125			0.012 ± 0.000	1					waxy, soapy, floral, citrus,
Ethyl benzenepropanoate	1358	1353		0.037 ± 0.001								spicy
Eugenol	1367	1367	0.006	0.073 ± 0.000	12					0.007 ± 0.001	1	sweet, spicy, woody

(Continued)

TABLE 5.5 (Continued)
Volatile Compounds of the Fresh Sapota Pulp and Its Dehydrated Powder

Compounds	LRI^E	LRI^L	OT	Juice	OAV	SM	OAV	SG	OAV	SMG	OAV	Odor Descriptors
Isoeugenol	1367	1396						0.006 ± 0.000				sweet, spicy, floral
4-methoxyphenethyl alcohol	1377	1374						0.044 ± 0.001				-
Farnesane	1382	1379								0.016 ± 3.001		-
Methyl eugenol	1412	1411	0.775	0.018 ± 0.000	<1					0.025 ± 0.001	<1	sweet, fresh, spicy
Benzophenone	1642	1644				0.003 ± 0.000						balsam, rose, geranium

LRI^E, Linear retention index obtained experimentally; LRI^L, Linear retention index from literature; OT, Odor threshold values reported: Pino and Mesa (2006), De Jesus et al. (2023), Neta et al. (2019). OAV, Odor activity value; SM, Rehydrated dried powder of Sapota juice with maltodextrin; SG, Rehydrated dried powder of Sapota juice with gum Arabic; SMG, Rehydrated dried powder of Sapota juice with maltodextrin and gum Arabic and nutritional properties. In this chapter, we have discussed the pathways for the formation of volatile compounds, the more recent methods for aroma analysis, and the definition of volatile markers and compounds with significant odor impact value (OAV). Moreover, the more recent works involve the drying process (lyophilization, spray drying, and hot air drying) and their impact on volatile compounds that are responsible for the aroma characteristics of fresh fruits, such as mango, sapota, tomato, and soursop.

REFERENCES

Ahmed, M., Akter, M. S., Lee, J. C., Eun, J. B. (2010). Encapsulation by spray drying of bioactive components, physicochemical and morphological properties from purple sweet potato. *LWT – Food Science and Technology*, v. 43, n. 9, p. 1307–1312. DOI: 10.1016/j.lwt.2010.05.014

Alahmad, W., Kaya, S. I., Cetinkaya, A., Varanusupakul, P., Ozkan, S. A. (2023). Green chemistry methods for food analysis: Overview of sample preparation and determination. *Advances in Sample Preparation*, p. 100053. DOI: 10.1016/j.sampre.2023.100053

Alcarde, A. R., Souza, P. A., Belluco, A. E. S. (2011). Chemical profile of sugarcane spirits produced by double distillation methodologies in rectifying still. *Ciência e Tecnologia de Alimentos*, v. 31, p. 355–360. DOI: 10.1590/S0101-20612011000200012

Andrade, E. H. A., Maia, J. G. S., Zoghbi, M. G. B. (2000). Aroma volatile constituents of Brazilian varieties of mango fruit. Journal of Food Composition and Analysis, v. 13(1), p. 27–33. http://dx.doi. org/10.1006/jfca.1999.0841.

Araujo, H. C. S., Jesus, M. S., Sandes, R. D. D., Nogueira, J. P., Leite Neta, M. T. S., Narain, N. (2023) Evaluation of performance of maltodextrin and gum Arabic usage on volatiles profile of Spray-dried powders of sapota (Manilkara zapota) fruit. *CIÊNCIA E TECNOLOGIA DE ALIMENTOS (ONLINE)*, v. 43, p. 1–1.

Bala, S., Kumar, J., Beniway, B. S. (2017). Changes in ADF and NDF content in different stages during, storage, drying in pulp and peel of sapota (*Manilkara zapota* L.) *Annals of Applied Biology*, v. 33, p. 215–219: 0970-0153.

Baysal, T., Deminrdoven, A. (2007). Lipoxygenase in fruits and vegetables: A review. *Enzyme Microbiology Technology*, v. 40, p. 491–496. DOI: 10.1016/j.enzmictec.2006.11.025

Berk, Z. (2013). Freeze drying (Lyophilization) and freeze concentration. *Food Process Engineering and Technology*, p. 567–581. DOI: 10.1016/b978-0-12-415923-5.00023-x

Bicas, J. L., Silva, J. C., Dionísio, A. P., Pastore, G. M. (2010). Biotechnological production of bioflavors and functional sugars. *Ciência e Tecnologia De Alimentos*, v. 30, p. 7–18. DOI: 10.1590/S0101-20612010000100002

Bonneau, A., Boulanger, R., Lebrun, M., Maraval, I., Gunata, Z. (2016). Aroma compounds in fresh and dried mango fruit (*Mangifera indica* L. cv. Kent): Impact of drying on volatile composition. *International Journal of Food Science and Technology*, v. 51, p. 789–800. DOI: 10.1111/ijfs.13038

Canuto, K.M., Souza Neto, M.A., Garruti, D.S. (2009) Composição Química Volátil, Em Diferentes Estádios De Maturação, De Manga 'Tommy Atkins' Produzida No Vale Do São Francisco. *Revista Química Nova*. v. 32, p. 2377–2381.

Capozzi, L. C., Trout, B. L., Pisano, R. (2019). From batch to continuous: Freeze-drying of suspended vials for pharmaceuticals in unit-doses. *Industrial & Engineering Chemistry Research*, v. 58, p. 1635–1649. DOI: 10.1021/acs.iecr.8b02886

Celestino, S. M. C. (2010). Princípio de Secagem em Alimentos. Embrapa Cerrados. Planatina, DF. Available in: https://ainfo.cnptia.embrapa.br/digital/bitstream/item/77765/1/doc-276.pdf. Accessed in: 01 of May of 2023.

Chen, J. L., Zhang, J., Song, L., Jiang, Y., Wu, J., Song Hu, X. (2010) Changes in microorganism, enzyme, aroma of hami melon (*Cucumis melo* L.) juice treated with dense phase carbon dioxide and stored at 4°C. *Innovative Food Science and Emerging Technologies*, v. 11. p. 623–629. DOI: 10.1016/j.ifset.2010.05.008.

Chen, X., Ting, J. L. H., Peng, Y., Tangjaidee, P., Zhu, Y., Li, Q., ..., Quek, S. Y. (2021). Comparing three types of mandarin powders prepared via microfluidic-jet spray drying: Physical properties, phenolic retention and volatile profiling. *Foods*, *10*(1), 123.

Chen, Y., Siewers, V., Nielsen, J. (2012). Profiling of cytosolic and peroxisomal acetyl-CoA metabolism in *Saccharomyces cerevisiae*. *PLoS One*, v. 7, p. 42475. DOI: 10.1371/journal.pone.0042475

Chen, Y., Song, C., Li, Z., Chen, H., Jin, G. (2019). Effects of hot air and combined microwave and hot air drying on the quality attributes of celery stalk slices. *Journal of Food Processing and Preservation*. v. 44. 10.1111/jfpp.14310.

Cheong, K. W., Tan, C. P., Mirhosseini, H., Hamid, N. S. A., Osman, A., Basri, M. (2010). Equilibrium headspace analysis of volatile flavor compounds extracted from soursop (*Annona muricata*) using solid-phase microextraction. *Food Research International*, v. 43, n. 5, p. 1267–1276. DOI: 10.1016/j.foodres.2010.03.001

Cheong, K. W., Tan, C. P., Mirhosseini, H., Joanne-Kam, W. Y., Sheikh Abdul Hamid, N., Osman, A., Basri, M. (2014). The effect of prime emulsion components as a function of equilibrium headspace concentration of soursop flavor compounds. *Chemistry Central Journal*, v. 8, n. 1, p. 23. DOI: 10.1186/1752-153X-8-23

Choia, Y., Leeb, S. M., Leea, H. J., Kimb, Y.-S. (2018). Characterization of aroma-active compounds in Chinese quince (*Pseudocydonia sinensis* Schneid) by aroma dilution analyses. *Food Research International*, v. 105, p. 828–835. DOI: 10.1016/j.foodres.2017.12.015

Chong, S. Y., Wong, C. W. (2015). Production of spray-dried sapodilla (*Manilkara zapota*) powder from enzyme-aided liquefied puree. *Journal of Food Processing and Preservation*, v. 39, n. 6, p. 2604–2611. DOI: 10.1111/jfpp.12510

Christensen, L. P., Endelenbos, M. and Kreautzman. (2007) Fruits and vegetables of moderate climate. In *Flavours and fragrances*. Berlin, Heidelberg: Springe, v.1, p. 135–187. DOI: https://doi.org/10.1007/978-3-540-49339-6

Cudmore, R. S., Dons, A. (2000). Environmental Standard for Industrial Odour Effects: A Recomended Approach. *Aurora Envinmental*.

Dagli, M. M., Yealdam, P. A., Sevindik, O., Glucu, G., Kelebek, H., Selli, S. (2022). Investigation of aroma compounds of Queen Anne's Pocket Melon (Cucumis melo L. ssp. dudaim) Juice. *Journal of Raw Material to Processed Foods*. v. 3, n. 2, p 74–82. DOI: https://doi.org/10.57252/jrpfoods.2022.10

Dalal, K. B., Salunkhe, D. K., Olson, L. E., Do, J. Y. (1968). Volatile components of developing tomato fruit grown under filed and greenhouse conditions. *Plant and Cell Physiology*, v. 9, p. 389–400. DOI: 10.1093/oxfordjournals.pcp.a079355

de Jesus, M. S., Araujo, H. C. S., Denadai, M., Dutra Sandes, R. D., Nogueira, J. P., Leite Neta, M. T., Narain, N. (2023). Effect of different drying methods on the phenolic and volatile compounds of persimmon (*Diospyros kaki* L.). *Journal of Food Measurement and Characterization*, v. 17, p. 2576–2594. DOI: 10.1007/s11694-022-01803-6

Degenhardt, J., Kollner, T. G., Gershenzon, J. (2009). Monoterpene and sesquiterpene synthases and the origin of terpene skeletal diversity in plants. *Phytochemistry*, v. 70, p. 1621–1637. DOI: 10.1016/j.phytochem.2009.07.030

Diep, T. T., Yoo, M. J. Y., Pook, C., Sadooghy-Saraby, S., Gite, A., Rush, E. (2021). Volatile components and preliminary antibacterial activity of tamarillo (Solanum betaceum Cav.). *Foods*, v. *10*(9), 2212. DOI: 10.3390/foods10092212

Diez-Simon, C., Ammerlaan, B., van den Berg, M., van Duynhoven, J., Jacobs, D., Mumm, R., Hall, R. D. (2020). Comparison of volatile trapping techniques for the comprehensive analysis of food flavourings by gas chromatography-mass spectrometry. *Journal of Chromatography A*, v. 1624, p. 461191. DOI: 10.1016/j.chroma.2020.461191

Dried Fruits Global Market Report 2023 – By type (Apricots, Dates, Berries, Other Types), By Category (Conventional, Organic), By Application (Confectionaries, Dairy Products, Bakery Products, Snack and Bards, Desserts, Cerals, Other Applications – Market Size, Trends and Global Forecast 2023-2032. Available in: https://www.thebusinessresearchcompany.com/report/dried-fruits-global-market-report. Accessed in: 25 of July of 2023.

Eker, T., Darıcı, M., Cabaroglu, T. (2023). The impact of seed size on volatile composition in roasted peanuts extracted by the purge and trap method. *Journal of Food Measurement and Characterization*, v. 17, n. 1, p. 211–223. DOI: 10.1007/s11694-022-01583-z

El Hadi, M. A. M., Zhang, F.-J., Wu, F.-F., Zhou, C.-H., Tao, J. (2013). Advances in fruit aroma volatile research. *Molecules*, v. 18, p. 8200–8229. DOI: 10.3390/molecules18078200

Fennema, O. R., Damodaran, S., Parkin, K. L. (2010). Química de Alimentos de Fennema (4th ed.). Artmed: São Paulo.

Feron, G., Bonnarme, P., Durand, A. (1996). Prospects for the microbial production of food flavours. *Trends in Food Science and Technology*, v. 7, p. 285–293. DOI: 10.1016/0924-2244(96)10032-7

Feussner, I., Wasternak, C. (2002). The lipoxygenase pathway. *Annual Review of Plant Biology*, v. 53, p. 275–297. DOI: 10.1146/annurev.arplant.53.100301.135248

Fisher, C.F., Scott, T.R., Taylor, A. Mottram D.S. (1997). Food flavours: Biology and chemistry. Cambridge, UK: The Royal Society of Chemistry, 15–55. DOI: 10.1039/9781847550866

Fongin, S., Kawai, K., Harnkarnsujarit, N., Hagura, Y. (2017). Effects of water and maltodextrin on the glass transition temperature of freeze-dried mango pulp and an empirical model to predict plasticizing effect of water on dried fruits. *Journal of Food Engineering*, v. 210, p. 91–97. DOI: 10.1016/j.jfoodeng.2017.04.025

Forero, D. P., Orrego, C. E., Peterson, D. G., Osorio, C. (2015). Chemical and sensory comparison of fresh and dried lulo (*Solanum quitoense* Lam.) fruit aroma. *Food Chemistry*, v. 169, p. 85–91. DOI: 10.1016/j.foodchem.2014.07.111

Frabetti, A. C. C., de Moraes, J. O., Porto, A. S., da Silva Simão, R., Laurindo, J. B. (2021). Strawberry-hydrocolloids dried by continuous cast-tape drying to produce leather and powder. *Food Hydrocolloids*, *121*, 107041. DOI: 10.1016/j.foodhyd.2021.107041

Franchina, F. A., Zanella, D., Lazzari, E., Stefanuto, P. H., Focant, J. F. (2020). Investigating aroma diversity combining purge-and-trap, comprehensive two-dimensional gas chromatography, and mass spectrometry. *Journal of Separation Science*, v. 43, p. 1790–1799. DOI: 10.1002/jssc.201900902

Fratianni, A., Adiletta, G., Di Matteo, M., Panfili, G., Niro, S., Gentile, C., ..., Corona, O. (2020). Evolution of carotenoid content, antioxidant activity and volatiles compounds in dried mango fruits (Mangifera indica L.). *Foods*, *9*(10), 1424. DOI: 10.3390/foods9101424.

Fredes, A., Roselló, S., Beltrán, J., Cebolla-Cornejo, J., Pérez-de-Castro, A., Gisbert, C., Picó, M. B. (2017). Fruit quality assessment of watermelons grafted onto citron melon rootstock. *Journal of the Science of Food and Agriculture*, *97*(5), 1646–1655. DOI: 10.1002/jsfa.7915

Ganjyal, G. M., Hanna, M. A., Devadattam, D. S. K. (2003). Processing of sapota (Sapodilla): Drying. *Food Engineering and Physical Properties*, v. 68, n. 2, p. 517–520. DOI: 10.1111/j.1365-2621.2003.tb05704.x

Gatfield, I. L., (1996) Biotechnological production of flavour-active lactones in Biotechnology of Aroma Compounds, Ed. R. G. Berger. SpringerVerlag, Berlin, Heidelberg, New York, v. 1, p. 221–238.

Ghani, A. (2003). Medicinal plants of Bangladesh: Chemical constituents & uses. *Dhaka: Asiatic Society of Bangladesh* v.1, p. 1–16.

Gonzalez-Aguero, M., Troncoso, S., Gudenschwager, O., Campos-Vargas, R., Moya-Leon, M. A., Defilippi, B. G. (2009). Differential expression levels of aroma-related genes during ripening of apricot (*Prunus armeniaca* L.). *Plant Physiology and Biochemistry*, v. 47(5), p. 435–440. DOI: 10.1016/j.plaphy.2009.01.002

Gou, M., Chen, Q., Qiao, Y., Jin, X., Zhang, J., Yang, H., Fauconnier, M., Bi, J. (2023). Key aroma-active compounds identification of *Ziziphus jujuba* cv. Huizao: Effect of pilot scale freeze-drying. *Journal of Food Composition and Analysis*, v. 116. DOI: 10.1016/j.jfca.2022.105072

Grosch, W. (2001). Chemistry III: Volatile compounds. In: Coffee: Recent Developments. Blackwell Science, p. 68–89. DOI: 10.1002/9780470690499.ch3

Guadagni, D., Buttery, R., Turnbaugh, J. G. (1972). Odour thresholds and similarity ratings of some potato chip components. *Journal of the Science of Food and Agriculture*, v. 23, n. 12, p. 1435–1444. DOI: 10.1002/jsfa.2740231207

Guadagni, D. G., Okano, S., Buttery, R. G., Burr, H. K. (1966). Correlation of sensory and gas-liquid chromatographic measurements of apple volatiles. *Food Technology*, v. 20, p. 166–169.

Heath, H. B., Reineccius, G. (1986). Flavor Chemistry and Technology, Biogenesis of Flavor in Fruits and Vegetables. Van Nostrand Reinhold Company, p. 43–54.

Hemmerlin, A., Harwood, J. L., Bach, T. J. (2012). A raison d'être for two distinct pathways in the early steps of plant isoprenoid biosynthesis?. *Progress in Lipid Research*, v. 51, p. 95–148. DOI: 10.1016/j.plipres.2011.12.001

Hou, X., Jiang, J., Luo, C., Rehman, L., Li, X., Xie, X. (2023). Advances in detecting fruit aroma compounds by combining chromatography and spectrometry. *Journal of the Science of Food and Agriculture*. v. 103, n. 10, p. 4755–4766. DOI: 10.1002/jsfa.12498

Hwang, I. S., Kim, M. K. (2020). Influence of processing conditions on the flavor profiles of mulberry (Morus alba Linn) fruits using instrumental flavor analysis and descriptive sensory analysis. *Foods*, *9*(5), 581.DOI: 10.3390/foods9050581

Jahurul, M. H. A., Zaidul, I. S. M., Ghafoor, K., Al-Juhaimi, F. Y., Nyam, K. L., Norulaini, N. A. N., Sahena, F., Oma, A. K. M. (2015). Mango (*Mangifera indica* L.) by-products and their valuable components: A review. *Food Chemistry*, v. 183, p. 173–180. DOI: 10.1016/j.foodchem.2015.03.046

Jangam, S. V., Joshi, V. S., Mujumdar, A. S., Thorat, B. N. (2008). Studies on dehydration of sapota (Achras zapota). *Drying Technology*, 26(3), 369–377. DOI: 10.1080/07373930801898190

Janssens, L., De Pooter, H. L., Schamp, N. M., Vandamme, E. J. (1992). Production of flavours by microorganisms. *Process Biochemistry*, v. 27, p. 195–215. DOI: 10.1016/0032-9592(92)80020-4

Jeyaprakash, S., Heffernan, J. E., Driscoll, R. H., Frank, D. C. (2020). Impact of drying technologies on tomato flavor composition and sensory quality. LWT – Food Sci. Technology, 120, 108888 DOI: 10.1016/j.lwt.2019.108888.

Jiang, G. H., Ameer, K., Eun, J. B. (2020). Encapsulation of hot air-dried asian peaR powders using rice bran dietary fiber. *Food Bioscience*, 38. DOI: 10.1016/j.fbio.2020.100742

Jong, S. C., Birgmingham, J. M. (1993). Mushrooms as a Source of Natural Flavor and Aroma Compounds. The Chinese University Press.

Jubaer, H., Afshar, S., Le Maout, G., Mejean, S., Selomulya, C., Xiao, J., Chen, X. D., Jeantet, R., Woo, M. W. (2020). The impact of self-sustained oscillations on particle residence time in a commercial scale spray dryer. *Powder Technology*, v. 360, p. 1177–1191. DOI: 10.1016/j.powtec.2019.11.023

Karimkhani, M. M., Nasrollahzadeh, M., Maham, M., Jamshidi, A., Kharazmi, M. S., Dehnad, D., Jafari, S. M. (2022). Extraction and purification of α-pinene: A comprehensive review. *Critical Reviews in Food Science and Nutrition*, v. 16, p. 1–26. DOI: 10.1080/10408398.2022.2140331

Katsuno, T., Kasuga, H., Kusano, Y., Yaguchi, Y., Tomomura, M., Cui, J., Yang, Z., Baldermann, S., Nakamura, Y., Ohnishi, T., Mase, N., Watanabe, N., (2014). Characterisation of odorant compounds and their biochemical formation in green tea with a low temperature storage process. *Food Chemistry*. v. 148, 388–395. https://doi.org/ 10.1016/j.foodchem.2013.10.069.

Kempler, G. M. (1983). Production of flavor compounds by microorganisms. *Advances in Applied Microbiology*, v. 29, p. 29–51. DOI: 10.1016/S0065-2164(08)70353-8

Kumar, P., Chatli, M. K., Mehta, N., Malav, O. P., Verma, A. K., Kumar, D., Rathour, M. (2018). Antioxidant and antimicrobial efficacy of sapota powder in pork patties stored under different packaging conditions. *Korean Journal for Food Science of Animal Resources*, v. 38, n. 3, p. 593–605. DOI: 10.5851/kosfa.2018.38.3.593

Laohakunjit, N., Kerdchoechuen, O., Matta, F. B., Silva, J. L., Holmes, W. E. (2007). Postharvest survey of volatile compounds in five tropical fruits using headspace-solid phase microextraction (HSSPME). *HortScience*, v. 42, n. 2, p. 309–314. DOI: 10.21273/HORTSCI.42.2.309

Lasekan, O., Hashim, N., Lasekan, A. (2022). Flavour chemistry of dehydrated exotic fruits. *International Food Research Journal*, v. 29, n. 6. DOI: 10.47836/ifrj.29.6.03

Lasekan, O., Yap, S. P. (2018). Characterization of the aroma compounds in fresh and dried sapodilla (*Manikara zapota*, L.) by the application of aroma extract dilution analysis. *CYTA: Journal of Food*, v. 16, n. 1, p. 801–806. DOI: 10.1080/19476337.2018.1485748

LE CLOIREC, P., FANLO, J. L., DEGORGE-DUMAS, J. R. (1991) Odeurs et désodorisation industrielles. Ecole des Mines D'Alès.

LEITE NETA, M. T. S., DE SANTANA, K.L. ; DOS SANTOS, R.A.R., GALVÃO, M. S., FONTES, A. S., NARAIN, N. (2018). Fatty acids profile of pulp and peel of mango (Mangifera indica Linn.) cultivar –Espada from Brazilian northeast region. *Acta Horticulturae*, v. 1, p. 197–198, DOI: 10.17660/ActaHortic.2018.1198.31.

Leite Neta, M. T. S., de Jesus, M. S., da Silva, J. L. A., Araujo, H. C. S., Sandes, R. D., Shanmugam, D. S., Narain, N. (2019). Effect of spray drying on bioactive and volatile compounds in soursop (*Annona muricata*) fruit pulp. *Food Research International*, v. 124, p. 70–77. DOI: 10.1016/j.foodres.2018.09.039

Liang, Z., Zhang, P., Fang, Z. (2022). Modern technologies for extraction of aroma compounds from fruit peels: A review. *Critical Reviews in Food Science and Nutrition*, v. 62, n. 5, p. 1284–1307. DOI: 10.1080/10408398.2020.1840333

Lin, X., Xu, J.-L., Sun, D.-W. (2020). Comparison of moisture uniformity between microwave-vacuum and hot-air dried ginger slices using hyperspectral information combined with semivariogram. *Drying Technology*, v. 39 p. 1044–1058 DOI: 10.1080/07373937.2020.1741006

Link, J. V., Tribuzi, G., Laurindo, J. B. (2017). Improving quality of dried fruits: A comparison between conductive multi-flash and traditional drying methods. *LWT – Food Science and Technology*, v. 84, p. 717–725. DOI: 10.1016/j.lwt.2017.06.045

Liu, Y., Liao, Y., Guo, M., Zhang, W., Sang, Y., Wang, H., ..., Chen, G. (2022). Comparative elucidation of bioactive and volatile components in dry mature jujube fruit (Ziziphus jujuba Mill.) subjected to different drying methods. *Food Chemistry: X*, *14*, 100311. DOI: 10.1016/j.fochx.2022.100311

Lomascolo, A., Stentelaire, C., Asther, M., Lesage-Meessen, L. (1999). Basidiomycetes as new biotechnological tools to generate natural aromatic flavours for the food industry. *Trends in Biotechnology*, v. 17, p. 282–289. DOI: 10.1016/S0167-7799(99)01313-X

MacLeod, A. J., Troconis, N. G. (1983). Volatile flavor components of sapodilla fruit (*Achras sapota* L.). *Journal of Agricultural and Food Chemistry*, v. 30, n. 3, p. 515–517. DOI: 10.1021/jf00111a026

Mahmoud, S. S., Croteau, R. B. (2002). Strategies for transgenic manipulation of monoterpene biosynthesis in plants. *Trends in Plant Science*, v. 7, n. 8, p. 366–373. DOI: 10.1016/S1360-1385(02)02303-8

Marín-San Román, S., Rubio-Bretón, P., Pérez-Álvarez, E. P., Garde-Cerdán, T. (2020). Advancement in analytical techniques for the extraction of grape and wine volatile compounds. *Food Research International*, v. 137, p. 109712. DOI: 10.1016/j.foodres.2020.109712

Martina, M.; Tikunov, Y., Portis, E., Bovy, A. G. (2021). The Genetic Basis of Tomato Aroma. *Genes*, v. 12, n. 226., p. 1–26. DOI: 10.3390/ genes12020226

Matheyambath, A. C., Subramanian, J., Paliyath, G. (2016). Mangoes. Reference module in food science. *Encyclopedia of Food and Health*, v. 5, p. 641–645.

Monteiro, S. S., Ribeiro, S. R., Soquetta, M. B., Pires, F. J., Wagner, R., Da Rosa, C. S. (2018). Evaluation of the chemical, sensory and volatile composition of sapota-do Solimões pulp at different ripening stages. *Food Research International*, v. 109, p. 159–167. DOI: 10.1016/j.foodres.2018.04.033. PMID: 29803438.

Murray, K. E., & Whitfield, F. B. (1975). The occurrence of 3-alkyl-2-methoxypyrazines in raw vegetables. *Journal of the Science of Food and Agriculture*, v. 26, n. 7, 973–986. https://doi.org/10.1002/jsfa.2740260714

National Horticulture Board – NHB. (2021). Horticulture crops for 2019-20, India (1st advance estimates). Available in: https://nhb.gov.in/StatisticsViewer.aspx?enc=MWoUJibk35dW2g36TUJWAoZqESmAYFi7h2irlsmjlINTcFl1rG/kLbq8ZQbWUvuM. Accessed in: 17 of July of 2023.

Neta, M. T. S. L., de Jesus, M. S., da Silva, J. L. A., Araujo, H. C. S., Sandes, R. D. D., Shanmugam, S., Narain, N. (2019). Effect of spray drying on bioactive and volatile compounds in soursop (*Annona muricata*) fruit pulp. *Food Research International*, v. 124, p. 70–77. DOI: 10.1016/j.foodres.2018.09.039

Nogueira, J. P., de Siqueira, A. C. P., Sandes, R. D. D., de Sousa Galvão, M., Neta, M. T. S. L., Narain, N. (2018). An insight into key volatile compounds in acerola (*Malpighia emarginata* DC.) pulp based on their odour activity values and chemometric evaluation. *Analytical Methods*, v. 10, n. 48, p. 5851–5866. DOI: 10.1039/C8AY01427B

Nykänen, L. (1986). Formation and occurrence of flavour compounds in wine and distilled alcoholic beverages. *American Journal of Enology and Viticulture*, v. 37, p. 84–96. DOI: 10.5344/ajev.1986.37.1.84

Ojediran, J. O., Okonkwo, C. E., Adeyi, A. J., Adeyi, O., Olaniran, A. F., George, N. E., Olayanju, A. T. (2020). Drying characteristics of yam slices (*Dioscorea rotundata*) in a convective hot air dryer: Application of ANFIS in the prediction of drying kinetics. *Heliyon*, v. 6, p. e03555. DOI: 10.1016/j.heliyon.2020.e03555

Oliveira, V. S., Afonso, M. R. A., Costa, J. M. C. (2011). Caracterização físicoquímica e comportamento higroscópico de sapoti liofilizado. *Revista Ciência Agronômica*, v. 42, p. 342–348. DOI: 10.1590/S1806-66902011000200012

Oms-Oliub, G., Hertoga, M. L. A. T. M., Van de Poela, B., Ampofo-Asiamaa, J., Geeraerda, A. H., Nicolaı, B. M. (2011). Metabolic characterization of tomato fruit during preharvest development, ripening, and postharvest shelf-life. *Postharvest Biology and Technology*, v. 62, p. 7–16. DOI: 10.1016/j.postharvbio.2011.04.010

Onwude, D. I., Hashim, N., Chen, G. (2016). Recent advances of novel thermal combined hot air drying of agricultural crops. *Trends in Food Science & Technology*, v. 57, p. 132–145. DOI: 10.1016/j.tifs.2016.09.012

Pandit, S. S., Chidley, H. G., Kulkarni, R. S., Pujari, K. H., Giri, A. P., Gupta, V. S. (2009). Cultivar relationships in mango based on fruit volatile profiles. *Food Chemistry*, v. 114, p. 363–372. DOI: 10.1016/j.foodchem.2008.09.107

Parker, J. K. (2015). Introduction to aroma compounds in foods: In Flavour Development, Analysis and Perception in Food and Beverages. Woodhead Publishing Series in Food Science, p. 3–30. ISBN 9781782421030. DOI: 10.1016/B978-1-78242-103-0.00001-1

Perez, A. G., Sanz, C. (2008). Formation of fruit flavour. In: Fruit and Vegetable Flavour: Recent Advances and Future Prospects. Woodhead Publishing, p. 41–70. ISBN: 978-1-85573-467-8.

Petronilho, S., Coimbra, M. A., Rocha, S. M. (2014). A critical review on extraction techniques and gas chromatography based determination of grapevine derived sesquiterpenes. *Analytica Chimica Acta*, v. 846, p. 8–35. DOI: 10.1016/j.aca.2014.05.049

Pino, J. A., Marbot, R., Aguero, J. (2003). Volatile components of sapodilla fruit (*Manilkara achras* L.). *Journal of Essential Oil Research*, v. 15, p. 374–375. DOI: 10.1080/10412905.2003.9698614

Pino, J. A., Mesa, J. (2006). Contribution of volatile compounds to mango (*Mangifera indica* L.) aroma. *Flavour and Fragrance Journal*, v. 21, p. 207–213. DOI: 10.1002/ffj.1703

Pino, J. A., Mesa, J., Munoz, Y., Marti, M. P., Marbot, R. (2005). Volatile components from mango (*Mangifera indica* L.) cultivars. *Journal of Agriculture and Food Chemistry*, v. 53, p. 2213–2223. DOI: 10.1021/jf0402633

Prinsi, B., Negri, A. S., Fedeli, C., Morgutti, S., Negrini, N., Cocucci, M., Espen, L. (2011). Peach fruit ripening: A proteomic comparative analysis of the mesocarp of two cultivars with different flesh firmness at two ripening stages. *Phytochemistry*, v. 72, p. 1251–1262. DOI: 10.1016/j.phytochem.2011.01.012

Pu, Y.-Y., Sun, D.-W. (2017). Combined hot-air and microwave-vacuum drying for improving drying uniformity of mango slices based on hyperspectral imaging visualisation of moisture content distribution. *Biosystems Engineering*, v. 156, p. 108–119. DOI: 10.1016/j.biosystemseng.2017.01.006

Qiu, L., Zhang, M., Wang, Y., Liu, Y. (2019). Physicochemical and nutritional properties of wasabi (*Eutrema yunnanense*) dried by four different drying methods. *Drying Technology*, v. 37, n. 3, p. 363–372. DOI: 10.1080/07373937.2018.1458318

Rajkumar, G., Rajan, M., Araujo, H. C., de Jesus, M. S., Leite Neta, M. T. S., Sandes, R. D. D., Narain, N. (2021). Comparative evaluation of volatile profile of tomato subjected to hot air, freeze, and spray drying. *Drying Technology*, v. 39, n. 3, p. 383–391. DOI: 10.1080/07373937.2020.1842441

Rajkumar, G., Shanmugam, S., Galvão, M. S., Leite Neta, M. T. S., Sandes, R. D. D., Mujumdar, A. S., Narain, N. (2017). Comparative evaluation of physical properties and aroma profile of carrot slices subjected to hot air and freeze drying. *Drying Technology*, v. 35. DOI: 10.1080/07373937.2016.1206925

Raven, P. H., Evert, R. F., Eichhorn, S. E. (2001). A composição molecular das células vegetais. In: Biologia Vegetal. Guanabara Koogan, p. 17–39.

Ray, S., Raychaudhuri, U., Chakraborty, R. (2016). An overview of encapsulation of active compounds used in food products by drying technology. *Food Bioscience*, v. 13, p. 76–83. DOI: 10.1016/j.fbio.2015.12.009

Reeve, J., Smith, J. F., Carpenter-Boggs, L., Reganold, J. P. (2008). Soil-based cycling and differential uptake of amino acids by three species of strawberry (*Fragaria* spp.) plants. *Soil Biology & Biochemistry*, v. 40, p. 2547–2552. DOI: 10.1016/j.soilbio.2008.06.015

Reineccius, G. (2006). Flavour formation of fruits and vegetables. In: Reineccius, G. Flavour Chemistry and Technology. CRC Press: Boca Raton, FL, p. 73–101.

Rohmer, M. (1999). The discovery of a mevalonate-independent pathway for isoprenoid biosynthesis in bacteria, algae and higher plants. *Natural Product Reports*, v. 16, p. 565–574. DOI: 10.1039/A709175C

Rothe, M., Thomas, B. (1963). Aroma in bread. Evaluation of chemical taste analyses with the aid of threshold value. *Zeitschrift für Lebensmitteluntersuchung und Forschung*, v. 119, p. 302–310.

San, R. I., Alonso, M., Bartolomé, L., Alonso, R. M. (2014). Headspace gas chromatography methods and their potential industrial applications. *Current Chromatography*, v. 1. DOI: 10.2174/2213240601666140131235643

Santana, K. L., Galvão, M. S., Jesus, M. S., Nogueira, J. P., Narain, N. (2017). HS-SPME optimization and extraction of volatile compounds from soursop (*Annona muricata* L.) pulp with emphasis on their characteristic impact compounds. *Food Science and Technology*, v. 37, p. 250–260. DOI: 10.1590/1678-457X.20916

Schwab, W. (1998). Application of stable isotope ratio analysis explaining the bioformation of 2,5-dimethyl-4-hydroxy-3(2H)-furanone in plants by a biological Maillard reaction. *Journal of Agricultural and Food Chemistry*, v. 46, p. 2266–2269. DOI: 10.1021/jf9710037

Schwab, W., Davidovichi-Rikanati, R., Lewinsohn, E., (2008). Biosynthesis of plant-derived flavor compounds. *The Plant Journal*, v. 54, p. 712–732. DOI: 10.1111/j.1365-313X.2008.03446.x

Schweiggert, U., Hofmann, S., Reichel, M., Schieber, A., Carle, R. (2008). Enzyme-assisted liquefaction of ginger rhizomes (Zingiber officinale Rosc.) for the production of spray-dried and paste-like ginger condiments. *Journal of Food Engineering*, 84(1), 28–38. DOI: 10.1016/j.jfoodeng.2007.04.013

Shishir, M. R. I., Chen, W. (2017). Trends of spray drying: A critical review on drying of fruit and vegetable juices. *Trends in Food Science & Technology*, v. 65, p. 49–67. DOI: 10.1016/j.tifs.2017.05.006

Siddiqui, M. W., Longkumer, M., Ahmad, M. S., Barman, K., Thakur, P. K., Kabir, J. (2014). Postharvest biology and technology of sapota: A concise review. *Acta Physiologiae Plantarum*, v. 36, n. 12, p. 3115–3122. DOI: 10.1007/s11738-014-1696-4

Siegmund, B. (2015). Biogenesis of aroma compounds: Flavour formation in fruits and vegetables: In Woodhead Publishing Series in Food Science, Technology and Nutrition, Flavour Development, Analysis and Perception in Food and Beverages. Woodhead Publishing, p. 127–149, ISBN 9781782421030. DOI: 10.1016/B978-1-78242-103-0.00007-2

Slaughter, J. C. (1999). The naturally occurring furanones: Formation and function from pheromone to food. *Biological Reviews*, v. 74, p. 259–276.

Smogrovicova, D., Domény, Z., Gemeiner, P., Malovikova, A., Sturdik, E. (1999). Reactors for continuous primary beer fermentation using immobilised yeast, *Biotechnology Techniques*, v. 11, p. 261–264. DOI: 10.1023/A:1018494606777

Srikanth, K. S., Sharanagat, V. S., Kumar, Y., Bhadra, R., Singh, L., Nema, P. K., Kumar, V. (2019). Convective drying and quality attributes of elephant foot yam (*Amorphophallus paeoniifolius*). *LWT – Food Science and Technology*, v. 99, p. 8–16. DOI: 10.1016/j.lwt.2018.09.049

Stephan, A., Bücking, M., Steinhart, H. (2000). Novel analytical tools for food flavours. *Food Research International*, v. 33, n. 3–4, p. 199–209. DOI: 10.1016/S0963-9969(00)00035-1

Tandon, K. S., Baldwin, E. A., Scott, J. W., Shewfelt, R. L. (2003). Linking sensory descriptors to volatile and non-volatile components of fresh tomato flavor. *Journal. Food Science*, v. 68, p. 2366–2237. DOI: 10.1111/j.1365-2621.2003.tb05774.x

Tang, X. C., Pikal, M. J. (2004). Design of freeze-drying processes for pharmaceuticals: Practical advice xiaolin. *Pharmaceutical Research*, v. 21, p. 191–200. DOI: 10.1023/b:pham.0000016234.73023.75

Tholl, D. (2006). Terpene synthases and the regulation, diversity and biological roles of terpene metabolism. *Current Opinion in Plant Biology*, v. 9, p. 297–304. DOI: 10.1016/j.pbi.2006.03.014

Uekane, T. M., Nicolotti, L., Griglione, A., Bizzo, H. R., Rubiolo, P., Bicchi, C., Rocha-Leão, M. H., Rezende, C. M. (2017). Studies on the volatile fraction composition of three native Amazonian Brazilian fruits: Murici (*Byrsonima crassifolia* L., Malpighiaceae), bacuri (*Platonia insignis* M., Clusiaceae), and sapodilla (*Manilkara sapota* L., Sapotaceae). *Food Chemistry*, v. 219, p. 13–22. DOI: 10.1016/j.foodchem.2016.09.098. PMID: 27765208.

Varzakas, T., Tzia, C. (Eds.). (2014). Food Engineering Handbook: Food Process Engineering. CRC Press.

Wandjou, J. G. N., Sut, S., Giuliani, C., Fico, G., Papa, F., Ferraro, S., ..., Dall'Acqua, S. (2019). Characterization of nutrients, polyphenols and volatile components of the ancient apple cultivar 'Mela Rosa Dei Monti Sibillini'from Marche region, central Italy. *International Journal of Food Sciences and Nutrition*, *70*(7), 796–812.

Wang, B., Liu, F., Xiang, J., He, Y., Zhang, Z., Cheng, Z., Liu, W., Tan, S. (2021). A critical review of spray-dried amorphous pharmaceuticals: synthesis, analysis and application. *International Journal of Pharmaceutics*, v. 594, 120165. DOI:10.1016/j.ijpharm.2020.120165.

Wang, C.; Zhang, W.; Li, H.; Mao, J.; Guo, C.; Ding, R.; Wang, Y.; Fang, L.; Chen, Z.; Yang, G. (2019). Analysis of volatile compounds in pears by HS-SPME-GC×GC-TOFMS. *Molecules*, v. 24, 1795. DOI:10.3390/molecules24091795

Xiao, Z., Jiang, X., Niu, Y. (2022). Study on the interaction of sweet protein (thaumatin) with key aroma compounds in passion fruit juice using electronic nose, ultraviolet spectrum, thermodynamics, and molecular docking. *Lwt*, *162*, 113463. DOI: 10.1016/j.lwt.2022.113463

Xinyu, Z., Yi, W., Xue, H., Zixi, Y., Weiqiong, Y., Zhaolin, L. (2020). Characterization of volatile compounds in five blueberry varieties using purge and trap coupled to gas chromatography-mass spectrometry. *Italian Journal of Food Science*, v. 32, n. 2. DOI: https://doi.org/10.14674/IJFS-1675

Yan, Q., Wang, H., Zhu, H., Wu, Y., Zhou, F., Wang, Z., Wang.renlei, Hua, C. (2019). Changes and migration of volatile flavor components in freeze-dried peaches. *Food & Machinery*, v. 35, p. 20–25, DOI: 10.13652/j.issn.1003-5788.2019.07.004

Zabetakis, I., Gramshaw, J. W., Robinson, D. S. (1999). 2,5-dimethyl-4-hydroxy-2H-furan-3-one and its derivatives: Analysis, synthesis and biosynthesis – A review. *Food Chemistry*, v. 65, p. 139–151. DOI: 10.1016/S0308-8146(98)00203-9

Zhu, J., Chen, F., Wang, L., Niu, Y., Chen, H., Wang, H., Xiao, Z. (2016). Characterization of the key aroma volatile compounds in cranberry (*Vaccinium macrocarpon* Ait.) using gas Chromatography–Olfactometry (GC-o) and odor activity value (OAV). *Journal of Agricultural and Food Chemistry*, v. 64, n. 24, p. 4990–4999. DOI: 10.1021/acs.jafc.6b01150

6 Instrumentation Techniques and Analyses of Bioactive Compounds in Different Dried Fruit Products

Roberto Moscetti[1], *Alessandro Benelli*[1], *Pietro Ibba*[2], *Cinzia Mannozzi*[3], *Francesca Melini*[4], *and Riccardo Massantini*[1]

[1]Department of Sciences and Technology for Agriculture, Forest, Environment and Energy, University of Tuscia, Viterbo, Italy

[2]Faculty of Engineering, University of Bozen, Bolzano, Italy

[3]School of Pharmacy, University of Camerino, Camerino, Italy

[4]Council for Agricultural Research and Economics, Rome, Italy

6.1 BIOACTIVE COMPOUNDS AND NUTRITIONAL QUALITY OF DRIED FRUITS

Fruits are an excellent source of bioactive compounds, such as dietary fiber, minerals, water-soluble vitamins, provitamin A, and phenolic compounds (PCs), which can produce physiological effects. Over the last decades, epidemiological studies have shown an inverse correlation between a high intake of fruits (and vegetables) and the incidence of non-transmissible chronic diseases, including cardiovascular diseases, cancer, diabetes, obesity, etc. (de la Rosa et al., 2019; Ellong et al., 2015; Kris-Etherton et al., 2002; Thakur et al., 2020).

Among the plethora of bioactive compounds found in fruits, PCs have attracted great attention due to their antioxidant, antimicrobial, antidiabetic, antihypertensive, antithrombotic, and immunomodulatory activity (Ellong et al., 2015; Thakur et al., 2020). Polyphenols are a heterogeneous class of compounds synthesized by plants through either the shikimate or acetate pathway under normal and stressful conditions (Melini et al., 2020b). Based on the number of phenol rings and the structural elements binding the rings one to another, they can be classified into simple phenols, phenolic acids, flavonoids, xanthones, stilbenes, and lignans

DOI: 10.1201/9781003402916-6

(Fernando et al., 2020). According to a broader classification, PCs are grouped into flavonoids and non-flavonoids.

Flavonoids are the most abundant PCs in fruits and are typically classified into seven subclasses: flavonols, flavones, isoflavones, anthocyanidins, flavanones, flavanols, and chalcones. This classification is based on structural differences (Melini et al., 2020a, 2020b), specifically, the oxidation degree of the central heterocycle.

Flavan-3-ols (i.e., catechin and epicatechin) are present in apples (Kschonsek et al., 2018), cherries, and cranberries (de la Rosa et al., 2019). Flavanones (e.g., hesperetin, naringenin, and eriodictyol) are mainly present in citrus fruits (de la Rosa et al., 2019). Anthocyanidins (e.g., cyanidin, delphinidin, malvidin, pelargonidin, peonidin, and petunidin) are mainly available in red/purple-colored fruits, such as cherries, blueberries, elderberries, raspberries, blackberries, and cranberries, as well as in apples, red grapefruit, plums, strawberries, and watermelon (de la Rosa et al., 2019). The two flavonols quercetin and rutin are among the most abundant flavonoids in fruits and vegetables. Flavones (i.e., apigenin and luteolin) are present only in very small quantities in fruits (de la Rosa et al., 2019). Flavonoids generally have a significant impact on fruit quality attributes, such as color, flavor, and bitterness, and have been associated with beneficial biochemical effects (Williamson et al., 2018).

As regards the non-flavonoids group, phenolic acids are the predominant class of compounds in fruits. They present very diverse chemical structures, which are generally smaller and simpler than flavonoids (de la Rosa et al., 2019). Chemically, they contain a single phenyl group, substituted by one carboxylic group and one or more -OH groups. Phenolic acids consist of derivatives of hydroxybenzoic acid and hydroxycinnamic acid available in free, conjugated-soluble, and insoluble-bound forms. Phenolic acids are major components of berries; hydroxybenzoic acid is abundant in cranberry; ferulic acid in cranberries, blueberries, and bilberries; and hydroxycinnamic acid, *p*-hydroxybenzoic, and *p*-coumaric in strawberries (Williamson et al., 2018). Chlorogenic acid is found in plums, nectarines, peaches, apples, and pears (de la Rosa et al., 2019; Kschonsek et al., 2018).

Besides polyphenols, carotenoids are another class of bioactive compounds widely available in fruits and are generally responsible for red, orange, and yellow colors. Like other bioactive compounds, carotenoids are essential for human health, providing protection against cardiovascular diseases, cancer, cataracts, and macular degeneration (Arscott, 2013). Cultivated fruits, such as citrus fruits, apricots, peaches, mangoes, and papaya, are important sources of carotenoids (Meléndez-Martínez et al., 2022). Apricots and orange fruits are rich in β-carotene; watermelons and pink grapefruits in lycopene; and mandarins and some orange varieties in β-cryptoxanthin (Meléndez-Martínez et al., 2022). A high dietary intake of phytochemicals is ensured by the consumption of fruits in their fresh form; however, many fruits are seasonal, highly perishable, and not available year-round.

The recent interest in dried fruits and their increasing consumption follows up on the necessity to preserve seasonal and healthy fruits, together with convenience and portability. Therefore, it is nevertheless a primary concern to assess the phytochemical content and profile of dried fruits and understand how processing can affect them.

6.2 EFFECTS OF DRYING PROCESSING CONDITIONS ON BIOACTIVE COMPOUNDS

The phenolic content and composition of fruits are determined by genetic and environmental factors but may also be modified during processing and storage (Sadler et al., 2019). Processing, such as thermal treatments, can liberate PCs from the food matrix. Phenolic acids are, in fact, metabolic intermediates accumulating in the vacuoles, and high temperatures can break down the cellular constituents and release the bound phenolics (McSweeney and Seetharaman, 2015; Zhu et al., 2023). Following thermal degradation, change and reduction of their antioxidant potential can also be observed (Devahastin and Niamnuy, 2010). Degradation can be either enzymatic or non-enzymatic: the enzymatic pathway implies the involvement of the enzymes polyphenol oxidase (PPO), lipoxygenase, and peroxidase, while non-enzymatic degradation includes the occurrence of the Maillard reaction, which then leads to phenol oxidation (McSweeney and Seetharaman, 2015). High temperatures can, however, inactivate the enzymes and slow down enzyme activity and polyphenol degradation (McSweeney and Seetharaman, 2015). Within this framework, the knowledge of the effect of pre-drying treatments, drying techniques, and process parameters thereof on the retention and/or degradation of polyphenols is crucial to guaranteeing the production of dried fruits rich in phytochemicals.

Hot air-drying (HAD) is the most commonly used dehydration technique (Ju et al., 2018) due to its affordability and simplicity of equipment when compared to other drying techniques. In HAD, energy from an energy source (i.e., a continuous flow of hot air) is transferred to a food item, and then moisture is transferred from the food to the atmosphere by diffusion or capillary mechanism. The concentration gradient is the driving force of this process. Temperature, humidity, air velocity, and air exchange, but also food product geometry and product thickness, are crucial process parameters.

A balanced setting of all parameters is pivotal for polyphenol retention in HAD. Evidence showed, on the one hand, that the hotter the air is, the higher the drying rate is; but, on the other hand, elevated temperatures or long drying times can have negative effects on the color and nutrient content of the dried fruit. PCs and carotenoids are susceptible to degradation when exposed to high temperatures, oxygen, humidity, and light (Lu et al., 2020). High temperatures can cause chemical, physical, and biological changes in foods, and, upon chemical and enzymatic reactions, the polyphenol content and composition of fruits can be affected (McSweeney and Seetharaman, 2015). The application of HAD to Asian pears, at 50°C and 2 m/s, showed, for instance, a decrease in the content of some PCs, namely, 3,4-dimethoxybenzaldehyde, 4-hydroxybenzoic acid, catechin, chlorogenic acid, chrysin, epicatechin, and trans-cinnamic acid compared to the fresh product, as well as to the pears dried with ultrasound-assisted vacuum drying (USVD) and freeze-drying (FD) (Gutiérrez-Cárdenas et al., 2022). In HAD, the setting of temperatures is also important for PCs retention, as observed in pomegranate arils. Good retention of total polyphenols and total tannins was observed at 75°C compared to the other drying temperatures (55°C and 65°C). However, the best preservation of total monomeric anthocyanins was observed at 55°C (Süfer and Palazoğlu, 2019). To overcome the limitations of HAD, pretreatments have been explored to reduce the drying time and apply lower temperatures. As regards the application of pretreatments, blanching is one of the most widely used and allows for the inactivation of enzymes, the expulsion of intracellular air, the reduction of color loss and organoleptic properties, and the improvement of the drying rate (Llavata et al., 2020). A combination of blanching and HAD (at 60°C) can increase the release of PCs, such as individual anthocyanins, procyanidins, and flavonoids, compared to unblanched foods dried in open-air or in a hot-air oven (Kessy et al., 2016).

Osmotic dehydration is another conventional pretreatment and consists of the introduction of the food matrix into a hypertonic solution. It implies a partial loss of water and offers some advantages, such as a shorter drying time (energy saving), a reduction of both solid gain and heat-induced damages, and subsequently better organoleptic or functional properties (Llavata et al., 2020). The application of vacuum at the beginning of the process has also been explored, as it can promote an increase in the rate of mass transfer (Llavata et al., 2020). The pressure gradient generates an increase in the rate of water-related weight loss and solid gain. Pulsed vacuum osmotic dehydration used as a pretreatment allowed for higher retention of PCs content in cranberries (*Vaccinium macrocarpon*) and blueberries (*Vaccinium corymbosum*) (Liu et al., 2020). On the other hand, it did not affect the decrease of total monomeric anthocyanins; this class of compounds is, in fact, particularly sensitive to thermal degradation and losses due to leaching into osmotic solution.

Ultrasound, microwaves, or pulsed electric fields (PEF) have also been explored as drying pretreatments since they allow a reduction of the drying time and energy consumption, as well as an improvement of the final food quality (Llavata et al., 2020).

When ultrasound is applied in hyper- or hypotonic solutions, microchannels are generated, and a major water movement occurs inside the solid food matrix. In addition, the breaking down of the cellular structure and the formation of microcracks enable diffusion in the structure, as shown for apples (Magalhães et al., 2017), blueberries (Celejewska et al., 2018), and kiwifruits (Wang et al., 2019). This change in the food matrix implies that lower drying temperatures and shorter drying times are required. The application of ultrasound to fruits also allows for a better retention of color. In fact, upon application of ultrasound in hypertonic media, solids are uptaken, and gas is released from the food. A good retention of pigments after ultrasound pretreatment has been observed in strawberries (Prosapio and Norton, 2018) and plums (Dehghannya et al., 2017). However, following the higher mass transfer brought by ultrasound, a loss in bioactive compounds, such as vitamin C, can occur; the dilution thereof in the hypertonic solution is eased by the soluble character of the component and the microchannels formed in the food matrix. This behavior has been observed in strawberry (Buratti et al., 2006; Noshad et al., 2018), kiwifruit (Wang et al., 2019), pomegranate (Allahdad et al., 2019), and persimmon fruit (Bozkir et al., 2019).

Microwave drying (MWD) is another pretreatment, which allows rapid moisture removal and is claimed, together with ultrasound, to positively affect the texture and nutritional value of the final product and preserve bioactive compounds in fruits (Liu et al., 2020; Tylewicz et al., 2022; Wang et al., 2018). MWD is characterized by more rapid and uniform heating due to the penetration of microwaves into the product (Chatzilia et al., 2023; Liu et al., 2020). The rapid heating offers the advantage of preventing tissue structure shrinkage and enhancing the rehydration ability. As regards the effect of the MWD pretreatment on phenolics, it retained the highest content of bioactive compounds in wild blueberries compared to other pretreatments such as ultrasound water-bath and ultrasound probe (Sharif et al., 2018). The application of MWD to pineapple slices also allowed higher retention of PCs than HAD and FD (Izli et al., 2018). It is possible that bound PCs were released from the cellular constituents, leading to the formation of free ones. In addition, higher cell disruption may occur in MWD and subsequently more PCs are released (Izli et al., 2018). However, MWD also presents some drawbacks. Food morphology and thickness affect the uniformity of drying, and there is a risk of surface overheating and textural damage. As a result, MWD is often combined with other drying techniques in a two-step procedure which relies on the advantages of each method. For instance, an initial HAD step followed by MWD has been reported to be an efficient approach to improve product quality while significantly reducing time and energy requirements (Chatzilia et al., 2023). HAD removes surface moisture from the product surface, while microwave energy targets moisture within the inner part of the food.

PEF pretreatment consists of applying short pulses at high voltage to a biological material placed between two electrodes. Its main effect is electroporation, which means induction and development of pores in the cell membrane, thus promoting the subsequent mass transfer through them (Llavata et al., 2020). However, no clear trend has been found about the effect of the PEF pretreatment on the retention of the bioactive compounds. Antioxidant degradation and formation of new PCs have been observed in apples pretreated with PEF, while β-carotene content increased in dried apricots (Lammerskitten et al., 2019).

As regards the availability of drying techniques other than HAD, FD is a three-step drying process, such as (i) a fruit is first frozen and ice crystals are formed, (ii) then sublimation occurs, and (iii) finally water is removed by desorption. This technique is a time-consuming and energy-intensive process, but has the advantage of preserving the nutritional properties of fruits to a greater extent compared to HAD. This technique is, in fact, effective in minimizing the formation of large ice crystals and reducing drip losses during freezing. Generally, FD allows preserving polyphenol content more than HAD, due to the low applied temperatures and the vacuum pressure, as observed in different fruits, such as jujube fruits (McSweeney and Seetharaman, 2015; Wojdyło et al., 2019), *Citrus* × *sinensis* (L.), and Osbeck fruits (Kumar et al., 2022). It also prevents polyphenol degradation by enzymatic reactions, thanks to the freezing step. Phenolic composition of Asian pears, apples, kiwi, kaki, and goji fruits is also well-retained by FD (Donno et al., 2019; Kayacan et al., 2022): catechins, flavonols, and phenolic acid in apples; chlorogenic and ρ-coumaric acids, (−)-epicatechin in kiwi and kaki; ferulic acid and catechins in goji dried fruits; β-carotene in kiwi, kaki, and goji fruits; lutein in goji fruits (Donno et al., 2019).

Microwave-vacuum drying (MWVD) is a relatively new technique that combines the benefits of vacuum and microwave heating. It utilizes rapid heating from microwave energy, and the low boiling temperature of water is determined by the reduced-pressure environment (Monteiro et al., 2020). The MWVD technique helps to reduce undesired sensory changes and nutrient loss that can occur due to prolonged drying times or high surface temperatures. The application of this technique also has positive effects on PCs composition and content. When jujube fruits and chokeberries were treated with a pre-drying treatment by HAD and a second step by MWVD, phenolics were retained more effectively than with HAD (Samoticha et al., 2016; Wojdyło et al., 2019). MWVD, with a relatively low microwave power (i.e., 100–300 W), also allowed the highest content of bioactive compounds and levels of antioxidant activity in cranberries (*Vaccinium macrocarpon* L.) (Zielinska and Zielinska, 2019). However, it is important to note that MWVD may result in uneven heating and drying due to a nonuniform field distribution in the drying chamber and irregular shape and size of the product. In addition, its effect on PCs composition is not univocal. Total phenolic content (TPC) decreased when poniol (*Flacourtia jangomas*) fruit was treated at 600 W and a vacuum of 450 mm Hg (Dash et al., 2023). This decrease may be attributed to the higher temperature generated by increased microwave power, which can enhance the activity of PPO. On the other hand, when a higher vacuum was applied, in combination with citric acid, an increase in TPC was observed.

Intermittent air drying (IAD) has also been explored as a technique which may potentially decrease the operation cost and increase the quality of fruits. The principle of this methodology is the transfer, by temperature gradient, of moisture from the inner part of a food to the surface. In addition, the temperature drop allows reducing nutrient degradation. The application of IAD to litchi fruit (*Litchi Chinensis* Sonn.) allowed higher retention of vitamin C and polyphenols compared to other drying methods (Cao et al., 2020). A possible reason for polyphenol retention might be the shorter time of exposure to oxygen for phenolic decay in IAD.

The choice of the most suitable drying technology depends on the specific characteristics of the raw materials and the desired attributes of the dried products. Selecting the proper drying process is crucial to obtain high-quality food products with unique nutritional properties.

6.3 TECHNIQUES FOR MONITORING BIOACTIVE COMPOUNDS IN DRIED PRODUCTS

Analysis of the drying process includes chemical and physical characterization of raw materials, intermediates, and end products, as well as monitoring of drying process parameters. Psychrometric properties of air are measured at various locations within the aeraulic circuit. Monitoring and controlling air velocity can help to ensure that the product is properly dehydrated without causing quality issues (W. P. Zhang et al., 2021). Exhaust air recirculation is also essential to maintain low energy consumption without negatively affecting the dehydration velocity (Zohrabi et al., 2019). In addition, by keeping track of the weight of the product, it is possible to ensure that the product is not under-/over-dried (Chakravartula et al., 2023). In addition to these parameters, other monitoring systems may also be utilized, such as computer vision (CV), single-point spectroscopy, and hyperspectral imaging (HSI). CV can provide valuable information about the appearance of the product, such as color, texture, shape, and size, as indicators of sensorial and nutritional quality

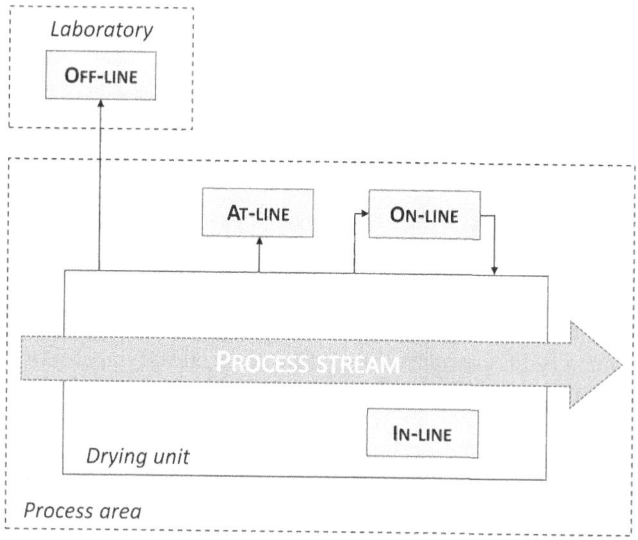

FIGURE 6.1 Analytical implementations for a drying process unit.

(Aghbashlo et al., 2014; Chakravartula et al., 2023). Single-point spectroscopy and HSI can be used to analyze the composition of the product during the drying process (Moscetti et al., 2017, 2018a, 2018b). Both techniques can be potentially used to predict bioactive compounds directly and indirectly, during and after drying.

Methods for process analysis are generally classified based on their spatial and temporal proximity to the process unit in off-line, at-line, on-line, and in-line. The food industry is transitioning from a reactive approach to food safety and quality management to a more proactive and predictive approach, where on-line and in-line methods play a crucial role (Figure 6.1). The reactive approach is often associated with batch processes and involves monitoring intermediate and end products using off-/at-line methods. The shift toward a proactive and predictive approach reflects the industry's desire to improve food safety and quality by preventing problems before they occur using on-line and in-line methods (Kourti, 2007).

One advantage of off-line analysis in the food industry, particularly for monitoring the quality of dehydrated foods, is the availability of high levels of expert knowledge in the central laboratory. This laboratory can adjust marginal conditions, such as sample preparation and equipment configuration, to suit the method of analysis and the types of food being examined. This results in a high level of flexibility and, as a rule, full utilization of the analysis equipment. However, off-line analysis is slow, and there are often several days between sampling and the availability of the analysis results, making direct process control impossible, which proves to be a disadvantage if deviations or faults occur in the production.

At-line analysis is optimized for the specific method of analysis and allows for a shorter time span between sampling and availability of results compared to off-line analysis. This is particularly advantageous in food drying processes as it enables the rapid detection of changes that could impact the quality of the final product. However, at-line analysis has some disadvantages, including the need for manual sampling that is not compatible with drying.

On-line analysis is an automated process analysis method that provides quick results, allowing for prompt corrective measures to be taken if there are deviations from target values in the process control system. It also provides feedback for comparison with target values, and planned values can be linked with expected instantaneous data for feedforward. This method can be used for process and cost optimization. However, the need for automatic sampling from the drying unit does not ensure full compatibility with all drying systems.

In-line analysis is a process analysis method that differs from other methods in that the equipment is directly placed above the material stream, eliminating the need for sampling or preparation. This allows for continuous analysis without restricting the analysis to a part of the material stream. As a result, in-line analysis fulfills the requirements for continuous data flow between the analysis equipment and process control, enabling direct process control. Both on-/in-line solutions require robust analytical equipment and trained personnel to manage them.

In the context of analyzing bioactive compounds in dried products, chromatographic techniques are employed as off-line and at-line methods. In-line solutions, such as CV, spectrophotometric technologies, and microsensors, are either already implemented or currently under development for scientific research. On-line solutions are typically employed to monitor and quantify product weight loss during the drying process.

6.4 CHROMATOGRAPHIC TECHNIQUES

The choice of methodology for the quantitative and qualitative determination of polyphenols in dried fruits, as well as in any plant-origin food, is very complex, and there is no single and standard method. In addition, the first step for reliable quantification by colorimetric assays and profiling by chromatographic techniques is related to the identification of the most effective sample preparation and extraction methods. This ensures that chemical alteration of the target compound is avoided, and accurate determination of polyphenols is possible.

Liquid samples are generally filtered or centrifuged before extraction, while solid samples require pretreatment processes, such as drying, FD, milling, or grinding, sieving, and homogenization (Pyrzynska and Sentkowska, 2019). Sample particle size is also an aspect that needs to be considered in sample preparation. Grinding enables, in fact, a better accessibility of polyphenols to solvent extraction compared with ungrounded samples (Prasedya et al., 2021); in combination with defatting where applicable, it is advised prior to extraction for accurate sample preparation.

The isolation of the polyphenols is mainly carried out by solvent extraction. The nature of the sample and the chemical properties of the phenolics, including molecular structure, polarity, concentration, number of aromatic rings, and hydroxyl groups, deeply affect the solubility and separation properties of polyphenols and must be hence considered in the preparation and extraction procedures (Alara et al., 2021). Generally, PCs are considered polar to medium-polar compounds, susceptible to degradation after exposure to high temperature, oxidation, and radiation. The presence of compounds with similar physicochemical properties can also affect polyphenol extraction. Cleaning methods, such as solid-phase extraction, droplet counter-current chromatography, and column chromatography, are additional steps used to remove unwanted phenolics and other interfering substances, such as pigments, terpenes, fats, and wax (Alara et al., 2021). In addition, polyphenols can form complexes with proteins, carbohydrates, or other molecules, which thus hinder the complete extraction of some phenolics (Alara et al., 2021). Aqueous solutions of methanol, ethanol, and acetone are generally used as solvents. Adjustment of aqueous-organic solvents to acidic pH is made for anthocyanins. To release polymeric polyphenols from the residue of extraction, acid, alkaline, or enzymatic hydrolysis treatments are also performed. Extraction efficiency is influenced not only by solvent choice and composition but also by other process parameters, such as temperature, extraction time, and solvent-to-sample ratio (Melini et al., 2023; Melini and Melini, 2021a).

Crude extracts can be obtained using conventional extraction approaches (e.g., maceration, Soxhlet extraction, and shaking), as well as with other methods such as microwave-assisted extraction (MAW), ultrasound-assisted extraction (UAE), and supercritical fluid extraction (SFE) in different formats. The application of these techniques has been increasingly preferred because they allow for the use of environmentally friendly solvents, reduce extraction time, decrease solvent consumption, and result in lower energy consumption (Melini and Melini, 2021b).

In dried fruits, polyphenols have mostly been extracted using conventional extraction techniques, with (i) varying solvent-to-sample ratios, ranging from 2.5 to 20 mL g^{-1}; (ii) different volumes (1–25 mL) of extraction solution (mostly methanol); (iii) variable ratios of methanol: water (80:20 to 100:00 v/v); and (iv) different acidifying solvents (e.g., HCl and/or ascorbic acid) (Donno et al., 2019; Izli et al., 2018; Sharif et al., 2018; Süfer and Palazoğlu, 2019; Wojdyło et al., 2019). The lack of a standard and unique method for the determination of polyphenol content and profiling of single compounds thus makes it difficult to compare different dried fruits, as well as different drying methods.

Quantification of polyphenols in extracts is generally carried out by colorimetric assays, such as the Folin-Ciocalteu reagent (FCR), the aluminum chloride ($AlCl_3$) assay, and the pH differential method. These methods are quick, easy to perform, and cost-effective, allowing for the determination of TPC, total flavonoid content (TFC), and total monomeric anthocyanin concentration, respectively. These assays have been mostly used to quantify polyphenols in dried fruits. The Folin-Ciocâlteu assay is based on the reduction of the FCR in alkaline medium (sodium carbonate), resulting in the formation of blue tungsten and molybdenum oxides, which absorb light at 760 nm. The intensity of the blue color is proportional to the amount of reactive PCs in the sample. Gallic acid is the most common reference standard for their quantification. However, when applying this assay to dried fruits, various modifications are available. These modifications include (i) the total assay volume, which can range from at least 1 up to 10 mL; (ii) the amount of FCR, ranging from 90 μL up to 2.5 mL per sample; (iii) the order of the reagent addition; (iv) the sodium carbonate concentration; (v) the reading wavelength (from 725 up to 765 nm); (vi) and the reaction time (30–120 min) (Dash et al., 2023; Donno et al., 2019; Izli et al., 2018;

Liu et al., 2020; Sharif et al., 2018; Süfer and Palazoğlu, 2019; Wojdyło et al., 2019). It is important to note that accurate quantification using this method is challenging because other components present in the food extract (e.g., ascorbic acid, amino acids, or sugars) can behave as reducing agents, leading to potential interference in the results.

TFC is commonly quantified using the $AlCl_3$ colorimetric assay, where Al (III) acts as a complexing agent. The assay relies on the affinity of flavonoids to bind metal ions such as Al (III), due to their oxo and hydroxyl groups, depending on experimental conditions, including pH value. Experimentally, upon the addition of $AlCl_3$ in the absence of $NaNO_2$, yellow-colored Al (III)-flavonoid complexes are formed, and their absorbance is subsequently measured at 410–440 nm. TFC is quantified on reference flavonoid standards, such as quercetin, catechin, and rutin. As regards dried fruits, TFC has been investigated only in one study on wild blueberries treated by osmotic dehydration after microwave pretreatment (Sharif et al., 2018).

Identification of individual PCs is not possible with colorimetric methods. High performance liquid chromatography (HPLC) and gas chromatography (GC), or their combination with mass spectrometry (MS), are the two most applied chromatographic techniques used to quantify them (Alara et al., 2021).

HPLC is preferred for the separation and quantification of polyphenols in fruits. The chromatographic conditions of the HPLC methods generally include the use of a reversed-phase C18 column, Ultraviolet-Visible (UV-Vis) diode array detector (DAD), and a gradient mobile phase consisting of two or more reagents, which are typically a polar organic solvent (e.g., acetonitrile or methanol) and an aqueous solution of a weak acid (e.g., acetic acid or phosphoric), while anthocyanins, more highly acidic mobile phases (e.g., trifluoroacetic acid), are used. Detection methods, other than UV-Vis, are also available, e.g., Fluorescence, Refractive Index, Evaporative Light Scattering, Electrochemical, Charged Aerosol, and MS. However, they have been used to a lesser extent or in no study on the determination of polyphenols in dried fruits.

The general elution order of polyphenols on RP-LC columns is usually hydroxybenzoic acids < cinnamic acids < *O*-glycosides < flavonoid aglycones. When non-highly acidic mobile phases are used, anthocyanins elute close to flavanols. The analytes are identified and quantified by comparison against standard solutions of phenolic acids, flavonoids, and anthocyanins.

However, peak separation can become complex, resulting in amalgamated peaks, which can subsequently hamper an exact determination of the compound concentration. In such cases, HPLC may not be suitable for a completely accurate and reliable characterization and quantification of PCs or flavonoids. In some foods, the extract may also contain molecules that are chemically related to polyphenols or have similar polarity. Therefore, a purification stage, where potential interfering components are removed, must be foreseen (liquid–liquid partitioning or solid–phase extraction are possible solutions).

The identification of PCs and flavonoids in dried fruits has been carried out using various methods, including HPLC coupled with a diode array (HPLC-DAD) (Kayacan et al., 2022); UV-Vis and FLD (HPLC-UV-FLD) (Srednicka-Tober et al., 2020; Yasir et al., 2016), and Ultra Performance Liquid Chromatography – tandem mass spectrometry (UPLC-MS/MS) (Oszmiański and Lachowicz, 2016).

The determination of PCs (e.g., 4-OH-Benzoic acid, chlorogenic acid, and trans-cinnamic acid) and flavonoids (e.g., catechin, epicatechin, and galangin) in dried Asian pear (*Pyrus pyrifolia*) was achieved using HPLC-DAD (Kayacan et al., 2022). In a separate study, an HPLC-UV-FLD method was used to determine PCs and flavonoids in four different apple cultivars processed with heat pump drying (Serni et al., 2020). The mentioned study addressed the challenge of simultaneous determination of 13 different PCs in apple extract without the need for fractionation steps or other preliminary sample treatments. Various classes of PCs were successfully detected: hydroxybenzoic acids (i.e., chlorogenic acid and gallic acid); dihydrochalcones (i.e., phloridzin); flavan-3-ols (i.e., catechin and procyanidin); and flavonols (i.e., quercetin-3-*O*-glucoside, quercetin-3-*O*-xyloside, quercetin-3-*O*-arabinoside, quercetin-3-*O*-rhamnoside, and quercetin-3-*O*-galactoside) (Serni et al., 2020). The application of a more sensitive and selective detector, such as a mass spectrometer, to an LC system (LC–MS) or tandem mass spectrometry (LC–MS/MS), offers significant advantages in overcoming various limitations associated with polyphenol determination by HPLC. These limitations include the reliance on compound identification solely through retention time, Ultraviolet (UV) spectra, the need of standards for quantification, and the challenges of detecting and quantifying polyphenols in complex samples. The literature shows that only a few studies have explored the use of LC–MS techniques for the identification and quantification of bioactive compounds on dried fruits (Oszmiański and Lachowicz, 2016; Yasir et al., 2016). However, LC-MS methods are increasingly becoming a prevalent approach due to their ability to identify and predict the exact molecular weight of compounds from small amounts of food sample extracts, even at low concentrations (López-fernández et al., 2020).

PCs in dried fruits have been successfully identified using MS techniques with electrospray ionization source operating in negative and/or positive ionization mode. The use of high-resolution mass spectrometers allows for precise identification of analytes based on their retention time and the MS and MS/MS spectra with respect to those present in the libraries (Oszmiański and Lachowicz, 2016; Yasir et al., 2016). In particular, polyphenols in dried fruits of chokeberry (*Aronia melanocarpa* L.) have been quantitatively analyzed by using UPLC-MS/MS (time of flight Q-TOF/MS detector). This detector allows the selection of specific ions in the quadrupole to be fragmented and analyzed in the TOF according to the arrival time of the mass fragments.

6.5 SINGLE-POINT SPECTROSCOPY

Rapid and/or nondestructive assessment of fruit quality attributes by using vibrational spectroscopic methods is widely reported in recent scientific literature (Beghi et al., 2017; Bureau et al., 2019; Johnson et al., 2023; Manzoor et al., 2021, 2022; Sultanbawa et al., 2021; Tahir et al., 2019; X. Zhang et al., 2021). Optical techniques combined with chemometrics have been utilized to identify correlations with the physical/chemical properties of the fruit sample, enabling their quantification. Included among the optical analytical techniques, vibrational spectroscopy, which relies on the interaction between matter and electromagnetic radiation, has become more accessible over the years, thanks to advancements in data processing capabilities and calibration procedures (Ruiz-Altisent et al., 2010).

Vibrational spectroscopic techniques include mid-infrared (MIR), NIR, Raman, and fluorescence spectroscopy. The absorption of light by a molecule can be explained by quantum mechanics, where the absorption results in discrete changes in the energy levels (quantum levels) of the molecule. When an intramolecular bond absorbs an energy equal to the difference between two adjacent quantum levels, fundamental vibrations between atoms are generated. These vibrations are characterized by specific vibrational wavelengths that fall within the range of *MIR spectroscopy* (4000–400 cm^{-1}, equivalent to 2500–25000 nm) (Delwiche, 2015). Furthermore, the vibrations cause a change in the dipole moment of the molecule. By approximating the single bond to a simple diatomic harmonic oscillator, as described by classical mechanics, it can be observed that features associated with specific bonds appear in specific regions. Since molecules, excluding optical isomers, differ from each other, the IR spectrum of each molecule is characteristic. As a result, IR spectra are suitable for qualitative analysis by identifying functional groups in key spectral ranges, which allows for tracing the structural features of unknown molecules (Bureau et al., 2019).

NIR spectroscopy is mainly based on the vibration of hydrogen bonds in the functional groups of organic molecules, such as C–H, O–H, and N–H bonds. These vibrations exhibit anharmonicity, which allows for overtone transitions and combination bands, falling within the NIR spectral range (780–2500 nm). Vibrations of intrachain bonds, such as C–C bonds, are not active in NIR spectroscopy.

Raman spectroscopy provides a structural fingerprint for identifying molecules. It relies on inelastic scattering of photons and provides narrow, well-resolved bands that result from the vibration of covalent bonds, such as relative neutral intramolecular C–C, C–H, and C=C bonds. Raman spectroscopy eliminates water interferences and can be used to characterize the orientation of anisotropic crystal structures. The technique adopts lasers as light sources, covering spectral ranges from UV to NIR (Mosca et al., 2021; Zhang, 2017).

Fluorescence spectroscopy focuses on fluorophores, which are specific organic compounds that emit unique fluorescence when excited by UV-Vis radiation. When reverting to the ground state by emitting photons, fluorophores lose some of their absorbed energy through vibrational relaxation of the excited state. As a result, a shift in the fluorescence spectrum toward longer wavelengths (hence lower energy) compared to the absorption spectrum occurs, as described by Stokes' rule (Hussain et al., 2018; Sikorska et al., 2019). Fluorophores are mainly polyaromatic hydrocarbons, heterocyclic compounds, and some highly unsaturated aliphatic compounds. They often exhibit conjugated systems of C=C double bonds, aromaticity, and rigid molecular structure. This includes various PCs such as gallic acid, ferulic acid, and monomeric flavan-3-ols ((+)-catechin, (–)-epicatechin). Furthermore, fluorescence is influenced by the nature and position of substituents (Sikorska et al., 2019).

The spectroscopic measurement mode can be selected based on the optical properties of the samples. Transparent samples can be measured in transmittance mode, while turbid liquids, semisolids, and solids require the use of diffuse transmittance, transflectance, or diffuse reflectance modes, depending on the absorption and scattering characteristics of the analyzed material (Figure 6.2) (Reich, 2005).

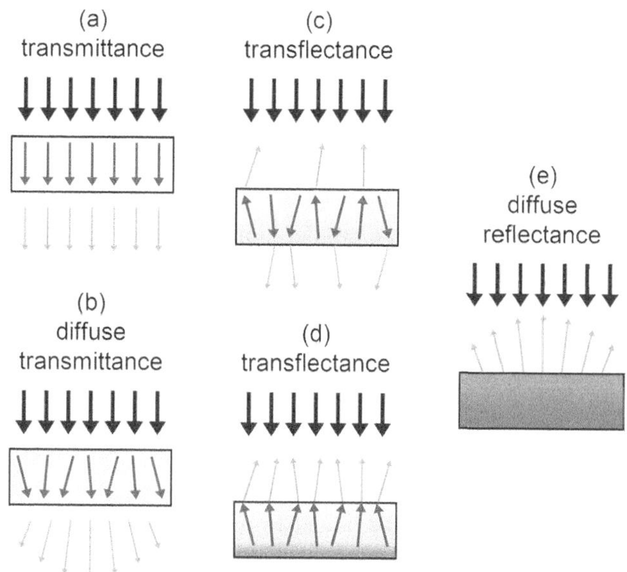

FIGURE 6.2 NIR Measuring modes: (a) transmittance, (b) diffuse transmittance, (c/d) transflectance, and (e) diffuse reflectance. The black arrows represent the light coming from the light source, the dark gray arrows the light interacting with the analyzed material, and the light gray arrows the light reaching the NIR sensor.

Each of these techniques has advantages and drawbacks. The weakness of the absorption bands in the NIR region can eliminate the need for sample dilution, while their broadness and overlapping peaks make them suitable for quantitative analysis of substances with specific functional groups. However, NIR spectroscopy is not suitable for qualitative analysis due to peak overlap (Delwiche, 2015). The nondestructive analysis of whole fruits using NIR spectroscopy depends on the physical structure of the fruit, resulting in specific limitations for each type of fruit (de Oliveira et al., 2014a). On the other hand, MIR spectroscopy, due to its high sensitivity to chemical composition, is a technique best suited for high-throughput quali-quantitative analysis of the main constituents of fruit by-products, such as juices, homogenates, and extracts (Bureau et al., 2012). However, the sensitivity of NIR or MIR spectroscopy may not be adequate for the quantification of minor food components in those foods, such as fruits, that are primarily composed of water. On the contrary, the high sensitivity and selectivity of fluorescence spectroscopy make it well-suited for studying minor and trace components in complex food matrices. It is important to note that fluorescence spectroscopy targets fluorescent analytes, although nonfluorescent ones could indirectly provide additional information through their influence on fluorophores. Finally, Raman spectroscopy provides complementary information to MIR spectroscopy, offering the advantage of eliminating water interference. However, it is characterized by low signal intensity and is subject to interference caused by sample fluorescence (Sikorska et al., 2019).

Among the minor constituents of fruit, bioactive compounds are of great interest because of their health-promoting properties. The spectroscopic techniques introduced above can be employed for the qualitative and quantitative determination of the bioactive compounds, especially in dried fruit, where water interference on the spectroscopic data, which affects NIR and MIR techniques, is not a concern. In this regard, Table 6.1 shows various applications of NIR, MIR, and Raman spectroscopy on powders obtained from dried fruit. Prior to spectroscopic analyses of minor components, the dried samples are often ground to minimize scattering of the incident electromagnetic radiation.

To highlight the strengths and weaknesses of the works in Table 6.1, a brief discussion follows. *Apple* polyphenols are susceptible to degradation by PPOs, leading to enzymatic browning. To mitigate this, Bureau et al. (2012) used FD as a preparation method for MIR analysis of apple samples (Table 6.1). The best results from the partial least squares (PLS) model for total polyphenols were obtained for the unoxidized samples, with a root-mean-square error of prediction (RMSEP) of 9.0%. However, it is worth noting that the analysis was conducted on a limited number of samples (55, of which 39 in calibration and 16 in validation), which may affect the robustness of the prediction model (Bureau et al., 2012).

Despite the relatively small number of samples (38 in the calibration subset and 27 in the validation subset), Gardana et al. (2018) predicted the anthocyanin content of dry powdered bilberry extract using PLS, which showed an excellent RMSEP of 0.30% w/w (Table 6.1). The authors analyzed anthocyanin content using Fourier Transform Near-Infrared (FT-NIR) and developed a method to detect fraud, specifically the substitution with cheaper alternatives. The study revealed that anthocyanins accounted for 18–34% w/w DW. Furthermore, FT-NIR successfully differentiated the authentic from counterfeit samples (six) (Gardana et al., 2018).

The physical structure of passion fruit hinders the determination of qualitative attributes using vibrational spectroscopy on intact fruit (de Oliveira et al., 2014a). To overcome these challenges, freeze-dried passion fruit pulp was used in the study of de Oliveira et al. (2014b) (Table 6.1). The determination of organic acids (i.e., malic and citric acids) yielded the best result for citric acid using MIR spectroscopy, with an RMSEP of 4.52%, probably influenced by the median concentration of citric acid, 5.57 times higher than that of malic acid. Regarding carotenoids, β-carotene is the primary one present in passion fruit, albeit in very low concentrations. Once again, vibrational spectroscopy, NIR in this case, did not perform well, as evidenced by an RMSEP of 25.51% (de Oliveira et al., 2014b).

The study conducted by McIntyre et al. (2020) focused on freeze-dried plums (Table 6.1). The analysis was affected by the rather low concentration of the analytes: phenols, although being the most represented, exhibited a narrow range of measurements below the uncertainty level, making them unsuitable for quantitative regression. Similarly, due to their skewed distribution and their multi-chemical nature, antioxidants were also deemed unsuitable for quantitative regression. Vitamin C and anthocyanins, although quantitatively less abundant than phenols, were successfully determined using FT-Raman, MIR, and NIR spectroscopy combined with PLS models. The best results were achieved using FT-Raman; the RMSEP, probably affected by the low concentrations of the analytes, was 12 and 0.69 mg/100 g DW for

TABLE 6.1
Applications of Spectroscopic Techniques to the Analysis of Powdered Dried Fruit

Food Matrix, Sample Type	Analytes	Sample Size (cal/val)	Spectral Range	Optical Geometry	Statistical Method	Cross-validation	Test Set	Accuracy	Error	Notes	Reference
Apple (*Malus domestica* Borkh.), freeze-dried powder	**Polyphenols**: **flavan-3-ols** (including **procyanidins**), **dihydrochalcones**, **hydrocinnamic acids** malic acid, citric acid	39/16 nonoxidized samples	MIR: 1568–1010 cm^{-1}	Reflectance	PLS	NS ("full cross-validation")	Randomly distributed (one-third of the samples)	R^2_P: flavan-3-ols: 0.95; procyanidins: 0.94; dihydrochalcones: 0.90; hydroxycinnamic acids: 0.48; total polyphenols: 0.96	RMSEP%: flavan-3-ols: 9.3; procyanidins: 9.8; dihydrochalcones: 19.2; hydroxycinnamic acids: 49.3; total polyphenols: 9.0	All sample data (oxidized and not) models gave very high RMSEP%	(Bureau et al., 2012)
Bilberry (*Vaccinium myrtillus* L.), dried powder	**Anthocyanins**	38/27	FT-NIR: 1064–1640 nm, 1833–2354 nm	Reflectance	PLS	LOO	Dependent test set (randomly selected samples)	R^2_P: 0.995	RMSEP: 0.30% w/w	NIR analysis could identify counterfeit bilberry samples	(Gardana et al., 2018)
Passion fruit (*Passiflora edulis* Sims), freeze-dried (lyophilized) pulp	citric acid, malic acid, **carotenoids** (**β-carotene**)	56 (80% calibration/20% validation)	NIR: 800–2500 nm; MIR:4000–600 cm^{-1}	Reflectance	PLS	LOO	PCA to define calibration and validation subsets; not specified how	R^2_P: MIR: citric acid: 0.913; NIR: malic acid: 0.506; total acids: 0.903; β-carotene: 0.772	RMSEP%: MIR: citric acid: 4.518; NIR: malic acid: 25.383; total acids: 5.885; β-carotene: 25.509	Limitations due to the physical structure of the fruit	(de Oliveira et al. 2014a, 2014b)

(Continued)

TABLE 6.1 *(Continued)*
Applications of Spectroscopic Techniques to the Analysis of Powdered Dried Fruit

Food Matrix, Sample Type	Analytes	Sample Size (cal/val)	Spectral Range	Optical Geometry	Statistical Method	Cross-validation	Test Set	Accuracy	Error	Notes	Reference
Plum (*Prunus salicina*), ground to powder and freeze-dried	**Anthocyanins, vitamin C, phenolic content**	24 samples: 4 replicates per sample, 3 for calibration and 1 for validation	FT-Raman: 200–4000 cm^{-1}; MIR: 450–4000 cm^{-1}; NIR: 920–1670 nm	FT-Raman: Raman intensity; MIR and NIR: absorbance	PLS	LOO	One biological replicate kept separate to act as an independent test set	R^2_P: Anthocyanin, FT-Raman: 0.91; MIR: 0.74; NIR: 0.79; Vitamin C, FT-Raman: 0.68; MIR: 0.68; NIR: 0.51	RMSEP (mg/100g): Anthocyanin, FT-Raman: 12; MIR: 16; NIR: 17; Vitamin C, FT-Raman: 0.69; MIR: 0.71; NIR: 0.90	4 varieties from 2 locations during a single growing season. Low number of samples.	(McIntyre et al., 2020)
Plum, kakadu (*Terminalia ferdinandiana* Exell), powder	**Vitamin C**	80/NS (CV); 80/5 (Prediction)	MIR: 4000–400 cm^{-1}; FT-NIR: 11 550–3950 cm^{-1}	Reflectance	PLS	NS	Obtained in 2020 and sourced from a commercial supplier	R^2_{CV}: MIR: 0.93; NIR; 0.91 R^2_P: MIR: 0.65; NIR: 0.73	SECV (mg/100 g DW): MIR: 1811; NIR: 1839 SEP (mg/100 g DW): MIR: 2367; NIR: 4773	Low number of samples; very low number of samples as validation set	(Cozzolino et al., 2021)
Ziziphus jujuba (*Z. Jujuba* Mill.), puff dried ground to powder	**Phenolics** and **flavonoids**	52/26	NIR: 900–1700 nm	Reflectance	Si-ACO-PLS	–	*Z. jujuba* samples were systematically separated into two subsets in the 2:1 ratio	R^2_P: Gallic acid: 0.88; Caffeic acid: 0.89; L-epicatechin: 0.91; Phloridzin: 0.86; Cianidanol: 0.8360	RMSE (μg/g) Gallic acid: 3.06; Caffeic acid: 6.04; L-epicatechin: 16.30; Phloridzin: 0.84; Cianidanol: 16.01	Fresh samples from five geographical origin of China. Low number of samples	(Arslan et al., 2020)

Note: cal/val, calibration/validation; NS, not specified; CV, cross-validation; MIR, mid-infrared; FT-NIR, Fourier transform near-infrared; NIR, near-infrared; FT-Raman, Fourier transform Raman; PLS, partial least squares; Si-ACO-PLS, synergy interval with ant colony optimization partial least square; LOO, leave-one-out; R^2_P, coefficient of determination of prediction; R^2_{CV}, coefficient of determination of cross-validation; RMSEP, root-mean-square error of prediction; SECV, standard error of cross-validation; DW, dry weight; SEP, standard error of prediction; RMSE, root-mean-square error.

anthocyanins and vitamin C, respectively (corresponding to 14.3 and 16.4% of the measured range, respectively) (McIntyre et al., 2020).

Vitamin C was also determined on Kakadu plum powder samples by Cozzolino et al. (2021) using MIR and NIR spectroscopy (Table 6.1). The average vitamin C content was approximately 14% of DW, significantly higher than the average reported by McIntyre et al. (2020). The PLS models were validated with only five samples, which may have affected the robustness and accuracy of the predictions. The standard error of prediction (SEP) values obtained were 2367 and 4773 mg/100 g DW for MIR and NIR spectroscopy, respectively, corresponding to 33 and 67% of the range of vitamin C content of the prediction samples, respectively.

Finally, Arslan et al. (2020) predicted phenolics and flavonoids content of puff-dried *Ziziphus jujuba* samples (Table 6.1). A fairly low number of samples were considered (52 in calibration and 26 in validation), as already pointed out by Bureau et al. (2012) and Gardana et al. (2018). Furthermore, the concentrations of the bioactive compounds under study are in the order of μg/g, low enough to affect the accuracy of prediction models. The ratio of RMSEP (from the best PLS models) to the respective concentration range of the validation subset, expressed as a percentage, ranges from 11.34% for phlorizin to 16.69% for cianidanol (Arslan et al., 2020). From the studies reported, it was observed that the use of vibrational spectroscopy techniques for determining bioactive compounds in powdered fruit was mainly affected by the low concentration of analytes. Nevertheless, it would be interesting to investigate the potential application of vibrational spectroscopy in at-line, in-line, or on-line process control systems for monitoring bioactive compounds, especially in whole or powdered dried fruit that contains an adequate amount of these compounds. If this condition is met, MIR or NIR spectroscopy could be a suitable choice.

6.6 COMPUTER VISION SYSTEMS

CV aims to replicate human visual perception in machines, enabling the automatic extraction and understanding of valuable information from images using algorithms. It has found extensive application in various fields, including food production quality assessment, robotics, medical diagnosis, and industrial systems. A CV setup typically consists of five elements: illumination system, sensor or camera, digitizer (if needed), computer, and image-processing software. In the food industry, CV has rapidly grown in popularity due to its ease of implementation, nondestructive product inspection, and high inspection rate (Raponi et al., 2017).

CV systems are an integral component of Process Analytical Technology (PAT) tools used for monitoring and controlling smart drying processes, i.e., advanced and automated drying techniques that utilize various sensors, data analysis techniques, and control systems to optimize and monitor the drying conditions. In detail, CV systems offer cost-effective, noninvasive, and real-time solutions to monitor attributes like appearance and texture, which are quantifiable into relevant quality parameters (e.g., color, size, shape, etc.). In other words, CV can be implemented as in-line data acquisition tool for quality monitoring, multivariate modeling, and feedback, providing comprehensive and precise information throughout the drying process. CV systems can be classified into two main categories: classical image analysis and HSI.

In classical image analysis, algorithms are used to process and interpret 2D images, extracting color features and patterns for various applications. On the other hand, HSI involves capturing and analyzing 3D images in multiple narrow and contiguous spectral bands, enabling the identification and characterization of materials based on their unique spectral signatures. Both approaches play crucial roles in modern CV applications, offering valuable insights and solutions across different industries and fields (Figure 6.3).

For classical image analysis, most machine vision systems are equipped with a red, green, and blue color space (RGB) camera, which is commonly used for external inspection of products. This RGB camera captures images by acquiring three wavelengths in the visible (Vis) spectral region using filters centered at 700 nm (red), 560.1 nm (green), and 465 nm (blue) (Wegener et al., 2017). As a result of light absorption, delocalization of electrons along the chromophore results in an excited state of molecules and the manifestation of colors (Sant'Anna et al., 2013). RGB is one of many existing color spaces, and converting its data to others (e.g., CIELab, HSV, CIELUV, etc.) can provide more accurate and standardized color information to facilitate various analyses, such as color comparison across different lighting conditions or devices, quality evaluation, and defect detection. Indeed, the RGB space lacks perceptual uniformity and does not accurately represent the colors perceived naturally by humans. For these reasons, the RGB to CIELab conversion is widely used in food industry applications. The CIELab color space consists of a psychometric lightness index (L*) and two-color coordinates (a* and b*), representing the luminosity, green to red, and yellow to blue aspects of colors, respectively. It also offers a more comprehensive coverage of the visible spectrum. CIELab coordinates provide the necessary information to compute color indexes such as hue angle (h), chroma (C*), and color difference (ΔE*) (Botondi et al., 2016). In addition to them, various color parameters from CIELab, such as browning, whiteness, and yellowness indexes, as well as ratios/differences between coordinates, have been developed for monitoring the color of specific fresh and processed fruits. Modeling the color degradation kinetics of food, including fruits, is a complex task. Several experimental studies have reported zero-order or first-order degradation kinetics of pigments and color changes in fruits during heat exposure. Pathare et al. (2012) conducted a comprehensive review, listing numerous methods suitable for monitoring color changes in fruits exposed to thermal processing and drying. A color degradation kinetics model can play an important role in revealing valuable insights into pigment (bioactive) compounds, including carotenoids, anthocyanins, betalains, and chlorophylls, abundantly found in fruits. In addition, several color indexes, particularly the browning index, can be used to assess (i) the oxidation of PCs caused by the action of the PPO enzyme (enzymatic browning) and (ii) the Maillard reaction (non-enzymatic browning), the rate of which is accelerated by the decrease of water activity. Below are some examples of color analysis successfully used to predict bioactive compounds.

PCs are crucial for sensory characteristics like color, astringency, and bitterness, as well as for providing antioxidants in the human diet. Their degradation is influenced by factors such as temperature, enzymatic activity, organic acids, sugar concentration, pH, and light and oxygen exposure. Colorimetric techniques can be used

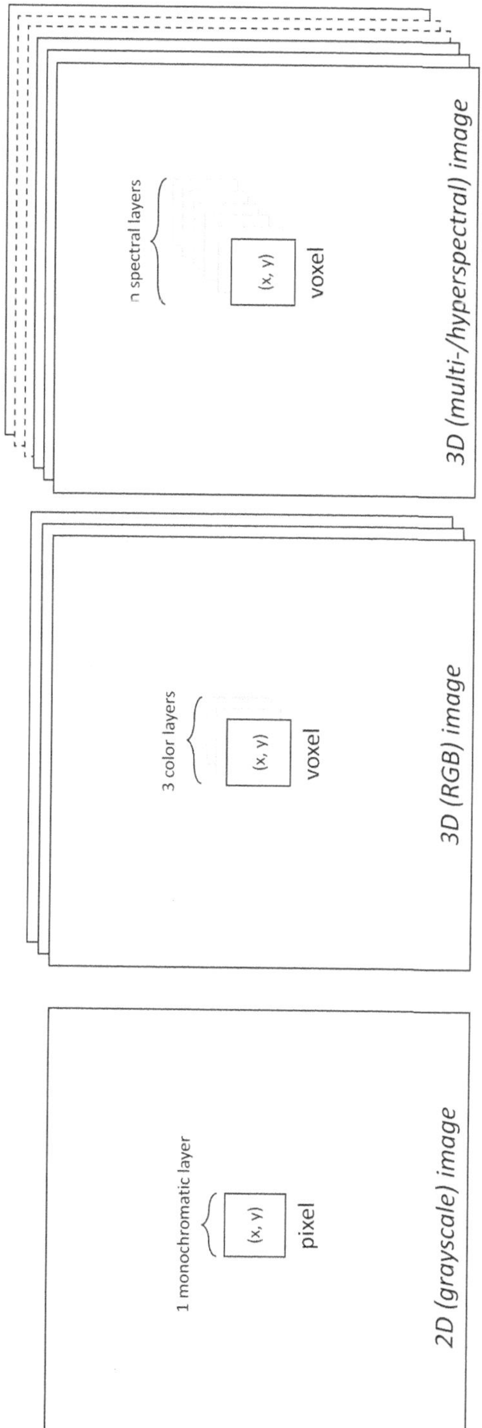

FIGURE 6.3 Data layers in monochromatic, RGB, and multi/hyperspectral images.

to assess the degradation of polyphenols, and there may be a relationship between the color profile and residual TPC in the dried product. Specifically, the TPC during drying exhibited a positive correlation with lightness (L*) in red grapes (R = 0.97) (Larrauri et al., 1997) and blueberries (R = 0.77) (Jiménez-Aguilar et al., 2011), and a negative relationship with CIELab color difference (ΔE*) in apple slices (R > 0.83) (Vega-Gálvez et al., 2012). Both authors also reported a negative correlation between antioxidant activity (AA) and hue angle (h) in red grapes and a* value in blueberries. Luminance (L*) and redness (a*) coordinates decrease and increase, respectively, during drying of apricots because of browning development (García-Martínez et al., 2013). Anthocyanins, which are sensitive to heat exposure, can undergo degradation mainly affecting the redness (a*+) and the hue angle (h) of the product (Sant'Anna et al., 2013). Zhao et al. (2023) found from good to very good relationships (R = 0.82–0.95) between color parameters and TPC, TAC, TFC, and DPPH.

Carotenoids are important fat-soluble pigments found in fruits and vegetables, known for their color and potential antioxidant capacity due to their vitamin A precursor properties and antioxidant capabilities. Carotenoids are susceptible to degradation caused by external factors such as heat, low pH, oxygen, and light exposure. The degradation can lead to color changes and the formation of various oxidation products with potential biological implications. Many studies have demonstrated the capability of color analysis in quantifying and characterizing carotenoids. Redness (a*+), yellowness (b*+), hue angle (h), chroma (C*), and color difference (ΔE*) were successfully applied to determine carotenoids cashew apple (Zepka et al., 2009), and orange fruit (Lee and Coates, 2003). The L*/b* ratio is also a good indicator of both browning development and carotenoids content according to Moscetti et al. (2018b).

Chlorophylls are present in tissues capable of photosynthesis, with fruits containing only "a" and "b" types of chlorophylls. The molecule is highly susceptible to chemical and/or biochemical degradation during drying, leading to changes in the food color. Chemical reactions involve the formation of pheophytin, which may occur during heat exposure. Strong heat treatments may also lead to the conversion of pheophytin to pyropheophytin due to decarbomethoxylation. A major chlorophyll degradation was observed in herbs with drying temperature higher than 60°C (Thamkaew et al., 2020), with possible implications also in fruit drying. In the context of the CIELab color space, the degradation of chlorophylls can be successfully monitored through the a* parameter, which is related to the greenish color of the product (Akar and Barutçu Mazı, 2019). However, several authors have found a correlation between chlorophyll content and luminance (L*) (Radicetti et al., 2016; Sanmartín et al., 2011), as well as the successful application of multiple linear regression using L*, chroma (C*), and hue angle (h) as predictors (Sant'Anna et al., 2013).

Betalains are water-soluble nitrogen-containing pigments synthesized in the cytoplasm and stored in the vacuoles of many flowers and fruits. These pigments are classified as yellow or red-purple betacyanins based on their chromophore. Sant'Anna et al. (2013) provided some examples of the correlation between color parameters (L*, a*, and C*) and Betalain content.

Despite an RGB camera being effective in monitoring pigment compounds, as well as enzymatic and non-enzymatic browning development during drying, the analysis is limited to the product surface. This limitation frequently makes the technology unsuitable for internal investigations of the product (Sun, 2010). As a result, if analytes are not visible in the visible region or are not directly related to color, they may remain undetected by the system. To address these challenges, the integration of HSI could provide valuable insights for detecting analytes invisible to RGB cameras and optimizing the drying process.

HSI systems are valuable tools due to their ability to provide both spectroscopic and spatial information (Mollazade et al., 2012; Moscetti et al., 2015). The data structure of a hyperspectral image is known as a hypercube. In the context of food drying analysis, HSI devices operate in the Vis/NIR region, utilizing absorption spectroscopy (Menesatti et al., 2009). The main advantage of HSI is its capability to monitor both external and internal quality parameters. With this technology, it becomes feasible to distinguish and analyze objects with similar color, shape, size, and overlapping spectra. However, HSI does have certain drawbacks, including the long acquisition time, a high amount of redundant information, and a large volume of acquired data, which limits its applicability in real-time processing (Raghavan et al., 2005). Additionally, the computational time required to develop prediction models increases significantly. To address these limitations, consideration can be given to the development of a Multispectral Imaging (MSI) system. MSI enables the selection of the most relevant wavelengths (typically 3–15) for predicting specific attributes of interest (Zhang et al., 2014). Compared to HSI, MSI offers advantages such as a faster scan rate, feasibility for on-line and in-line applications in the food processing industry, and reduced demand for computer memory during image acquisition and processing (Mahesh et al., 2015). However, MSI devices lack flexibility and are often custom-built by researchers for specific imaging tasks. Additionally, the use of an HSI system may be necessary before the development of an MSI device.

HSI is similar to single-point spectroscopy (Section 6.5) in terms of their application and underlying principles because both techniques involve the analysis of spectral information from a wide range of wavelengths. They share common principles of capturing spectral information, and their applications often overlap, making them powerful tools for in-line monitoring in industrial processes, fruit drying included. The HSI has the advantage of combining spectral and spatial information, allowing for comprehensive analysis of a sample's chemical composition and spatial distribution in a single imaging system. This capability sets it apart from single-point spectroscopy, which can only provide spectral data at specific points on the sample (Moscetti et al., 2015).

6.7 EMBEDDED SENSING TECHNOLOGIES (INTERNET OF THINGS)

There are various types of dryers and drying methods used in fruit drying, each with its own advantages and suitability for specific fruit and final desired product, both at a laboratory and industrial scale. The various approaches include "natural" methods like sun and air drying, as well as "artificial" ones such as oven, freeze, and MWD (Radojčin et al., 2021). While the main concept is shared, i.e., the removal of moisture from fresh fruits to preserve them for extended periods, the different technological approaches can have a great influence on other aspects of the final product quality. In fact, the drying process not only helps to preserve the product but also has the potential to improve the overall quality of the materials and to produce value-added compounds (Bustos et al., 2018).

During the drying process, the fruit sample undergoes a series of interconnected transient processes involving heat, mass, and momentum transfer, as well as physical, chemical, and phase changes (Hasan et al., 2019). In this context, the selection and precise monitoring of drying conditions are of paramount importance for preserving the valuable nutritional characteristics of the final products (Nakagawa et al., 2021). Here, the end product quality, affected by factors such as drying method, pre- and post-treatments, and process parameters, can be enhanced by estimating its quality attributes during the drying process, both in a direct and indirect way, i.e., through the use of sensing solutions (Section 6.3).

6.7.1 Sensing Solutions for Fruit Drying Monitoring

A sensor is a device capable of measuring a physical quantity related to the sample or a phenomenon related to it and transducing it into information (i.e., data) (Miranda et al., 2019). This information can then be used to finely control the fruit drying process using appropriate algorithms. Depending on the specific requirements of the process, fruit dryers can implement several types of sensors, ranging in complexity from controlling process parameters (i.e., T, relative humidity, and airflow) to those employed for direct monitoring of sample quality (e.g., weight and gas sensors). The selection of sensors to be embedded in fruit dryers depends on factors such as the required accuracy, precision, range, drift, and response time, as well as the unique characteristics of the drying process. This section provides a summary of some of the most commonly used sensors, along with their working principles and applications in fruit drying.

Temperature sensors are essential for monitoring the temperature inside the drying chamber and ensuring that it remains within the optimal range for drying fruits throughout all the phases of the process. Accurate temperature monitoring plays a critical role in preventing nonoptimal outcomes during the process, such as over- or under-drying of the fruits, which can result in a loss in the final product's sensory and nutritional quality. Among such sensors, the most commonly used are: (i) Thermocouples, consisting of the recording of a voltage proportional to the T difference between two dissimilar metals. Thermocouples are known for their durability and capability to withstand high temperatures (Heichal et al., 2005); (ii) Resistance Temperature Detectors (RTDs), based on the principle that the electrical resistance of certain metals changes with temperature. RTDs offer high accuracy, stability, and repeatability, making them suitable for applications that require precise temperature control (Santos and Vasconcelos, 2008); (iii) Infrared (IR) sensors, which detect the IR radiation emitted by objects, allowing for noncontact temperature measurements. IR sensors can directly and rapidly measure the surface temperature of a drying fruit are required (Santos and Vasconcelos, 2008).

Humidity sensors are employed to monitor and control the humidity levels during the drying process, as Relative Humidity (RH) is a key parameter to ensure the desired equilibrium moisture content. Various commonly used humidity sensors are employed in fruit drying: (i) Capacitive sensors, which are the most widely used humidity sensors. These sensors operate based on the principle that the dielectric constant of a material changes with humidity. Capacitive sensors offer good accuracy, reliability, and response time (Hassanzadeh et al., 2008); (ii) Resistive sensors, also known as hygroscopic polymer sensors, are based on the principle that certain materials change their electrical resistance with humidity. They are less employed in dryers, but are relatively inexpensive and have a fast response time (Drumea and Svasta, 2010); (iii) Thermal Conductivity Sensors, which measure the changes in the thermal conductivity of a humidity-sensitive material as humidity levels vary. They provide good accuracy and stability, particularly in low humidity ranges (Mendonça et al., 2005).

Airflow sensors are essential for optimizing the airflow and ensuring that the fruits are drying evenly. Various commonly implemented airflow sensors include: (i) anemometers, which directly measure the velocity of air and can be based on different working principles, such as hot-wire anemometers, which measure air velocity based on the cooling effect of the air on a heated wire (Karim and Hawlader, 2005); (ii) differential pressure flowmeters determine airflow by measuring the pressure losses over fixed dimension paths or calculating the pressure difference between two points in the airflow path, which are then used to infer airflow velocity and direction (O'Sullivan et al., 2014).

Weight sensors are a class of sensors that indirectly monitor moisture loss of a sample by correlating it with the change in weight. The most commonly used approaches for weight sensors are as follows: (i) load cells, consisting of strain gauges or piezoelectric elements that deform in response to the applied force, generating a proportional electrical signal. Load cells are widely used due to their accuracy and wide range of measurable weights they offer (Muller et al., 2010); (ii) piezoelectric sensors, which utilize the piezoelectric effect, where certain materials generate an electrical charge when subjected to mechanical stress. These sensors commonly provide a rapid response time (Kim et al., 2021); (iii) capacitive sensors, which operate based on changes in capacitance, caused by a change in the distance between the capacitor plates, as a result of the applied force or weight loss of the samples during the drying process (Mishra et al., 2021).

Gas sensors and electronics nose represent a novel approach, nowadays employed more at a laboratory scale rather than directly integrated into fruit dryers. They can be used to detect and analyze odors and volatile compounds emitted by fruit during the drying process, giving insights not only on the kinetics of the process but also on the development and retention of desirable quality characteristics. Gas sensors can utilize different approaches for volatile compound detection, such as metal oxide (MOX) sensors operating based on the principle of changing electrical resistance when the specific MOX is exposed to the volatile compound of interest (Tan and Ang, 2021). While they can be sensitive to a wide range of odors, they lack specificity to a single gas. For this reason, they are often used in combination with other gas sensors in electronic nose systems or coupled with membranes to enhance their selectivity (Graunke et al., 2016); quartz crystal microbalance (QCM) sensors use a quartz crystal that vibrates at a specific frequency. When exposed to odorous compounds, the mass on the crystal surface changes, causing a shift in the resonance frequency. This shift can be detected and correlated to the concentration of the gaseous molecule of interest (Liu and Zhang, 2021). QCM gas sensors offer a high sensitivity and allow tailored selectivity to the specific gas of interest, depending on the materials and coating technologies employed in their fabrication (Muñoz-Aguirre et al., 2007).

The use of sensing technologies for the real-time monitoring of processing parameters, especially when carried out concurrently with the monitoring of product quality attributes (e.g., color, shrinkage, and water activity), allows for more efficient modulation of the dryer, resulting in energy savings and a higher quality dried product (Raponi et al., 2022). However, the current state of this technology is limited by several factors. First, the accuracy and reliability of sensors still have drawbacks and are also considered uneconomical, making it challenging to properly distribute them within the drying chamber (Chen et al., 2020). This aspect is especially important for the drying of fruit, which typically exhibits nonuniform drying kinetics due to uneven heat and humidity distribution within the drying chamber. These challenges could be overcome through improved modeling, design, and building of the instrumentation, as well as widespread sensor positioning in the drying chamber to finely monitor and optimize the process. In this regard, commercially available sensors are not an ideal solution due to the limitations in cost, size, and required circuitry. Therefore, a viable and promising alternative is the use of printing technologies to manufacture custom-made small sensors.

6.7.2 A New Approach: Printed Sensors

Printed electronics and printed sensors are a rapidly expanding sector within sensing technologies consisting of the patterning of conductive structures over a wide range of functional materials on flexible substrates (Wang et al., 2022). They offer low fabrication cost, high throughput, and the ability to print several small patterns over large areas, making them very attractive for potential use in the fruit drying field. Thanks to such advantages, printed sensors can be integrated with objects with complex shapes, such as dryers, with application-specific functionalities, and in multisensor arrays (Angeli et al., 2020).

In the context of fruit drying, the latter is especially advantageous, since a widespread distribution within a drying chamber, considering each of the tray layers, would allow a monitoring that accounts for the complexity of the drying process. In this context, the sensor output can be exploited, allowing for a better evaluation of the multidimensional distribution of the physicochemical and structural responses as well as for the multi-phased transport mechanisms (e.g., solid-solid, solid-liquid, and liquid-gas) occurring during drying (Welsh et al., 2021). While the integration of such innovative solutions has not yet been considered in smart dryers, the technology could be easily adapted from other fields. For instance, fully printed sensor arrays have been developed for large throughput and precision sensing of temperature, RH (Bücher et al., 2022), and force (Sapra et al., 2023), mainly for robotics and healthcare-related applications.

In addition, the employment of sensor arrays yielding high-throughput data, precisely mapping the entirety of the food dryers, can be beneficial for both a more precise optimization of the systems and a higher quality of the data analysis. In fact, the multiscale data generated by such arrays can be employed in multivariate analysis and machine learning modeling (Khan et al., 2020) and for the validation and fine-tuning of the multiphysics modeling of the fruit drying process, considering the parameters of interests (e.g., T and RH) distribution within the drying chamber (Mutuli et al., 2020).

6.8 PROCESS IMPROVEMENT THROUGH DATA-DRIVEN APPROACHES

A widely used approach for the optimization of the drying process relies on the use of modeling. Such an approach, while not allowing the in-line monitoring and process tuning capabilities offered by sensors, offers several other advantages and has been employed as a preliminary step for the design of drying systems, for the optimization of the drying process, as well as for the characterization of the drying behavior of fruit (Onwude et al., 2016). The latter approach is particularly relevant in this context as it allows us to analyze the fruit drying behavior, describe the complex kinetics, and adapt the process parameters to model each particular fruit product, widely varying in, e.g., structure, water distribution among tissues, and size. By modeling the heat and mass transfer rates, it is in fact possible to optimize the drying conditions, such as airflow rates, T profiles, and operating time, to minimize energy consumption while maximizing drying efficiency and final product quality.

The main aspects related to the drying setup and process which can be modeled are:

Fruit Geometry, namely the shape, size, and surface characteristics to understand how they affect the drying behavior. This can include considering the fruit's external and internal structure, such as the presence of peel, pulp, and seeds.

Heat Transfer, crucial for understanding the temperature distribution within the fruit and the drying chamber. It involves considering conduction within the fruit, convection between the fruit surface and the surrounding air, and radiation if applicable.

Mass Transfer, employed to simulate the movement of moisture from the fruit to the surrounding air. This includes the diffusion of moisture within the fruit and its subsequent evaporation at the fruit surface. Factors like humidity gradients, air velocity, and moisture content gradients affect the drying rate.

Airflow and Fluid Dynamics within the drying chamber, which can significantly impact the drying process. Modeling airflow patterns, air velocity, and pressure distribution helps understand how heat and moisture are transported within the system.

In addition to the parameters to model, a crucial input to define in each modeling task are the boundary conditions and the material properties, essential for an accurate simulation. The former includes setting the initial conditions, such as the temperature and moisture content of the fruit and the surrounding air, as well as specifying the boundary conditions at the fruit-air interface, such as convective heat and mass transfer coefficients. The latter include the properties of the fruit and the surrounding air, such as thermal conductivity, diffusivity, specific heat capacity, density, and viscosity. These properties can vary with temperature, moisture content,

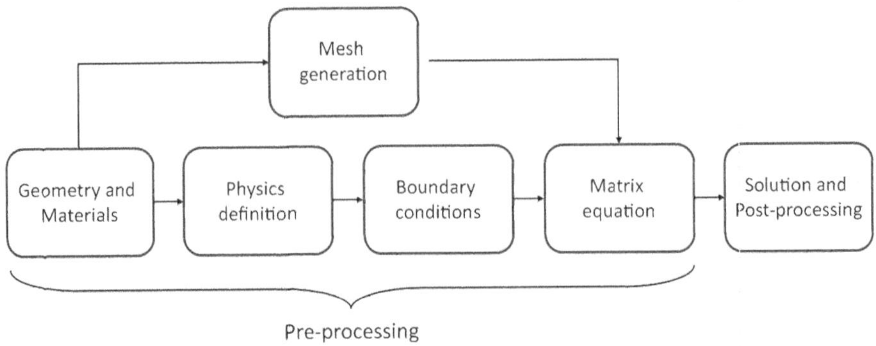

FIGURE 6.4 Basic workflow of a simulation process.

and other factors and should be carefully considered. In this context, the two most commonly used approaches are represented by two related but distinct concepts in the field of computational modeling and simulation, namely finite element modeling (FEM) and Multiphysics simulations.

Finite Element Modeling. FEM is a numerical method used to solve complex problems by dividing the domain into smaller, and thus simpler regions called finite elements. It discretizes the domain and solves the governing equations using finite element analysis (Akter et al., 2022). In the context of fruit drying, it is widely used to analyze and simulate a wide range of physical processes, including structural mechanics, heat, mass transfer, and fluid flow (Castro et al., 2018). FEM involves the following key steps, briefly summarized in Figure 6.4: (i) Geometry and mesh generation, starting from a simplified representation of the fruit slice. This step allows for the creation of a computational domain and for the discretization into finite elements; (ii) Formulation of the governing equations for the specific physical phenomena being modeled (e.g., mass transfer equation); (iii) Definition and assignment of the material properties to each of the predefined finite element; (iv) Specification of the appropriate boundary conditions representing the system being modeled, in this case the fruit slice; (v) Solution of the predefined finite element equations using numerical methods; and (vi) Analysis of the simulation results.

FEM has been extensively used in the context of fruit drying optimization for the prediction of several quality and process-related attributes. For instance, Janjai et al. (2008) employed it to model and predict the moisture content profiles of mango fruit, giving an accurate representation of the movement of moisture inside the fruit during the drying process. The same authors modeled moisture contents in litchi fruit during drying, providing insights into the transport processes happening within its different components (i.e., flesh, seed coat, and seed stalk) (Janjai et al., 2010). FEM was also employed by Seyedabadi et al. (2017) to model the simultaneous heat and mass transfer in cylindrical banana samples during convective drying, considering both axial and radial shrinkage effects and satisfactorily predicting the moisture content and core temperature kinetics. While such studies support the effectiveness of FEM for the optimization of the fruit drying process, such approach also presents several drawbacks limiting its potential, such as the computational cost, the model simplification, and its validation with experimental data (Castro et al., 2018). Here, the main limitation is that FEM solves a single physical phenomenon within a problem domain, and while it might be modeled adequately, it doesn't consider the interaction and coupling of multiple physical interactions happening in a real drying scenario. Despite these limitations, FEM remains a widely used and effective approach for modeling and simulating single physical phenomena, providing useful insights into the drying process. However, when dealing with complex systems involving multiple interacting factors, other approaches such as Multiphysics simulations offer a more comprehensive and integrated approach to capture the interdependencies of the system.

Multiphysics simulations. Multiphysics refers to the simultaneous modeling and analysis of multiple physical phenomena that are coupled and interact with each other, simultaneously solving multiple sets of governing equations that describe them (Knoerzer et al., 2015). In the context of a fruit drying process, multiphysics simulation considers the interplay of phenomena such as heat transfer, mass transfer, fluid flow, and structural effects, while also considering the precise design and constraints introduced by the drying process. In fact, such an approach offers the possibility of considering both the size and shape of the fruit samples and their distribution and drying behavior within different parts of the dryer.

The simulation process itself is typically similar to the above-mentioned FEM one, with the inclusion of some additional elements. The main difference is the consideration of a more complex first step of geometry and mesh generation, including the fruit as well as the fruit drying system, the drying chamber, and any other relevant components. This model is then discretized into small elements or cells to create a mesh, which allows for a more comprehensive simulative process. In the second step, the governing equations related to heat and mass transfer, including convection, conduction, and evaporation, are defined, considering also their interplay. The boundary conditions, such as T and RH, are then defined not only as fixed parameters at the interface between the fruit and the air but also considering the position in the drying chamber (Madadi Avargani et al., 2023). The subsequent steps of material properties definition, numerical simulations, validation, and analysis are then proceeding in a similar way compared to FEM simulations.

Multiphysics simulations have been used to model several aspects of the fruit drying process at different scales. For instance, Prawiranto et al. (2018) developed a 3D model to quantify the impact of the changes in the cellular structure of apple cells during the entire dehydration process (i.e., from the turgid cell to its shrinkage, plasmolysis, and lysis) on the tissue sorption isotherm and water permeability. Considering a larger scale, Lemus-Mondaca et al. (2013) carried out an experimental and numerical study for the drying process of Chilean papaya slices in a range of air temperatures from 40° to 80°C, predicting with good accuracy both temperature and moisture distributions inside the sample with an unsteady 3D coupled heat conduction and mass diffusion mathematical model. In recent work, a mechanistic drying model was coupled with fruit quality decay models and weather data to create a digital twin of an apple fruit undergoing a solar drying process. Such a model was employed to predict the drying time and quality attributes while also providing spatial and temporal distribution of T and RH within the fruit during the process (Prawiranto et al., 2021).

The implementation of simulation steps in the design, prototyping, and testing phases of the fruit drying process avoids a time- and resource-consuming trial and error process. The correct employment of such approaches not only yields an optimized and safer final product but also helps in the retention of its desired organoleptic characteristics (Martins, 2006).

6.9 FUTURE PERSPECTIVES

The field of instrumentation techniques for analyzing bioactive compounds in dried fruit products is continuously evolving, and several exciting future perspectives are emerging. As researchers and technology developers continue to advance their knowledge and capabilities, the following areas are expected to shape the future of this field: (i) sensing technologies, development of more sophisticated and sensitive sensors with improved accuracy, higher resolution, and the ability to detect a broader range of compounds; (ii) HSI can play a prominent role in the analysis of bioactive compounds due to its nondestructive and comprehensive data collection capabilities; (iii) automation and robotics, automated sample handling and data acquisition can lead to faster and more reliable results; (iv) miniaturization and portability of instrumentation, portable instruments for on-site quality control and product assessment; (v) data analytics and artificial intelligence, AI-driven data processing can help identify patterns, correlations, and predictive models, leading to better process optimization and quality control; (vi) multi-modal techniques, integration of multiple analysis techniques (e.g., spectroscopy, chromatography, and CV) for a more comprehensive understanding of bioactive compounds in dried fruits; (vii) Internet of Things (IoT), real-time data monitoring, and decision-making; (viii) novel bioactive compound characterization, ongoing research in identifying and characterizing new bioactive compounds in dried fruits may lead to the discovery of additional health benefits and applications; (ix) product and process sustainability, efforts to reduce energy consumption and waste generation during analysis; and (x) standardization and guidelines, establishment of protocols and guidelines to ensure consistency and comparability of results across different studies.

Overall, the future perspectives in instrumentation techniques for analyzing bioactive compounds in dried fruit products hold significant promise for advancing food science and contributing to the improvement of the nutritional quality and safety of dried fruit products for consumers worldwide. Researchers, industry stakeholders, and policymakers must collaborate to drive innovation in this field and unlock new opportunities for the food industry.

6.10 CONCLUDING REMARKS

In conclusion, the techniques employed for the analysis of bioactive compounds in dried fruit products have significantly contributed to advancing our understanding of their nutritional quality and potential health benefits. The effects of various drying processing conditions on the preservation of these bioactive compounds have been investigated, providing valuable insights into optimizing drying methods to retain their nutritional content. Through the application of chromatographic, spectroscopic, CV, and sensing technologies, a vast array of instrumentation can identify and quantify essential bioactive compounds, such as antioxidants and vitamins. However, it is noteworthy that most of the analyses are currently conducted using off-line solutions. As we move forward, there is a pressing need to bridge the gap between laboratory research and industrial applications by developing and implementing real-time on-line and in-line solutions. Transferring these advanced techniques from the controlled lab environment to the dynamic industrial setting holds the key to enhancing process monitoring, optimizing production efficiency, and ensuring the consistent quality and safety of dried fruit products.

REFERENCES

Aghbashlo, M., Hosseinpour, S., Ghasemi-Varnamkhasti, M., 2014. Computer vision technology for real-time food quality assurance during drying process. Trends Food Sci. Technol. 39, 76–84. https://doi.org/10.1016/J.TIFS.2014.06.003

Akar, G., Barutçu Mazı, I., 2019. Color change, ascorbic acid degradation kinetics, and rehydration behavior of kiwifruit as affected by different drying methods. J. Food Process Eng. 42, e13011. https://doi.org/10.1111/JFPE.13011

Akter, F., Muhury, R., Sultana, A., Deb, U.K., 2022. A comprehensive review of mathematical modeling for drying processes of fruits and vegetables. Int. J. Food Sci. 2022. https://doi.org/10.1155/2022/6195257

Alara, O.R., Abdurahman, N.H., Ukaegbu, C.I., 2021. Extraction of phenolic compounds: A review. Curr. Res. Food Sci. 4, 200–214. https://doi.org/10.1016/J.CRFS.2021.03.011

Allahdad, Z., Nasiri, M., Varidi, M., Varidi, M.J., 2019. Effect of sonication on osmotic dehydration and subsequent air-drying of pomegranate arils. J. Food Eng. 244, 202–211. https://doi.org/10.1016/J.JFOODENG.2018.09.017

Angeli, M.A.C., Caronna, F., Cramer, T., Gastaldi, D., Magagnin, L., Fraboni, B., Vena, P., 2020. Strain mapping inkjet-printed resistive sensors array. IEEE Sens. J. 20, 4087–4095. https://doi.org/10.1109/JSEN.2019.2961747

Arscott, S.A., 2013. Food sources of carotenoids. Carotenoids Hum. Heal. 3–19. https://doi.org/10.1007/978-1-62703-203-2_1

Arslan, M., Xiaobo, Z., Shi, J., Elrasheid Tahir, H., Zareef, M., Rakha, A., Bilal, M., 2020. In situ prediction of phenolic compounds in puff dried Ziziphus jujuba Mill. Using hand-held spectral analytical system. Food Chem. 331, 127361. https://doi.org/10.1016/J.FOODCHEM.2020.127361

Beghi, R., Buratti, S., Giovenzana, V., Benedetti, S., Guidetti, R., 2017. Electronic nose and visible-near infrared spectroscopy in fruit and vegetable monitoring. Rev. Anal. Chem. 36. https://doi.org/10.1515/revac-2016-0016

Botondi, R., Moscetti, R., Massantini, R., 2016. A comparative study on the effectiveness of ozonated water and peracetic acid in the storability of packaged fresh-cut melon. J. Food Sci. Technol. 53, 2352–2360. https://doi.org/10.1007/s13197-016-2207-y

Bozkir, H., Rayman Ergün, A., Serdar, E., Metin, G., Baysal, T., 2019. Influence of ultrasound and osmotic dehydration pretreatments on drying and quality properties of persimmon fruit. Ultrason. Sonochem. 54, 135–141. https://doi.org/10.1016/J.ULTSONCH.2019.02.006

Bücher, T., Huber, R., Eschenbaum, C., Mertens, A., Lemmer, U., Amrouch, H., 2022. Printed temperature sensor array for high-resolution thermal mapping. Sci. Reports. 12, 14231. https://doi.org/10.1038/s41598-022-18321-6

Buratti, S., Rizzolo, A., Benedetti, S., Torreggiani, D., 2006. Electronic nose to detect strawberry aroma changes during osmotic dehydration. J. Food Sci. 71, E184–E189. https://doi.org/10.1111/J.1750-3841.2006.00007.X

Bureau, S., Cozzolino, D., Clark, C.J., 2019. Contributions of Fourier-transform mid infrared (FT-MIR) spectroscopy to the study of fruit and vegetables: A review. Postharvest Biol. Technol. 148, 1–14. https://doi.org/10.1016/J.POSTHARVBIO.2018.10.003

Bureau, S., Ścibisz, I., Le Bourvellec, C., Renard, C.M.G.C., 2012. Effect of sample preparation on the measurement of sugars, organic acids, and polyphenols in apple fruit by mid-infrared spectroscopy. J. Agric. Food Chem. 60, 3551–3563. https://doi.org/10.1021/JF204785W

Bustos, M.C., Rocha-Parra, D., Sampedro, I., De Pascual-Teresa, S., León, A.E., 2018. The influence of different air-drying conditions on bioactive compounds and antioxidant activity of berries. J. Agric. Food Chem. 66, 2714–2723. https://doi.org/10.1021/acs.jafc.7b05395

Cao, X., Islam, M.N., Xu, W., Chen, J., Chitrakar, B., Jia, X., Liu, X., Zhong, S., 2020. Energy consumption, colour, texture, antioxidants, odours, and taste qualities of litchi fruit dried by intermittent ohmic heating. Foods. 9, 425. https://doi.org/10.3390/FOODS9040425

Castro, A.M., Mayorga, E.Y., Moreno, F.L., 2018. Mathematical modelling of convective drying of fruits: A review. J. Food Eng. 223, 152–167. https://doi.org/10.1016/J.JFOODENG.2017.12.012

Celejewska, K., Mieszczakowska-Frąc, M., Konopacka, D., Krupa, T., 2018. The influence of ultrasound and cultivar selection on the biocompounds and physicochemical characteristics of dried blueberry (*Vaccinium corymbosum* L.) Snacks. J. Food Sci. 83, 2305–2316. https://doi.org/10.1111/1750-3841.14292

Chakravartula, S.S.N., Bandiera, A., Nardella, M., Bedini, G., Ibba, P., Massantini, R., Moscetti, R., 2023. Computer vision-based smart monitoring and control system for food drying: A study on carrot slices. Comput. Electron. Agric. 206, 107654. https://doi.org/10.1016/J.COMPAG.2023.107654

Chatzilia, T., Kaderides, K., Goula, A.M., 2023. Drying of peaches by a combination of convective and microwave methods. J. Food Process Eng. 46, e14296. https://doi.org/10.1111/JFPE.14296

Chen, J., Zhang, M., Xu, B., Sun, J., Mujumdar, A.S., 2020. Artificial intelligence assisted technologies for controlling the drying of fruits and vegetables using physical fields: A review. Trends Food Sci. Technol. 105, 251–260. https://doi.org/10.1016/J.TIFS.2020.08.015

Cozzolino, D., Phan, A.D.T., Netzel, M.E., Smyth, H., Sultanbawa, Y., 2021. The use of vibrational spectroscopy to predict vitamin C in Kakadu plum powders (*Terminalia ferdinandiana Exell, Combretaceae*). J. Sci. Food Agric. 101, 3208–3213. https://doi.org/10.1002/JSFA.10950

Dash, K.K., Boro, S., Bhagya Raj, G.V.S., 2023. Effect of ultrasound pretreatment and microwave vacuum drying in the production of dried poniol fruit. J. Food Process Eng. 46, e14301. https://doi.org/10.1111/JFPE.14301

de la Rosa, L.A., Moreno-Escamilla, J.O., Rodrigo-García, J., Alvarez-Parrilla, E., 2019. Phenolic compounds. Postharvest Physiol. Biochem. Fruits Veg. 253–271. https://doi.org/10.1016/B978-0-12-813278-4.00012-9

de Oliveira, G.A., Bureau, S., Renard, C.M.G.C., Pereira-Netto, A.B., De Castilhos, F., 2014a. Comparison of NIRS approach for prediction of internal quality traits in three fruit species. Food Chem. 143, 223–230. https://doi.org/10.1016/J.FOODCHEM.2013.07.122

de Oliveira, G.A., de Castilhos, F., Renard, C.M.G.C., Bureau, S., 2014b. Comparison of NIR and MIR spectroscopic methods for determination of individual sugars, organic acids and carotenoids in passion fruit. Food Res. Int. 60, 154–162. https://doi.org/10.1016/J.FOODRES.2013.10.051

Dehghannya, J., Gorbani, R., Ghanbarzadeh, B., 2017. Influence of combined pretreatments on color parameters during convective drying of Mirabelle plum (*Prunus domestica subsp. syriaca*). Heat Mass Transf. und Stoffuebertragung. 53, 2425–2433. https://doi.org/10.1007/s00231-017-1995-6

Delwiche, S.R., 2015. Basics of spectroscopic analysis. Food Eng. Ser. 57–79. https://doi.org/10.1007/978-1-4939-2836-1_3

Devahastin, S., Niamnuy, C., 2010. Invited review: Modelling quality changes of fruits and vegetables during drying: A review. Int. J. Food Sci. Technol. 45, 1755–1767. https://doi.org/10.1111/J.1365-2621.2010.02352.X

Donno, D., Mellano, M.G., Riondato, I., De Biaggi, M., Andriamaniraka, H., Gamba, G., Beccaro, G.L., 2019. Traditional and unconventional dried fruit snacks as a source of health-promoting compounds. Antioxidants. 8, 396. https://doi.org/10.3390/ANTIOX8090396

Drumea, A., Svasta, P., 2010. Wireless sensor for temperature and humidity measurement. Proc. SPIE. 485–490. https://doi.org/10.1117/12.882390

Ellong, E.N., Billard, C., Adenet, S., Rochefort, K., Ellong, E.N., Billard, C., Adenet, S., Rochefort, K., 2015. Polyphenols, carotenoids, vitamin C content in tropical fruits and vegetables and impact of processing methods. Food Nutr. Sci. 6, 299–313. https://doi.org/10.4236/FNS.2015.63030

Fernando, I.P.S., Dias, M.K.H.M., Madusanka, D.M.D., Han, E.J., Kim, M.J., Jeon, Y.J., Lee, K., Cheong, S.H., Han, Y.S., Park, S.R., Ahn, G., 2020. Human keratinocyte UVB-protective effects of a low molecular weight fucoidan from Sargassum horneri purified by step gradient ethanol precipitation. Antioxidants. 9, 340. https://doi.org/10.3390/ANTIOX9040340

García-Martínez, E., Igual, M., Martín-Esparza, M.E., Martínez-Navarrete, N., 2013. Assessment of the bioactive compounds, color, and mechanical properties of apricots as affected by drying treatment. Food Bioprocess Technol. 6, 3247–3255. https://doi.org/10.1007/S11947-012-0988-1

Gardana, C., Scialpi, A., Fachechi, C., Simonetti, P., 2018. Near-infrared spectroscopy and chemometrics for the routine detection of bilberry extract adulteration and quantitative determination of the anthocyanins. J. Spectrosc. 2018. https://doi.org/10.1155/2018/4751247

Graunke, T., Schmitt, K., Wöllenstein, J., 2016. Organic membranes for selectivity enhancement of metal oxide gas sensors. J. Sensors. 2016. https://doi.org/10.1155/2016/2435945

Gutiérrez-Cárdenas, M.G., Kaur, T., Corona-Chávez, A., Olvera-Cervantes, J.L., Gómez-Salazar, J.A., Rojas-Laguna, R., Sosa-Morales, M.E., 2022. Dielectric properties analysis of dough and further microwave baking for cookies. J. Food Process. Preserv. 46, e16940. https://doi.org/10.1111/JFPP.16940

Hasan, M.U., Malik, A.U., Ali, S., Imtiaz, A., Munir, A., Amjad, W., Anwar, R., 2019. Modern drying techniques in fruits and vegetables to overcome postharvest losses: A review. J. Food Process. Preserv. 43, e14280. https://doi.org/10.1111/JFPP.14280

Hassanzadeh, A., Borghei, A., Lindquist, R.G., 2008. Relative humidity measurement using capacitive sensors. Conf. Proc. – IEEE SOUTHEASTCON 396–398. https://doi.org/10.1109/SECON.2008.4494326

Heichal, Y., Chandra, S., Bordatchev, E., 2005. A fast-response thin film thermocouple to measure rapid surface temperature changes. Exp. Therm. Fluid Sci. 30, 153–159. https://doi.org/10.1016/J.EXPTHERMFLUSCI.2005.05.004

Hussain, A., Pu, H., Sun, D.W., 2018. Innovative nondestructive imaging techniques for ripening and maturity of fruits – A review of recent applications. Trends Food Sci. Technol. 72, 144–152. https://doi.org/10.1016/J.TIFS.2017.12.010

Izli, N., Izli, G., Taskin, O., 2018. Impact of different drying methods on the drying kinetics, color, total phenolic content and antioxidant capacity of pineapple. http://mc.manuscriptcentral.com/tcyt CyTA – Journal of Food. 16, 213–221. https://doi.org/10.1080/19476337.2017.1381174

Janjai, S., Lamlert, N., Intawee, P., Mahayothee, B., Haewsungcharern, M., Bala, B.K., Müller, J., 2008. Finite element simulation of drying of mango. Biosyst. Eng. 99, 523–531. https://doi.org/10.1016/J.BIOSYSTEMSENG.2007.12.010

Janjai, S., Mahayothee, B., Lamlert, N., Bala, B.K., Precoppe, M., Nagle, M., Müller, J., 2010. Diffusivity, shrinkage and simulated drying of Litchi fruit (*Litchi Chinensis* Sonn.). J. Food Eng. 96, 214–221. https://doi.org/10.1016/J.JFOODENG.2009.07.015

Jiménez-Aguilar, D.M., Ortega-Regules, A.E., Lozada-Ramírez, J.D., Pérez-Pérez, M.C.I., Vernon-Carter, E.J., Welti-Chanes, J., 2011. Color and chemical stability of spray-dried blueberry extract using mesquite gum as wall material. J. Food Compos. Anal. 24, 889–894. https://doi.org/10.1016/J.JFCA.2011.04.012

Johnson, J.B., Walsh, K.B., Naiker, M., Ameer, K., 2023. The use of infrared spectroscopy for the quantification of bioactive compounds in food: A review. Mol. 28, 3215. https://doi.org/10.3390/MOLECULES28073215

Ju, H.Y., Zhao, S.H., Mujumdar, A.S., Fang, X.M., Gao, Z.J., Zheng, Z.A., Xiao, H.W., 2018. Energy efficient improvements in hot air drying by controlling relative humidity based on weibull and bi-di models. Food Bioprod. Process. 111, 20–29. https://doi.org/10.1016/J.FBP.2018.06.002

Karim, M.A., Hawlader, M.N.A., 2005. Mathematical modelling and experimental investigation of tropical fruits drying. Int. J. Heat Mass Transf. 48, 4914–4925. https://doi.org/10.1016/J.IJHEATMASSTRANSFER.2005.04.035

Kayacan, S., Sagdic, O., Doymaz, I., Karasu, S., 2022. The effect of different drying methods on total bioactive properties, individual phenolic compounds, rehydration ability, color, and microstructural characteristics of Asian pear. J. Food Process. Preserv. 46, e16682. https://doi.org/10.1111/JFPP.16682

Kessy, H.N.E., Hu, Z., Zhao, L., Zhou, M., 2016. Effect of steam blanching and drying on phenolic compounds of litchi pericarp. Mol. 21, 729. https://doi.org/10.3390/MOLECULES21060729

Khan, M.I.H., Sablani, S.S., Joardder, M.U.H., Karim, M.A., 2020. Application of machine learning-based approach in food drying: Opportunities and challenges. Drying Technology. 40, 1051–1067. https://doi.org/10.1080/07373937.2020.1853152

Kim, K., Kim, J., Jiang, X., Kim, T., 2021. Static force measurement using Piezoelectric sensors. J. Sensors. 2021. https://doi.org/10.1155/2021/6664200

Knoerzer, K., Buckow, R., Trujillo, F.J., Juliano, P., 2015. Multiphysics simulation of innovative food processing technologies. Food Eng. Rev. 7, 64–81. https://doi.org/10.1007/s12393-014-9098-3

Kourti, T., 2007. Process analytical technology beyond real-time analyzers: The role of multivariate analysis. Crit. Rev. Anal. Chem. 36, 257–278. https://doi.org/10.1080/10408340600969957

Kris-Etherton, P.M., Hecker, K.D., Bonanome, A., Coval, S.M., Binkoski, A.E., Hilpert, K.F., Griel, A.E., Etherton, T.D., 2002. Bioactive compounds in foods: Their role in the prevention of cardiovascular disease and cancer. Am. J. Med. 113, 71–88. https://doi.org/10.1016/S0002-9343(01)00995-0

Kschonsek, J., Wolfram, T., Stöckl, A., Böhm, V., 2018. Polyphenolic compounds analysis of old and new apple cultivars and contribution of polyphenolic profile to the in vitro antioxidant capacity. Antioxidants. 7, 20. https://doi.org/10.3390/ANTIOX7010020

Kumar, D., Ladaniya, M.S., Gurjar, M., Kumar, S., 2022. Impact of drying methods on natural antioxidants, phenols and flavanones of immature dropped Citrus sinensis L. Osbeck fruits. Sci. Reports. 12, 1–12. https://doi.org/10.1038/s41598-022-10661-7

Lammerskitten, A., Wiktor, A., Siemer, C., Toepfl, S., Mykhailyk, V., Gondek, E., Rybak, K., Witrowa-Rajchert, D., Parniakov, O., 2019. The effects of pulsed electric fields on the quality parameters of freeze-dried apples. J. Food Eng. 252, 36–43. https://doi.org/10.1016/J.JFOODENG.2019.02.006

Larrauri, J.A., Rupérez, P., Saura-Calixto, F., 1997. Effect of drying temperature on the stability of polyphenols and antioxidant activity of red grape pomace peels. J. Agric. Food Chem. 45, 1390–1393. https://doi.org/10.1021/JF960282F

Lee, H.S., Coates, G.A., 2003. Effect of thermal pasteurization on Valencia orange juice color and pigments. LWT – Food Sci. Technol. 36, 153–156. https://doi.org/10.1016/S0023-6438(02)00087-7

Lemus-Mondaca, R.A., Zambra, C.E., Vega-Gálvez, A., Moraga, N.O., 2013. Coupled 3D heat and mass transfer model for numerical analysis of drying process in papaya slices. J. Food Eng. 116, 109–117. https://doi.org/10.1016/J.JFOODENG.2012.10.050

Liu, K., Zhang, C., 2021. Volatile organic compounds gas sensor based on quartz crystal microbalance for fruit freshness detection: A review. Food Chem. 334, 127615. https://doi.org/10.1016/J.FOODCHEM.2020.127615

Liu, Z.L., Staniszewska, I., Zielinska, D., Zhou, Y.H., Nowak, K.W., Xiao, H.W., Pan, Z., Zielinska, M., 2020. Combined hot air and microwave-vacuum drying of cranberries: Effects of pretreatments and pulsed vacuum osmotic dehydration on drying kinetics and physicochemical properties. Food Bioprocess Technol. 13, 1848–1856. https://doi.org/10.1007/s11947-020-02507-9

Llavata, B., García-Pérez, J.V., Simal, S., Cárcel, J.A., 2020. Innovative pre-treatments to enhance food drying: A current review. Curr. Opin. Food Sci. 35, 20–26. https://doi.org/10.1016/J.COFS.2019.12.001

López-fernández, O., Domínguez, R., Pateiro, M., Munekata, P.E.S., Rocchetti, G., Lorenzo, J.M., 2020. Determination of polyphenols using liquid chromatography–tandem mass spectrometry technique (LC–MS/MS): A review. Antioxidants. 9, 479. https://doi.org/10.3390/ANTIOX9060479

Lu, W., Maidannyk, V.A., Lim, A.S.L., 2020. Carotenoids degradation and precautions during processing. Carotenoids Prop. Process. Appl. 223–258. https://doi.org/10.1016/B978-0-12-817067-0.00007-5

Madadi Avargani, V., Abdlla Maarof, H., Zendehboudi, S., 2023. Multiphysics CFD modeling to assess performance of a perforated multi-plate indirect solar dryer with a V-corrugated absorber surface. Appl. Therm. Eng. 227, 120387. https://doi.org/10.1016/J.APPLTHERMALENG.2023.120387

Magalhães, M.L., Cartaxo, S.J.M., Gallão, M.I., García-Pérez, J.V., Cárcel, J.A., Rodrigues, S., Fernandes, F.A.N., 2017. Drying intensification combining ultrasound pre-treatment and ultrasound-assisted air drying. J. Food Eng. 215, 72–77. https://doi.org/10.1016/J.JFOODENG.2017.07.027

Mahesh, S., Jayas, D.S., Paliwal, J., White, N.D.G., 2015. Hyperspectral imaging to classify and monitor quality of agricultural materials. J. Stored Prod. Res. 61, 17–26. https://doi.org/10.1016/J.JSPR.2015.01.006

Manzoor, M.F., Hussain, A., Naumovski, N., Ranjha, M.M.A.N., Ahmad, N., Karrar, E., Xu, B., Ibrahim, S.A., 2022. A narrative review of recent advances in rapid assessment of anthocyanins in agricultural and food products. Front. Nutr. 9, 901342. https://doi.org/10.3389/fnut.2022.901342

Manzoor, M.F., Hussain, A., Sameen, A., Sahar, A., Khan, S., Siddique, R., Aadil, R.M., Xu, B., 2021. Novel extraction, rapid assessment and bioavailability improvement of quercetin: A review. Ultrason. Sonochem. 78, 105686. https://doi.org/10.1016/J.ULTSONCH.2021.105686

Martins, R.C., 2006. Simple finite volumes and finite elements procedures for food quality and safety simulations. J. Food Eng. 73, 327–338. https://doi.org/10.1016/J.JFOODENG.2005.01.033

McIntyre, S.M., Ma, Q., Burritt, D.J., Oey, I., Gordon, K.C., Fraser-Miller, S.J., 2020. Vibrational spectroscopy and chemometrics for quantifying key bioactive components of various plum cultivars grown in New Zealand. J. Raman Spectrosc. 51, 1138–1152. https://doi.org/10.1002/JRS.5867

McSweeney, M., Seetharaman, K., 2015. State of polyphenols in the drying process of fruits and vegetables. Crit. Rev. Food Sci. Nutr. 55, 660–669. https://doi.org/10.1080/10408398.2012.670673

Meléndez-Martínez, A.J., Mandić, A.I., Bantis, F., Böhm, V., Borge, G.I.A., Brnčić, M., Bysted, A., Cano, M.P., Dias, M.G., Elgersma, A., Fikselová, M., García-Alonso, J., Giuffrida, D., Gonçalves, V.S.S., Hornero-Méndez, D., Kljak, K., Lavelli, V., Manganaris, G.A., Mapelli-Brahm, P., Marounek, M., Olmedilla-Alonso, B., Periago-Castón, M.J., Pintea, A., Sheehan, J.J., Tumbas Šaponjac, V., Valšíková-Frey, M., Meulebroek, L., Van, O'Brien, N., 2022. A comprehensive review on carotenoids in foods and feeds: Status quo, applications, patents, and research needs. Crit. Rev. Food Sci. Nutr. 62, 1999–2049. https://doi.org/10.1080/10408398.2020.1867959

Melini, V., Melini, F., 2021a. Modelling and optimization of ultrasound-assisted extraction of phenolic compounds from black quinoa by response surface methodology. Mol. 26, 3616. https://doi.org/10.3390/MOLECULES26123616

Melini, V., Melini, F., 2021b. Compositional and nutritional analysis. Innov. Food Anal. 1–39. https://doi.org/10.1016/B978-0-12-819493-5.00001-7

Melini, V., Melini, F., Acquistucci, R., 2020a. Phenolic compounds and bioaccessibility thereof in functional pasta. Antioxidants. 9, 343. https://doi.org/10.3390/ANTIOX9040343

Melini, V., Melini, F., Comendador, F.J., 2023. Response surface methodology as an experimental strategy for ultrasound-assisted extraction of phenolic compounds from artichoke heads. Antioxidants. 12, 1360. https://doi.org/10.3390/ANTIOX12071360

Melini, V., Melini, F., Luziatelli, F., Ruzzi, M., 2020b. Functional ingredients from Agri-food waste: Effect of inclusion thereof on phenolic compound content and bioaccessibility in bakery products. Antioxidants. 9, 1216. https://doi.org/10.3390/ANTIOX9121216

Mendonça, S.L.R., Filho, C.R.B., Da Silva, Z.E., 2005. Transient conduction in spherical fruits: Method to estimate the thermal conductivity and volumetric thermal capacity. J. Food Eng. 67, 261–266. https://doi.org/10.1016/J.JFOODENG.2004.04.026

Menesatti, P., Zanella, A., D'Andrea, S., Costa, C., Paglia, G., Pallottino, F., 2009.Food Bioprocess Technol. 2, 308–314. https://doi.org/10.1007/s11947-008-0120-8

Miranda, J., Ponce, P., Molina, A., Wright, P., 2019. Sensing, smart and sustainable technologies for Agri-food 4.0. Comput. Ind. 108, 21–36. https://doi.org/10.1016/J.COMPIND.2019.02.002

Mishra, R.B., El-Atab, N., Hussain, A.M., Hussain, M.M., 2021. Recent progress on flexible capacitive pressure sensors: From design and materials to applications. Adv. Mater. Technol. 6, 2001023. https://doi.org/10.1002/ADMT.202001023

Mollazade, K., Omid, M., Tab, F.A., Mohtasebi, S.S., 2012. Principles and applications of light backscattering imaging in quality evaluation of agro-food products: A review. Food Bioprocess Technol. 5, 1465–1485. https://doi.org/10.1007/s11947-012-0821-x

Monteiro, R.L., Gomide, A.I., Link, J.V., Carciofi, B.A.M., Laurindo, J.B., 2020. Microwave vacuum drying of foods with temperature control by power modulation. Innov. Food Sci. Emerg. Technol. 65, 102473. https://doi.org/10.1016/J.IFSET.2020.102473

Mosca, S., Conti, C., Stone, N., Matousek, P., 2021. Spatially offset Raman spectroscopy. Nat. Rev. Methods Prim. 1, 1–16. https://doi.org/10.1038/s43586-021-00019-0

Moscetti, R., Haff, R.P., Ferri, S., Raponi, F., Monarca, D., Liang, P., Massantini, R., 2017. Real-time monitoring of organic carrot (var. Romance) during hot-air drying using near-infrared spectroscopy. Food Bioprocess Technol. 10, 2046–2059. https://doi.org/10.1007/s11947-017-1975-3

Moscetti, R., Raponi, F., Ferri, S., Colantoni, A., Monarca, D., Massantini, R., 2018a. Real-time monitoring of organic apple (var. Gala) during hot-air drying using near-infrared spectroscopy. J. Food Eng. 222, 139–150. https://doi.org/10.1016/J.JFOODENG.2017.11.023

Moscetti, R., Saeys, W., Keresztes, J.C., Goodarzi, M., Cecchini, M., Danilo, M., Massantini, R., 2015. Hazelnut quality sorting using high dynamic range short-wave infrared hyperspectral imaging. Food Bioprocess Technol. 8, 1593–1604. https://doi.org/10.1007/s11947-015-1503-2

Moscetti, R., Sturm, B., Crichton, S.O.J., Amjad, W., Massantini, R., 2018b. Postharvest monitoring of organic potato (cv. Anuschka) during hot-air drying using visible–NIR hyperspectral imaging. J. Sci. Food Agric. 98, 2507–2517. https://doi.org/10.1002/JSFA.8737

Muller, I., De Brito, R., Pereira, C.E., Brusamarello, V., 2010. Load cells in force sensing analysis – Theory and a novel application. IEEE Instrum. Meas. Mag. 13, 15–19. https://doi.org/10.1109/MIM.2010.5399212

Muñoz-Aguirre, S., Yoshino, A., Nakamoto, T., Moriizumi, T., 2007. Odor approximation of fruit flavors using a QCM odor sensing system. Sensors Actuators B Chem. 123, 1101–1106. https://doi.org/10.1016/J.SNB.2006.11.025

Mutuli, G.P., Gitau, A.N., Mbuge, D.O., 2020. Convective drying modeling approaches: A review for herbs, vegetables, and fruits. J. Biosyst. Eng. 45, 197–212. https://doi.org/10.1007/s42853-020-00056-9

Nakagawa, K., Horie, A., Nakabayashi, M., Nishimura, K., Yasunobu, T., 2021. Influence of processing conditions of atmospheric freeze-drying/low-temperature drying on the drying kinetics of sliced fruits and their vitamin c retention. J. Agric. Food Res. 6, 100231. https://doi.org/10.1016/J.JAFR.2021.100231

Noshad, M., Savari, M., Roueita, G., 2018. A hybrid AHP-TOPSIS method for prospectively modeling of ultrasound-assisted osmotic dehydration of strawberry. J. Food Process Eng. 41, e12928. https://doi.org/10.1111/JFPE.12928

O'Sullivan, J., Ferrua, M., Love, R., Verboven, P., Nicolaï, B., East, A., 2014. Airflow measurement techniques for the improvement of forced-air cooling, refrigeration and drying operations. J. Food Eng. 143, 90–101. https://doi.org/10.1016/J.JFOODENG.2014.06.041

Onwude, D.I., Hashim, N., Janius, R.B., Nawi, N.M., Abdan, K., 2016. Modeling the thin-layer drying of fruits and vegetables: A review. Compr. Rev. Food Sci. Food Saf. 15, 599–618. https://doi.org/10.1111/1541-4337.12196

Oszmiański, J., Lachowicz, S., 2016. Effect of the production of dried fruits and juice from chokeberry (*Aronia melanocarpa* L.) On the content and antioxidative activity of bioactive compounds. Mol. 21, 1098. https://doi.org/10.3390/molecules21081098

Pathare, P.B., Opara, U.L., Al-Said, F.A.J., 2012. Colour measurement and analysis in fresh and processed foods: A review. Food Bioprocess Technol. 6, 36–60. https://doi.org/10.1007/S11947-012-0867-9

Prasedya, E., Frediansyah, S., Martyasari, A., Ilhami, N.W.R., Abidin, B.K., Padmi, S., Fahrurrozi, H., Juanssilfero, A.B., Widyastuti, S., Sunarwidhi, A.L., 2021. Effect of particle size on phytochemical composition and antioxidant properties of Sargassum cristaefolium ethanol extract. Sci. Reports. 11, 1–9. https://doi.org/10.1038/s41598-021-95769-y

Prawiranto, K., Carmeliet, J., Defraeye, T., 2021. Physics-based digital twin identifies trade-offs between drying time, fruit quality, and energy use for solar drying. Front. Sustain. Food Syst. 4, 606845. https://doi.org/10.3389/fsufs.2020.606845

Prawiranto, K., Defraeye, T., Derome, D., Verboven, P., Nicolai, B., Carmeliet, J., 2018. New insights into the apple fruit dehydration process at the cellular scale by 3D continuum modeling. J. Food Eng. 239, 52–63. https://doi.org/10.1016/J.JFOODENG.2018.06.023

Prosapio, V., Norton, I., 2018. Simultaneous application of ultrasounds and firming agents to improve the quality properties of osmotic + freeze-dried foods. LWT. 96, 402–410. https://doi.org/10.1016/J.LWT.2018.05.068

Pyrzynska, K., Sentkowska, A., 2019. Chromatographic analysis of polyphenols. Polyphenols in Plants. 353–364. https://doi.org/10.1016/B978-0-12-813768-0.00021-9

Radicetti, E., Massantini, R., Campiglia, E., Mancinelli, R., Ferri, S., Moscetti, R., 2016. Yield and quality of eggplant (*Solanum melongena* L.) as affected by cover crop species and residue management. Sci. Hortic. (Amsterdam). 204, 161–171. https://doi.org/10.1016/J.SCIENTA.2016.04.005

Radojčin, M., Pavkov, I., Kovačević, D.B., Putnik, P., Wiktor, A., Stamenković, Z., Kešelj, K., Gere, A., 2021. Effect of selected drying methods and emerging drying intensification technologies on the quality of dried fruit: A review. Process. 9, 132. https://doi.org/10.3390/PR9010132

Raghavan, G.S.V., Rennie, T.J., Sunjka, P.S., Orsat, V., Phaphuangwittayakul, W., Terdtoon, P., 2005. Overview of new techniques for drying biological materials with emphasis on energy aspects. Brazilian J. Chem. Eng. 22, 195–201. https://doi.org/10.1590/S0104-66322005000200005

Raponi, F., Moscetti, R., Monarca, D., Colantoni, A., Massantini, R., 2017. Monitoring and optimization of the process of drying fruits and vegetables using computer vision: A review. Sustain. 9, 2009. https://doi.org/10.3390/SU9112009

Raponi, F., Moscetti, R., Nallan Chakravartula, S.S., Fidaleo, M., Massantini, R., 2022. Monitoring the hot-air drying process of organically grown apples (cv. Gala) using computer vision. Biosyst. Eng. 223, 1–13. https://doi.org/10.1016/J.BIOSYSTEMSENG.2021.07.005

Reich, G., 2005. Near-infrared spectroscopy and imaging: Basic principles and pharmaceutical applications. Adv. Drug Deliv. Rev. 57, 1109–1143. https://doi.org/10.1016/J.ADDR.2005.01.020

Ruiz-Altisent, M., Ruiz-Garcia, L., Moreda, G.P., Lu, R., Hernandez-Sanchez, N., Correa, E.C., Diezma, B., Nicolaï, B., García-Ramos, J., 2010. Sensors for product characterization and quality of specialty crops – A review. Comput. Electron. Agric. 74, 176–194. https://doi.org/10.1016/J.COMPAG.2010.07.002

Sadler, M.J., Gibson, S., Whelan, K., Ha, M.A., Lovegrove, J., Higgs, J., 2019. Dried fruit and public health – What does the evidence tell us?. Int. J. Fruit Sci. 70, 675–687. https://doi.org/10.1080/09637486.2019.1568398

Samoticha, J., Wojdyło, A., Lech, K., 2016. The influence of different the drying methods on chemical composition and antioxidant activity in chokeberries. LWT – Food Sci. Technol. 66, 484–489. https://doi.org/10.1016/J.LWT.2015.10.073

Sanmartín, P., Villa, F., Silva, B., Cappitelli, F., Prieto, B., 2011. Color measurements as a reliable method for estimating chlorophyll degradation to phaeopigments. Biodegradation. 22, 763–771. https://doi.org/10.1007/s10532-010-9402-8

Sant'Anna, V., Gurak, P.D., Ferreira Marczak, L.D., Tessaro, I.C., 2013. Tracking bioactive compounds with colour changes in foods – A review. Dye. Pigment. 98, 601–608. https://doi.org/10.1016/J.DYEPIG.2013.04.011

Santos, E.J.P., Vasconcelos, I.B., 2008. RTD-based smart temperature sensor: process development and circuit design. "2008 26th Int. Conf. Microelectron. Proceedings, MIEL 2008" 333–336. https://doi.org/10.1109/ICMEL.2008.4559289

Sapra, S., Chakraborthy, A., Nuthalapati, S., Nag, A., Inglis, D.W., Mukhopadhyay, S.C., Ercan Altinsoy, M., 2023. Printed, wearable e-skin force sensor array. Measurement. 206, 112348. https://doi.org/10.1016/J.MEASUREMENT.2022.112348

Serni, E., Venir, E., Romano, G., Guerra, W., Robatscher, P., 2020. Determination of major phenolics content in dried apples from three new cultivars (*Malus domestica Borkh.*) Using HPLC-UV-FL with pentafluorophenyl stationary phase. Food Anal. Methods. 13, 863–871. https://doi.org/10.1007/s12161-020-01703-9

Seyedabadi, E., Khojastehpour, M., Abbaspour-Fard, M.H., 2017. Convective drying simulation of banana slabs considering non-isotropic shrinkage using FEM with the arbitrary Lagrangian–Eulerian method. Int. J. Food Prop. 20, S36–S49. https://doi.org/10.1080/10942912.2017.1288134

Sharif, I., Adewale, P., Dalli, S.S., Rakshit, S., 2018. Microwave pretreatment and optimization of osmotic dehydration of wild blueberries using response surface methodology. Food Chem. 269, 300–310. https://doi.org/10.1016/J.FOODCHEM.2018.06.087

Sikorska, E., Khmelinskii, I., Sikorski, M., 2019. Fluorescence spectroscopy and imaging instruments for food quality evaluation. Eval. Technol. Food Qual. 491–533. https://doi.org/10.1016/B978-0-12-814217-2.00019-6

Srednicka-Tober, D., Kazimierczak, R., Ponder, A., Hallmann, E., 2020. Biologically active compounds in selected organic and conventionally produced dried fruits. Foods. 9, 1005. https://doi.org/10.3390/FOODS9081005

Süfer, Ö, Palazoğlu, T.K., 2019. A study on hot-air drying of pomegranate: Kinetics of dehydration, rehydration and effects on bioactive compounds. J. Therm. Anal. Calorim. 137, 1981–1990. https://doi.org/10.1007/s10973-019-08102-1

Sultanbawa, Y., Smyth, H.E., Truong, K., Chapman, J., Cozzolino, D., 2021. Insights on the role of chemometrics and vibrational spectroscopy in fruit metabolite analysis. Food Chem. Mol. Sci. 3, 100033. https://doi.org/10.1016/J.FOCHMS.2021.100033

Sun, D.W., 2010. Hyperspectral imaging for food quality analysis and control. Hyperspectral Imaging Food Qual. Anal. Control. 1–477. https://doi.org/10.1016/C2009-0-01853-4

Tahir, H.E., Xiaobo, Z., Jianbo, X., Mahunu, G.K., Jiyong, S., Xu, J.L., Sun, D.W., 2019. Recent progress in rapid analyses of vitamins, phenolic, and volatile compounds in foods using vibrational spectroscopy combined with chemometrics: A review. Food Anal. Methods. 12, 2361–2382. https://doi.org/10.1007/S12161-019-01573-W

Tan, W.C., Ang, K.W., 2021. Volatile organic compound sensors based on 2D materials. Adv. Electron. Mater. 7, 2001071. https://doi.org/10.1002/AELM.202001071

Thakur, M., Singh, K., Khedkar, R., 2020. Phytochemicals: Extraction process, safety assessment, toxicological evaluations, and regulatory issues. Funct. Preserv. Prop. Phytochem. 341–361. https://doi.org/10.1016/B978-0-12-818593-3.00011-7

Thamkaew, G., Sjöholm, I., Galindo, F.G., 2020. A review of drying methods for improving the quality of dried herbs. Crit. Rev. Food Sci. Nutr. 61, 1763–1786. https://doi.org/10.1080/10408398.2020.1765309

Tylewicz, U., Mannozzi, C., Castagnini, J.M., Genovese, J., Romani, S., Rocculi, P., Rosa, M.D., 2022. Application of PEF- and OD-assisted drying for kiwifruit waste valorisation. Innov. Food Sci. Emerg. Technol. 77, 102952. https://doi.org/10.1016/J.IFSET.2022.102952

Vega-Gálvez A., Ah-Hen K., Chacana M., Vergara J., Martínez-Monzó J., García-Segovia P., Lemus-Mondaca R., Di Scala K., 2012. Effect of temperature and air velocity on drying kinetics, antioxidant capacity, total phenolic content, colour, texture and microstructure of apple (var. Granny Smith) slices. Food Chem. 132(1), 51–59. doi: 10.1016/j.foodchem.2011.10.029

Wang, J., Xiao, H.W., Ye, J.H., Wang, J., Raghavan, V., 2019. Ultrasound pretreatment to enhance drying kinetics of kiwifruit (*Actinidia deliciosa*) slices: Pros and cons. Food Bioprocess Technol. 12, 865–876. https://doi.org/10.1007/s11947-019-02256-4

Wang, L., Xu, B., Wei, B., Zeng, R., 2018. Low frequency ultrasound pretreatment of carrot slices: Effect on the moisture migration and quality attributes by intermediate-wave infrared radiation drying. Ultrason. Sonochem. 40, 619–628. https://doi.org/10.1016/J.ULTSONCH.2017.08.005

Wang, X., Zhang, M., Zhang, L., Xu, J., Xiao, X., Zhang, X., 2022. Inkjet-printed flexible sensors: From function materials, manufacture process, and applications perspective. Mater. Today Commun. 31, 103263. https://doi.org/10.1016/J.MTCOMM.2022.103263

Wegener, S., Kaufmann, M., Kroh, L.W., 2017. Influence of l-pyroglutamic acid on the color formation process of non-enzymatic browning reactions. Food Chem. 232, 450–454. https://doi.org/10.1016/J.FOODCHEM.2017.04.046

Welsh, Z.G., Khan, M.I.H., Karim, M.A., 2021. Multiscale modeling for food drying: A homogenized diffusion approach. J. Food Eng. 292, 110252. https://doi.org/10.1016/J.JFOODENG.2020.110252

Williamson, G., Kay, C.D., Crozier, A., 2018. The bioavailability, transport, and bioactivity of dietary flavonoids: A review from a historical perspective. Compr. Rev. Food Sci. Food Saf. 17, 1054–1112. https://doi.org/10.1111/1541-4337.12351

Wojdyło, A., Lech, K., Nowicka, P., Hernandez, F., Figiel, A., Carbonell-Barrachina, A.A., 2019. Influence of different drying techniques on phenolic compounds, antioxidant capacity and colour of Ziziphus jujube mill. Fruits. Mol. 24, 2361. https://doi.org/10.3390/MOLECULES24132361

Yasir, M., Sultana, B., Nigam, P.S., Owusu-Apenten, R., 2016. Antioxidant and genoprotective activity of selected cucurbitaceae seed extracts and LC–ESIMS/MS identification of phenolic components. Food Chem. 199, 307–313. https://doi.org/10.1016/J.FOODCHEM.2015.11.138

Zepka, L.Q., Borsarelli, C.D., Da Silva, M.A.A.P., Mercadante, A.Z., 2009. Thermal degradation kinetics of carotenoids in a cashew apple juice model and its impact on the system color. J. Agric. Food Chem. 57, 7841–7845. https://doi.org/10.1021/jf900558a

Zhang, B., Huang, W., Li, J., Zhao, C., Fan, S., Wu, J., Liu, C., 2014. Principles, developments and applications of computer vision for external quality inspection of fruits and vegetables: A review. Food Res. Int. 62, 326–343. https://doi.org/10.1016/J.FOODRES.2014.03.012

Zhang, W.P., Yang, X.H., Mujumdar, A.S., Ju, H.Y., Xiao, H.W., 2021. The influence mechanism and control strategy of relative humidity on hot air drying of fruits and vegetables: A review. Drying Technology. 40, 2217–2234. https://doi.org/10.1080/07373937.2021.1943669

Zhang, X., Yang, J., Lin, T., Ying, Y., 2021. Food and agro-product quality evaluation based on spectroscopy and deep learning: A review. Trends Food Sci. Technol. 112, 431–441. https://doi.org/10.1016/J.TIFS.2021.04.008

Zhang, Z., 2017. Raman spectroscopic sensing in food safety and quality analysis. Food Chem. Funct. Anal. 1–16. https://doi.org/10.1039/9781788010528-00001

Zhao, S., An, N., Zhang, K., Li, D., Wang, L., Wang, Y., 2023. Evaluation of drying kinetics, physical properties, bioactive compounds, antioxidant activity and microstructure of Acanthopanax sessiliflorus fruits dried by microwave-assisted hot air drying method. J. Food Eng. 357, 111642. https://doi.org/10.1016/J.JFOODENG.2023.111642

Zhu, R., Shen, J., Law, C.L., Ma, X., Li, D., Han, Y., Kiani, H., Manickam, S., Tao, Y., 2023. Combined calcium pretreatment and ultrasonic/microwave drying to dehydrate black chokeberry: Novel mass transfer modeling and metabolic pathways of polyphenols. Innov. Food Sci. Emerg. Technol. 83, 103215. https://doi.org/10.1016/J.IFSET.2022.103215

Zielinska, M., Zielinska, D., 2019. Effects of freezing, convective and microwave-vacuum drying on the content of bioactive compounds and color of cranberries. LWT. 104, 202–209. https://doi.org/10.1016/J.LWT.2019.01.041

Zohrabi, S., Seiiedlou, S.S., Aghbashlo, M., Scaar, H., Mellmann, J., 2019. Enhancing the exergetic performance of a pilot-scale convective dryer by exhaust air recirculation. Drying Technology. 38, 518–533. https://doi.org/10.1080/07373937.2019.1587617

7 Storage and Quality Degradation of Dried Fruit Products

S. Savitha
Institute of Chemical Technology Mumbai
Maharashtra, India

7.1 INTRODUCTION

Drying fruits would certainly increase the shelf-life (SL) as the moisture content is reduced. Hence, the water activity (a_w) is reduced, which makes the microbial survival challenging. However, dried fruit has concentrated carbohydrates, vitamins, carotenoids, minerals, polyphenols, and flavonoids, which become the major reasons for early spoilage. During dehydration, once the sugar concentration reaches saturation, it turns into a glassy state. Oxidation of polyphenols and other bioactives may lead to off-flavours, off-odours, and colour changes.

7.1.1 Characteristics of Dried Fruit Products

The characteristics of dried fruits depend on the dehydration method. Water sorption, porosity, colour, and bioactives are affected by drying methods such as conventional dehydration, vacuum drying, microwave heating, and freeze-drying. Conventional drying involves high temperatures and long dehydration duration, leading to low porosity and high density. For instance, convective air heating often closes the pores of the food matrix, leading to case hardening and uneven dehydration. Case hardening is a phenomenon wherein the outer layer is dehydrated quickly and causes the closure of pores, making it tough for the moisture from the internal layer to be released into the surroundings. This keeps the core of the product moist. Apart from that, rehydration is inefficient as the capillaries are closed for water absorption. Therefore, rehydratability becomes one of the major characteristics of dried fruit.

While microwave drying provides uniform and faster drying. The high temperatures help in obtaining microbially safe products; however, a loss in bioactives and flavours have been witnessed. Vacuum drying gives better quality products by retaining the colour, aroma, and bioactives better than conventional drying. However, microbial inactivation may not be achieved. Freeze drying is an advanced technology that provides highly porous products with less shrinkage and retained colour, flavour, and bioactives (Tsami et al., 1998).

DOI: 10.1201/9781003402916-7

7.1.2 Factors to Care of during Storage

If storage of fresh fruits is considered, the nature of climacteric and non-climacteric fruits needs to be taken into account. Climacteric fruits are the ones that continue with ripening even after being detached from the parent plant. On the other hand, non-climacteric fruits need to be attached to the plant for ripening. During storage, respiration and transpiration of fruits continue, and after a certain time, the endogenous enzymes start taking over. The endogenous enzymes move into an autolytic mode and begin degrading the fruit skin. This modifies the texture by softening the skin, making it easier for microorganisms such as mould and fungi to feed on. This leads to spoilage of fruit after a certain time of storage. Enzymes such as polyphenol oxidases (PPO) and pectic methyl esterase cause enzymatic browning and texture degradation in fruits. Hence, controlling the enzyme activity and microbial growth is important to increase the SL. Water is a crucial molecule essential for biochemical pathways and reactions. On dehydration, moisture is removed, hampering these biochemical reactions. Lack of moisture halts the enzymatic activity involved and hence delays the spoilage-causing reactions. This is how dehydration helps increase the SL.

Additionally, it supports inactivation of spoilage microorganisms as a_w is reduced (Mansuri, n.d.). Water activity is the available moisture for the microorganisms to survive. Water activity, denoted by a_w, is the ratio of vapour pressure of water in the food matrix (P) to the vapour pressure of pure water (P_0) at the same temperature (Erkmen & Bozoglu, 2016).

For a dehydrated product, it is crucial to prevent moisture gain. Factors that might result in moisture gain are the relative humidity (RH) and, packaging material. Another major factor is the air composition, i.e., the oxygen concentration during storage. Other factors are temperature and the storage period. Oxygen can degrade the quality of the dried fruit by oxidizing lipid, degrading vitamin C, degrading proteins, and discolouration. It can also form harmful peroxides and aldehydes. Additionally, it also supports microbial growth, leading to spoilage, and causes non-enzymatic browning.

Hence, before storage, the packaging conditions such as the packaging material, headspace, and air composition are to be taken care of. The packaging material should be a good barrier to moisture, O_2, SO_2, and volatile compounds, depending on the fruit. Maintaining the headspace is also important to control the oxygen permeation into the packaging. Modified Atmosphere Packaging is a technology to maintain the desired air composition in the packaging. Active packaging is another technique in which oxygen scavengers, humidity absorbents, and antimicrobial materials are used. A third type of packaging technology is intelligent packaging, which uses biosensors, temperature-time, oxygen, carbon dioxide, and microbial growth indicators.

Glass is an inert and impermeable material and can be considered the best packaging material. However, due to its bulky, inflexible, and breakable nature, it might not be preferred for transport. Alternatively, packaging materials such as polyethylene and polypropylene are used, which are flexible, light, and unbreakable (Miranda et al., 2019). The internal headspace composition affects the biochemical reactions that alter the bioactive content, colour, and flavour quality.

7.1.3 Physical, Chemical, and Nutritional Quality

Conventional drying methods such as sun drying and tray drying are slow and may lead to inferior quality product. Novel methods such as microwave, freeze, and hybrid drying require shorter dehydration time and hence retain the product quality. The physical quality of the dried fruits refers to the surface colour and texture. Chemical quality refers to the characteristic compounds of the fruit that provide flavour, colour, and aroma. While the nutritional characteristics include the key vitamins, minerals, and antioxidants (Shah et al., 2022).

7.1.3.1 Quality of Traditionally Dried Fruits

Sun-dried products undergo a long dehydration time and are subject to degradation of light-sensitive compounds. Hence, some polyphenols may undergo photochemical reactions like oxidation, altering the nutritional or flavour profile. Browning and casehardening are observed to a large extent. Shade drying is employed as an alternative to avoid sun exposure and retain light-sensitive compounds. Osmotically dried fruits are widely accepted and enjoyed products. It retains volatile compounds and provides an improved texture. These have a pleasing colour and taste, often altered due to the salty or sugary layer caused by dehydration. Often, osmotic dehydration is combined with electric pulse and high pressure to enhance diffusion and have minimal effect on the quality.

7.1.3.2 Quality of Fruits Dehydrated Using Novel Technologies

Freeze-dried products undergo freezing and sublimation. The changes in bioactives such as phenols and flavonoids are found to be insignificant. This shows that oxidative reactions are negligible, and browning is avoided. Other nutritional quality aspects such as vitamins are also retained due to low temperature. Vacuum dehydration also requires low temperatures and less drying time as the dehydration is carried out under vacuum. The rehydration ratio is high as well as the shrinkage. However, the high pressure can cause darkening. Microwave drying requires shorter time for dehydration with volumetric heating. Good quality fruits have been achieved in terms of colour, shrinkage, and rehydration property (Review, 2006). There may be partial loss in aroma and other sensory aspects. Microwave has been combined with vacuum and osmotic dehydration for improved dehydration properties. Tray drying, on the other hand, provides products with an altered colour due to longer drying time, but a uniform dried product is obtained. Tunnel-dried products undergo high heat and hence face oxidation of compounds.

7.2 CHANGES OCCURRING DURING STORAGE

The physical, chemical, sensory, nutritional, and microbial changes that occur during dehydration vary as per the dehydration method. However, the changes occurring during storage depend on the storage conditions. If we compare the storage of fruits dehydrated by different methods stored under same conditions, the trend in changes in a particular attribute varies.

7.2.1 Physical Changes

Texture: The touch and feel of the dehydrated fruit are the textural aspects. The cellular structure of fresh fruit is made of rigid cell walls filled with liquid. On dehydration, as the fluid is removed, cell turgor pressure is lost, leading to collapse of the organized cell wall structure. Often, due to sudden dehydration, the structure collapses, leading to extreme shrinkage. High dehydration temperatures lead to case-hardening of the fruit surface, hampering the release of moisture from the fruit core, and leading to uneven dehydration. The microscopic and macroscopic levels of fruit structure and composition influence the structure. The texture can be analysed using a compression-extrusion test, in which force is applied to the fruit to flow through an orifice. The maximum force (N/cm^2) required for the extrusion is taken as an index of the textural quality. It has been observed that with a loss in moisture content, a higher maximum force is required (Miranda et al., 2014).

It is known that during fruit ripening, the firm texture and cell adhesion reduce and soften the fruit flesh. The pectic substances and the middle lamella aid cell adhesion and firmness to the matrix. During fruit ripening, degradation of pectic substances solubilize and depolymerize, causing loose walls. The middle lamella also disintegrates. The cell-wall degrading enzymes mainly cause the degradation of pectic substances. Some examples include enzymes such as polygalacturonase in peach, avocado; pectin methyl esterase, amylases, xylanases, pectinases, and laminarinases in banana.

7.2.2 Chemical Changes

pH: The pH of the fruit matrix is important for biochemical reactions. The moisture content and the a_w affect the biochemical changes and the pH. Dehydration changes the product's moisture content and a_w, which might change during storage due to moisture absorption. The pH also controls the microbial proliferation, as the bacterial and fungal species have specific optimal pH range for growth and proliferation. Bacteria prefer pH above 6.0, while yeasts and moulds thrive below pH 5.0 and can go until pH 2.0.

Sugar content: The sugar content varies in fruits ranging from as low as 0.4 g/100g in lime to 16 g/100 g in grapes. On dehydration, the sugar gets concentrated, and some of the examples are dried apricots (38.9 g/100 g), dried figs (62.3 g/100 g), dried mango (73 g/100g), etc. During storage at ambient temperatures, Aonla fruit underwent osmo-air drying, oven drying, solar drying, and indirect solar drying, which has shown a reduction of 6, 8, 5, and 10 g/100 g in reducing sugars after 90 days, respectively. While the total sugars were reduced by only 1 g/100 g. The reduction in total sugars is caused by non-specific hydrolysis of macromolecules, interconversions of sugars, and aggregation of monomers during storage. Dehydration causes the reducing sugars to convert to unsaturated sugars that are highly reactive. For example, hexose-reducing sugars convert to 2-furaldehyde and 5-hydroxymethyl-2-furaldehyde (Pragati et al., 2003). Sugar makes the oxygen unavailable for oxidation of bioactives such as vitamin C, hence preventing its degradation to some extent.

7.2.3 Nutrient Loss

Ascorbic acid content: Fruits containing high ascorbic acid (AA) like Aonla fruits, Plums, Cherries, Citrus fruits, etc. are desired for this key property. During storage of dehydrated fruits, the AA content has been seen to decrease. It also diminishes completely after 3 months of storage if the dehydration method utilized high temperatures. AA is heat sensitive. Apart from that, it oxidises easily during storage. Dehydration causes pore opening in the matrix as the moisture releases. This also gives more chance for oxidation. Exposure to heat for longer durations or at high temperatures causes higher degradation. For instance, strawberry dried by convective drying at 60°C degraded about 78% of L-ascorbic acid. On the other hand, freeze drying causes only 13% of degradation (Krzykowski et al., 2020). Indigenous enzymes such as AA oxidase, cytochrome oxidase, and peroxidase participate in AA degradation.

L-ascorbic acid degrades either aerobically or anaerobically (Giannakourou & Taoukis, 2021):

Aerobic pathway: L-ascorbic acid breakdown into dehydroascorbic acid (DHAA) reversibly in presence of oxygen. DHAA can further breakdown into other smaller compounds. While AA and DHAA show biological and antioxidant activity, further breakdown ceases the biological activity. For instance, DHAA can breakdown into 2,3-diketogluconic acid (DKG). The antioxidant activity of AA is five times higher than that of DHAA. To keep the oxidation of AA minimal, the dried fruit needs to be packaged with minimal oxygen. Which means the head space needs to be minimal. Apart from that, the packaging material permeability needs to be minimal to prevent entry of oxygen. During storage, a surge in DHAA is observed in the initial days due to the oxidation of AA and ceases gradually. This is the reaction-determining step. While AA content decreases during storage due to oxidation, as discussed above. Eventually, the headspace oxygen reduces as it is consumed due to oxidation.

Anaerobic pathway: L-ascorbic acid degrades in absence of oxygen; hence, it does not give DHAA. Rather, keto tautomeric forms of L-ascorbic acid are hydrolyzed to 2,3-diketogulonic acid. As products, reducones, furfural, and furancarboxylic acids are formed. These products may further get involved in browning reactions. Degradation of L-ascorbic acid through aerobic and anaerobic pathways has been graphically shown in Figure 7.1.

Tannin content: During storage, the tannin compounds undergo breakdown by enzymes such as PPO. Osmotically dried Aonla, Bael, and Pear fruits have shown a decrease in their tannin content during ambient storage conditions.

Phenols: The phenolic compounds differ in all fruits having a key characteristic phenolic specific for a fruit. The phenolic composition depends on the genetic and environmental factors. At the cellular level, lignin, some flavonoids, and ferulic acid esters are located at the cell wall. While soluble phenols are located in cell vacuoles majorly. These might be present in soluble, suspension, or colloidal forms or as a part of the cell wall component. The phenolic composition of the seeds, epidermal and subepidermal layers (peel), and the internal tissue (cortex) varies significantly at the tissue level. Soluble phenolics are majorly found in the peels than the internal tissue,

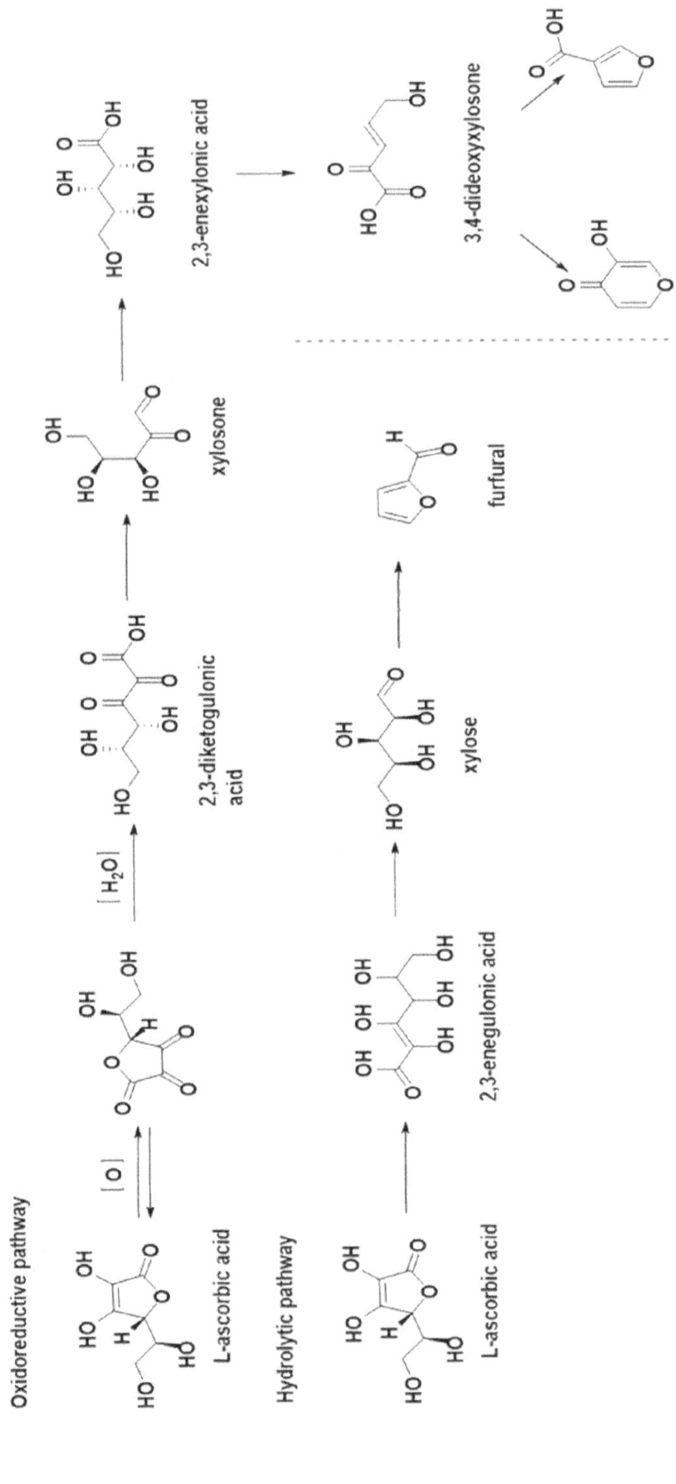

FIGURE 7.1 L-ascorbic acid degradation through aerobic and anaerobic pathways (Giannakourou & Taoukis, 2021).

Lipid oxidation by free radical chain propagation reactions	Termination of free radical chain propagation reaction:
$ROOH^{*} \longrightarrow ROO^{\cdot} + H^{\cdot}$	$ROO^{\cdot} + AH \longrightarrow ROOH + A^{\cdot}$
$ROOH \longrightarrow RO^{\cdot} + {}^{\cdot}OH$	$RO^{\cdot} + AH \longrightarrow ROH + A^{\cdot}$
$2ROOH \longrightarrow RO^{\cdot} + H_2O + ROO^{\cdot}$	

FIGURE 7.2 Antioxidant activity of phenols to prevent lipid oxidation.

e.g., flavonol glycosides. Anthocyanins are another type of phenol occurrence that is pH-dependent. In grapes, these are located in the skin in quinonoidal base form, while in strawberries and blueberries these are located in whole fruit in flavylium cationic form (Robards et al., 1999). Phenolic compounds have good antioxidant properties and subsequently cause browning reactions as well.

Antioxidant activity: Phenols undergo free radical scavenging, hydrogen donation, singlet oxygen quenching, metal-ion chelation, or can get attacked by superoxides. Phenols terminate the free radical chain reaction by donating hydrogen atoms to lipid radicals, for instance. Lipids are found in the seeds and in the pulp in small amounts. An illustration of the same is given in Figure 7.2.

Phenolic compounds as substrates for oxidation reactions: Naturally present phenols use oxygen, making them unavailable for AA oxidation until exhaustion. These are substrates for oxidoreductase enzymes like PPO and peroxidases (POD) are leading to enzymatic browning reactions. The site of reaction is in the vacuoles of the cytoplasm, and the products obtained are coloured quinones. Processing methods prior to drying such as chopping, crushing, and grinding can release such browning enzymes causing discolouration during dehydration and storage of the fruit. Phenols may also undergo non-enzymatic browning reactions in presence of metal ions. Metal-phenolics complexing in fruits may occur and have been seen in apples. Reactions of o-quinones with phenolic species are graphically shown in Figure 7.3.

7.2.4 Sensory Changes

Surface colour: The dehydration method decides the retention or degradation of the surface colour of the dehydrated fruit. Sun drying results in browning due to photo-oxidation. Indirect sun drying prevents browning due to photooxidation. However, other mechanisms such as enzymatic and non-enzymatic browning reactions may cause surface browning. During storage, enzymatic browning is caused by POD, PPO, and lipooxygenases. Non-enzymatic browning caused by oxidation also alters the surface colour. Depletion of antioxidants also causes browning. Antioxidants such as phenolic compounds and AA prevent oxidation of bioactives. Oxidation of

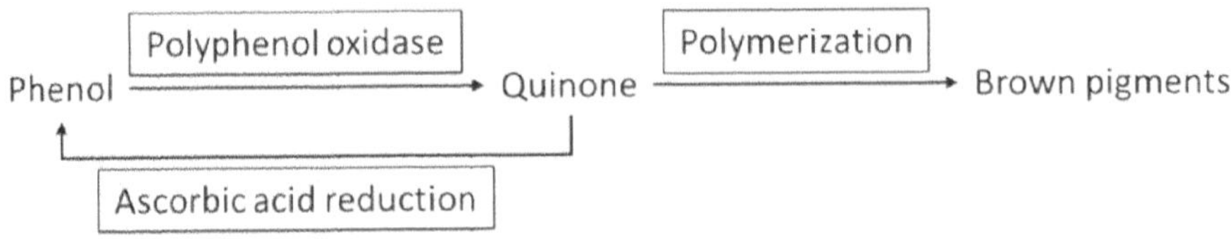

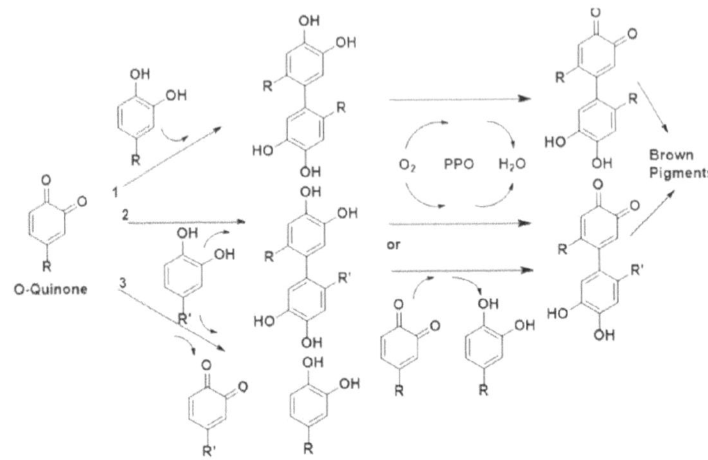

FIGURE 7.3 Reactions of o-quinones with phenolic species (Robards et al., 1999).

bioactives causes browning. Hence, when the antioxidant compounds start depleting during storage, the browning tendency of the fruit increases.

7.2.5 Microbial Changes

Dehydration is a step taken for preservation of the fruit. Water molecule is essential for biochemical reactions to take place. Removal of water from the matrix shuns these reactions by inactivating the essential enzymes involved. The pH is also altered to unfavourable limits as the moisture is removed. Removal of moisture and change in pH also prevent the microbial proliferation. It has been pointed out that pH levels below 6.0 and 5.0 hamper the bacterial and fungal growth, respectively. Water activity below 0.6 is suitable to prevent microbial growth. However, certain fungal and mould species may sustain at very low pH (<5.0) and a_w (<0.5). However, due to environmental changes, poor packing, and storage conditions, the product might gain moisture from the environment. This would start supporting the biochemical reactions and the enzymes in the fruit matrix. The fruit matrix is rich in sugar, minerals, and vitamins. All these factors provide a good medium for the microorganisms to flourish on. Dried Apricot, Fig, and other fruits have been found contaminated with pathogenic bacteria, coliforms, and fungi. Efforts on post-dehydration decontamination have been taken using thermal and non-thermal technologies such as infrared and ozone, respectively (Hussain et al., 2011; Öztekin et al., 2006).

7.3 STORAGE CONDITIONS REQUIRED BASED ON THE DRIED FRUIT PRODUCT

7.3.1 Packaging Material

The hygroscopic property of the product will attract moisture and increase the a_w, leading to spoilage. Air permeability would also affect the quality by oxidation of bioactives. Thus, packaging of the product becomes the final step and the most important step in processing and precedes storage and transport. It becomes essential to choose the right packaging material to protect from the storage and transport conditions faced by the product.

Dried fruits are preferred to be stored in glass jars and metallic containers as they have least permeability to air and moisture. However, these are bulky and consume space, making them expensive for storage and transport. They can also break easily, making the handling tough. Other acceptable packaging materials are plastic and paper bags. Paper bags are biodegradable but have poor barrier properties; hence, need to be layered with polymer material. On the other hand, plastic bags are fairly impermeable to air and moisture, are lightweighted, occupy less space, and are easy to handle. However, plastic is not environment friendly. Some examples are polyethylene, polypropylene, and polystyrene (Korese et al., 2022).

In case of dehydrated fruits, the packaging material should make a good barrier to moisture. To make it better, the polymer is layered with aluminium foil. Polyethylene has good thermal properties; hence, it is good for forming into a packaging material. It is often mixed with other polymers such as polyester and polypropylene. Polypropylene is made into the outer layer in case of dehydrated products as it has high density (0.92–0.94 g/mm^3) and can withstand high pressure. Polyester is a good moisture barrier and has good mechanical and chemical properties. Together, these provide good protection, strength, and moldability. Metalized foils are widely used due to their good barrier properties. Apart from being a good barrier, the tensile strength of the packaging material is important for the mobility and shaping. Layering different polymers for good barrier properties will affect the tensile strength of the packaging material (Aljilji et al., 2020).

7.3.2 Air Composition

The post-harvest management of fruits requires to maintain the air composition to slow down respiration. Continued respiration leads to ripening, and further ripening causes softening of the skin and supports spoilage. Low level of oxygen and slightly higher levels of CO_2 are preferred. For example, 20% of CO_2 retained the firmness of strawberry and reduced microbial spoilage. However, off-flavours were detected. Modified and controlled atmospheric packaging are technologies used to control post-harvest respiration. In the dehydrated fruit product, air in the packet may cause oxidation of the bioactives such as antioxidants, vitamin C, etc. Oxidative browning may also occur, as explained in the above sections.

7.3.3 Temperature

Temperature is an important factor that affects respiration. High temperature increases the respiration rate, leading to shorter storage time. High temperatures cause degradation of bioactives and antioxidants. Loss in colour, flavour, and aroma compounds occurs at high temperatures.

7.3.4 Humidity

Humidity is the most important factor for dehydrated fruits. Increase in the RH will increase the moisture content of the product, and hence the a_w. Increase in a_w supports bacterial and fungal growth, spoiling the product faster. Moisture gain increases the rate of quality degradation. As moisture is gained, the pH also changes. All these factors collectively lead to faster quality degradation of the product.

7.4 STORAGE AND SHELF-LIFE PREDICTION STUDIES CONDUCTED ON DIFFERENT DRIED FRUIT PRODUCTS

The SL of the dried fruit needs to be analysed to provide the customer with the right storage conditions and the storage time under which the product is safe to consume. The safe time to store the package before and after opening is essential to be informed. The conditions for storage such as temperature, RH, air composition, and packaging material can be assessed using SL studies. The same SL with the storage conditions needs to be declared on the packaging. Open dating is the duration in which the product will retain its quality and can be displayed on the store shelf, is a prerequisite in Europe, many South American and Arabic countries, Israel, and Taiwan. However, in many places in the United States, open dating is compulsory for infant formula and baby food. In about 20 US states, open dating is required for some food categories. The food producer is responsible for providing the date.

The SL is different from the safe life of the product. The SL period describes the storage period for which the product retains acceptable quality limits. While the safe-life relates to the microbial safety of the product. It is the storage period until which the product is safe to consume. Generally, the safe-life period is longer than the SL of the product. It is important to note that the ingredients used in the processing are introduced at different stages of the process, influencing the final product's SL. The quality of such a product may deteriorate faster after opening the package. Considering this, another concept called secondary shelf-life (SSL) comes into picture. The SSL informs about the period after opening the package during which the product quality is acceptable. This information is termed as "best within xx days of opening" on the food packet (Calligaris et al., 2019).

Traditional shelf-life analysis: The traditional method for SL analysis requires monitoring of the product during the storage period. It is real-time SL testing and is more suitable for a perishable food product as it undergoes quick quality changes. Real-time analysis is also suitable for SSL, as the food quality changes at a faster rate once it is opened and exposed to the environment. The storage conditions are preferred to be those on the market shelf, and these including temperature, RH, light exposure, and packaging material are experimented. The effect of such storage conditions is assessed based on the key characteristics of the product.

Accelerated shelf-life testing (ASLT): On the other hand, in the case of non-perishable foods such as dehydrated and frozen products, ASLT is practical since the physiological changes take longer in such products. Hence, storage conditions are set to extreme or harsh conditions to enhance the deterioration reactions. The environmental conditions allow faster biological changes, leading to faster quality degradation. ASLT requires to be applied systematically to compute the actual SL according to the market conditions.

An overlook at the SL testing protocol is described below:

i. *Preliminary conditions/deciding key parameters*:
   Different fruits have their characteristic qualities. Key parameters such as vitamin C, specific phenolic compounds, antioxidant activity, spoilage and browning enzymes, flavour compounds, and any other important bioactives particular to the fruit variety should be identified and studied during the storage period. Quality changes are categorized majorly into enzymatic, chemical, and physical changes. While microbiological changes influence the safety of the product. The mentioned changes occur simultaneously during storage. Chemical changes such as oxidation of flavour compounds, pigments, and non-enzymatic browning reactions may occur. Physical changes include structural collapse. As per the nutritional and safety regulatory requirements, the minimal limits of the specific quality parameters are aimed as acceptable limits. The storage time taken to reach the acceptable limit marks the SL of the product. In other words, the SL discriminates between acceptable and unacceptable limits.
ii. *Stability test/Experimentation:*
   After deciding the key parameters that define the quality of the fruit, the next step is to predict the deteriorative reaction kinetics as a function of different variables. The experimentation requires proper planning to assess the SL. The changes occurring in the indicator parameters are recorded at different time intervals during storage. Such changes mimic the changes that are expected to occur at the market shelf conditions. The rate of change in quality is a function of the intrinsic parameters (C_i), packaging factors (P_i), and environmental factors (E_i) as seen broadly. Hence, the reaction is affected by complex interactions between various factors. The function is stated as follows:

$$k = f(C_i, P_i, E_i)\ldots \tag{7.1}$$

   Generally, stable products such as dehydrated fruits are stored at 30–35°C. Changes in the storage conditions are inevitable at the market shelf and during transport.

   When ASLT is performed, the intrinsic and packaging variables are kept constant, while the environmental factor is varied to accelerated conditions to increase the reaction rate, causing quality degradation. Temperature, humidity, light intensity, air composition, and partial pressure are the parameters that could be played upon. The factors are selected independently, and their effect is seen on the degradation rate of the quality parameter selected. Then, the intensity of the factors is varied to see their effect on the quality parameter. The responses are recorded.

iii. *Modelling*:
Next, quality data collected at varying conditions is sorted to be taken as a function of storage time. Generally, the obtained data undergoes modelling as per the fundamental kinetic principles. The rate of changes in the quality parameters (I) as a storage time (t) is integrated as below,

$$\int_{I_0}^{I} \frac{dI}{I^n} = \int_{0}^{t} k dt \ldots \tag{7.2}$$

where k is the rate constant and n is the reaction order.

Based on the best-fit model, n is selected to find out the reaction order based on the determination coefficient of the linear regression (R^2). However, it does not give clear indication about the actual mechanism, but the apparent one. The rate constant is also the "apparent" rate constant. Generally, the apparent zero, first, and second order are used to describe the quality depletion. Linear regression analysis of I (zero order), ln I (first order), and 1/I (second order) as a function of storage time gives the rate constant. Once the reaction rate constant is determined, the SL is determined using the integrated form of the below equation:

$$SL = \frac{1}{k} \int_{I_0}^{I_{lim}} \frac{dI}{I^n} \ldots \tag{7.3}$$

Here, I_0 is the value of the critical indicator just after food production and I_{lim} is the critical indicator value corresponding to the previously defined acceptability limit.

So, for zero order reaction when $n = 0$:

$$SL = \frac{I_{lim} - I_0}{k} \ldots \tag{7.4}$$

for first order reaction when $n = 1$:

$$SL = \frac{\ln I_{lim} - \ln I_0}{k} \ldots \tag{7.5}$$

for second order reaction when $n = 2$:

$$SL = \frac{\frac{1}{I_{lim}} - \frac{1}{I_0}}{k} \ldots \tag{7.6}$$

The quality degradation reactions start from a lab phase, which is the induction period, followed by an exponential phase, which is a progressive increase, and finally the termination phase, where the reaction ceases.

Summarizing the ASLT protocol:

1. As a preliminary step, experimental plan including the sample size, sampling frequency, analytical tools, resources, and expenditure needs to be defined.
2. The next immediate step is to define and choose at least three important accelerating factors and their intensities. In case of dried fruits, the important conditions would be RH and temperature most importantly.
3. Planning the experimental process by anticipating the critical reactions and their rate is the next step.
4. Judging the effect of the condition on the chosen quality parameter and spacing the sampling time at the possible intervals is important. As the changes in the quality progress, one may adjust the intervals accordingly.
5. The data collected is modelled using a kinetic model to compute the reaction rate, k.
6. The estimated k is then applied to the below equation to relate with the accelerating factor (A).

$$\mathrm{k = f(A)} \ldots \tag{7.7}$$

7. Regression analysis to judge the accuracy of the model is conducted and the slope (m) is obtained. In case of a poor fit, more data points are required. The equation generated for SL prediction will be as follows:

$$\mathrm{k = mA + k_0} \ldots \tag{7.8}$$

This equation is used to compute the actual storage condition (A_{actual}) using the following equation:

$$\mathrm{SL} = \frac{\mathrm{I - I_0}}{\mathrm{k_{actual}}} = \frac{\mathrm{I - I_0}}{\mathrm{mA_{actual} + k_0}} \ldots \tag{7.9}$$

Here, k_{actual} is the rate of reaction and A_{actual} is the value of accelerating factor at the actual storage conditions desired. Hence, the SL of the product stored on the market shelf can be estimated.

7.5 FUTURE SCOPE

Dehydrated products are hygroscopic and can easily gain moisture from the environment. Hence, storage in high humid countries is challenging. Coming up with the right storage conditions for such high humidities is essential. Cheaper and more efficient packaging films need to be made. Post-dehydration decontamination technologies involving thermal effects are in the market. However, these affect the quality of the product. Alternatively, non-thermal technologies are under experimentation. However, these need to be made feasible for industrial applications. Training for accelerated SL tests is required at the industrial level.

7.6 CONCLUDING REMARKS

An overview of the characterization of dehydrated fruits and the storage conditions required for a longer SL of the product is highlighted in this chapter. The right storage conditions to retain the quality and safety are low moisture or humidity mainly. Computing the optimum storage conditions and the SL at the market shelf would help engineer the right packaging technology and storage conditions for the product.

REFERENCES

Aljilji, A., Mahmutović, O., Bašić, H., & Prazina, N. (2020). Mechanical properties of dried fruit packaging materials. *Periodicals of Engineering and Natural Sciences*, *8*(4), 2547–2552.

Calligaris, S., Manzocco, L., Anese, M., & Nicoli, M. C. (2019). Accelerated shelf life testing. In *Food Quality and Shelf Life*. Elsevier Inc. https://doi.org/10.1016/b978-0-12-817190-5.00012-4

Erkmen, O., & Bozoglu, T. F. (2016). Food Preservation by Reducing Water Activity. In *Food Microbiology: Principles into Practice* (pp. 44–58). https://doi.org/10.1002/9781119237860.ch30

Giannakourou, M. C., & Taoukis, P. S. (2021). Effect of alternative preservation steps and storage on vitamin C stability in fruit and vegetable products: Critical review and kinetic modelling approaches. *Foods*, *10*(11). https://doi.org/10.3390/FOODS10112630

Hussain, P. R., Meena, R. S., Dar, M. A., & Wani, A. M. (2011). Gamma irradiation of sun-dried apricots (*Prunus armeniaca* L.) for quality maintenance and quarantine purposes. *Radiation Physics and Chemistry*, *80*(7), 817–827. https://doi.org/10.1016/j.radphyschem.2011.02.029

Korese, J. K., Achaglinkame, M. A., & Adzitey, F. (2022). Effect of different packaging materials on storage stability of Gardenia erubescens stapf. & Hutch. Dried fruits and powder. *Applied Food Research*, *2*(2), 100143. https://doi.org/10.1016/j.afres.2022.100143

Krzykowski, A., Dziki, D., Rudy, S., Gawlik-Dziki, U., Janiszewska-Turak, E., & Biernacka, B. (2020). Wild strawberry *Fragaria vesca* L.: Kinetics of fruit drying and quality characteristics of the dried fruits. *Processes*, *8*(10), 1–13. https://doi.org/10.3390/pr8101265

Mansuri, S. M. (n.d.). *Storage of fresh fruits and vegetables.*

Miranda, G., Berna, À, González, R., & Mulet, A. (2014). The storage of dried apricots: The effect of packaging and temperature on the changes of texture and moisture. *Journal of Food Processing and Preservation*, *38*(1), 565–572. https://doi.org/10.1111/jfpp.12004

Miranda, G., Berna, A., & Mulet, A. (2019). Dried-fruit storage: An analysis of package headspace atmosphere changes. *Foods*, *8*(2), 1–11. https://doi.org/10.3390/foods8020056

Öztekin, S., Zorlugenç, B., & Zorlugenç, F. K. (2006). Effects of ozone treatment on microflora of dried figs. *Journal of Food Engineering*, *75*(3), 396–399. https://doi.org/10.1016/j.jfoodeng.2005.04.024

Pragati, Dahiya, S., & Dhawan, S. S. (2003). Effect of drying methods on nutritional composition of dehydrated aonla fruit (*Emblica officinalis* Garten) during storage. *Plant Foods for Human Nutrition*, *58*(3), 1–9. https://doi.org/10.1023/b:qual.0000040332.98384.c4

Review, S. P. (2006). Microwave drying of fruits and vegetables. *Stewart Postharvest Review*, *2*(6), 1–7. https://doi.org/10.2212/spr.2006.6.4

Robards, K., Prenzler, P. D., Tucker, G., Swatsitang, P., & Glover, W. (1999). Phenolic compounds and their role in oxidative processes in fruits. *Food Chemistry*, *66*(4), 401–436. https://doi.org/10.1016/S0308-8146(99)00093-X

Shah, A. S., Bhat, S. V., Muzaffar, K., Ibrahim, S. A., & Dar, B. N. (2022). Processing technology, chemical composition, microbial quality and health benefits of dried fruits. *Current Research in Nutrition and Food Science*, *10*(1), 71–84. https://doi.org/10.12944/CRNFSJ.10.1.06

Tsami, E., Krokida, M. K., & Drouzas, A. E. (1998). Effect of drying method on the sorption characteristics of model fruit powders. *Journal of Food Engineering*, *38*(4), 381–392. https://doi.org/10.1016/S0260-8774(98)00130-7

8 Emerging and Advanced Drying Technologies for Drying of Different Fruit Products

Michalska-Ciechanowska Anna
The Department of Fruit, Vegetable and Plant Nutraceutical Technology, the Faculty of Biotechnology and Food Science, Wrocław University of Environmental and Life Sciences
Wrocław, Poland

8.1 INTRODUCTION

Fruits have been perceived as nutritious, and today's consumers link them with a positive effect on health due to the presence of numerous biologically active compounds. Unfortunately, the consumption of fruits still does not meet the recommendations of the World Health Organization, even if the consciousness about their pro-health composition is continuously increasing. Another issue is linked to their relatively fast deterioration in the fresh form due to their high perishability, which results in a 40–50% market loss as these merchandise have a relatively high moisture content (Boateng 2022). A majority of fruits are produced on a seasonal basis, and their availability during a year is highly limited. What is more, the globalization of fruit production combined with the development of agricultural mechanization has caused a severe issue of local overproduction and unequal distribution of fruits around the world. This overproduction results in a huge amount of waste. Thus, their processing is an important sector of the economy. Consequently, fresh fruits are processed into dried forms. Chang et al. (2016) described dried fruits as vital healthful snacks being a concentrated form of their fresh counterpart (Alasalvar et al. 2023). The term 'dried fruit' is linked to a product that has a lower moisture content in comparison to its fresh form that has been removed by traditional open sun drying or various modern processing techniques, including industrial application of mechanical devices (Purohit et al. 2006; Alasalvar and Shahidi 2013). One of the promising approaches for making them a tradeable commodity is drying that includes diverse approaches with a range of different methods, either with a single application or combined (Sadler et al. 2019). Drying can be applied for dehydration of fruit and its fractions, including whole fruit, slices, halves, pomace (by-products/waste products), and also puree, juice, or extracts (to produce modern form of solubilized products, either from fruits or from pomace) (Raghavan et al. 2022):

- To maintain fruit-based products' presence on the market regardless of fruit seasonality (extent the shelf life),

 DOI: 10.1201/9781003402916-8

- To extent the availability of natural nutrients present in dried fruit,
- To extent the markets' offer in countries where particular fruits are not available,
- To decrease the packaging and shipping costs,
- To provide dried fruit-based products used as a sugar substitute,
- For production of modern fruit-based foodstuffs that may serve as alternative for sweets or salty snacks replacement,
- To obtain functional foods, natural additives, and colorants,
- To mitigate food waste and loss (FAO 2021).

Sadler et al. (2019) proposed a classification including conventional dried-fruit merchandises that can be extended by new forms of these products (Table 8.1). Dried fruit-based commodities can be grouped according to the way they are obtained. Traditional fruit-based products are usually produced by sun drying considered as the most environmental friendly technique for water removal from plants. Conventional dried fruits are produced without sugar addition. Due to the fact that such products are dehydrated under uncontrolled conditions, some modern approaches (pre-drying methods: addition of ascorbic acid or sulfites, steam blanching, etc.) were used to, e.g., inhibit the enzymatic reaction that may prevent color loss of dried merchandises being an important attribute for consumers (Chang et al. 2016).

Another group of dried fruit-based products are those gained by sugar infusion before drying. Such an approach is usually used to preserve the products from microbiological spoilage, and sugar can also be applied as a natural humectant to improve the structure of dried merchandises. In line with the current nutrition trends toward reduction of sugar in foods, the osmotic dehydration in fruit juices or concentrates has been used. Besides partial dehydration of fruits, additional fortification with bioactives can be made (Kinga Samborska et al. 2019; Masztalerz et al. 2021; Kowalska et al. 2023). This can serve as a new strategy for improving the bioactive potential of dried products and also enriching such products with bioactive compounds that are not present in raw materials. What is more, the fortification of fruit-based products with another fruit juice/concentrate as an osmotic agent may also alter the sensory properties of the products, making them more attractive to consumers (Cano-Lamadrid et al. 2017).

A very common form of dried fruits present on the market are candied dried fruit-based products that include candied dried fruit, crystallized, and glazed (Miletić et al. 2014). Basically, the sugar is used as a preservative for reduction of water activity that prolongs the shelf life of this form of foods being broadly applied in the bakery and confectionery products (Goldfein and Slavin 2015).

Recently, the dried fruit market has been experiencing a significant increase in value, especially due to the witnessed robust growth of snacking industry linked to the growing popularity of healthy and nutritious snacks. The rising trend among consumers to opt for ready-to-eat and tasty snacks that meet their daily nutritional needs is expected to drive demand for dried fruits. Additionally, the market is projected to expand further due to substantial investments in research and development of new types of dried fruit-based products that align with consumers' preferences for clean-label, healthy, organic, and plant-based snacking options (Global Report

TABLE 8.1
Classification of Dried Fruit-Based Products Including Modern Forms (Sadler et al. 2019)

Group of Fruit-based Products	Features	Processing Technique	Example (Material, Product type)
Traditional dried fruit-based products	*traditional:* no sugar addition *modern:* addition of SO_2, ascorbic acid, steam blanching	sun drying solar drying, air drying, smoking	prunes, grapes, figs, apricots, raisins, dates, apples, pears, etc.
Sugar-infused fruit-based products	*traditional*: sugar infusion before drying	sugar solution infusion, drying	cranberries, blueberries, cherries, mangoes, strawberries, pineapples, etc.
	modern: fruit juice/ concentrate infusion before drying	osmotic dehydration, drying	apples, pomegranate arils, strawberries
Candied dried fruit-based products	*traditional*: fruits or parts candied in sugar solution, crystallized, glazed	osmotic dehydration, drying	cherry, pineapple, cranberry, dates, plums, candied peel (citrus fruits)
Modern fruit-based products	whole fruit, fruit slice, puree, pomace *new approach*: improved texture by application of modern drying techniques	sun drying, oven drying microwave vacuum drying, combined drying	freeze-dried fruit, fruit snacks, fruit leathers, fruit pieces (apple, plum, cranberry, berries, mixed sources, etc.) fruit snacks chips (slices) (apple, mango, durian)
	powders whole fruit powders (partially resolubilized) *new approach:* juice/ concentrate/pomace extract powder products with functional carriers fruit/juice/pomace extracts	convective drying, freeze-drying spray drying in dehumidified air, vacuum drying, combined drying	plum, cranberry, blueberry, chokeberry, pear, apple, jaboticaba chokeberry, cranberry

of Dried Fruits 2022). In response to consumers' changing eating habits, numerous new forms of fruit-based products appear on the market. Currently, due to the popularity of freeze-dried fruits among consumers, a strong market demand from bakery and breakfast cereal industries has been observed (Global Report of Dried Fruits 2022). On the other side, the market growth of dried fruits has been experiencing a boost recently, thanks to the increasing trend of substituting salty or sweet snacks with fruit-based new product types, e.g., chips, fruit leather produced from puree (Nizamlioglu, Yasar, and Bulut 2022; Zhu et al. 2022; Thongcharoenpipat

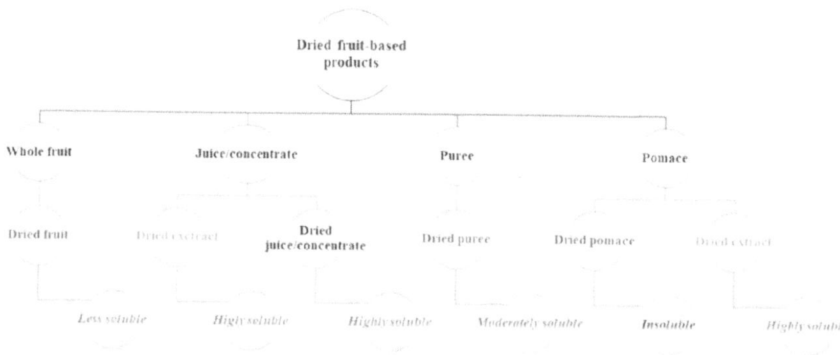

FIGURE 8.1 Solubility of different dried fruit-based products.

and Yamsaengsung 2023) or powders obtained from fruit (whole) (Gurak et al. 2014; Michalska et al. 2016) and its fractions (juice/concentrate, puree, and pomace) (Karam et al. 2016; Michalska et al. 2017; Michalska et al. 2018; Michalska-Ciechanowska et al. 2021; Katarzyna Samborska et al. 2022). The last-mentioned powders are easy-to-handle and convenient form of fruit-based products that can be broadly applied to other foodstuffs to improve their functionality (fortification with bioactive compounds) and color; however, their final properties in terms of solubility determine their application (Figure 8.1).

Currently, in line with sustainable development goals, the re-usage of nutritious fruit by-products, i.e., pomace, may promote sustainable resourcing of fruits. The management of fruit industrial by-products becomes a key issue in terms of recirculation of natural bioactives. Such a new approach can be successfully used for powder production that leads to the obtainment of highly soluble products (Michalska-Ciechanowska et al. 2021). However, the way of their extraction (solvent type) is a significant issue linked to their quality.

Nevertheless, the chemical composition of fruits poses a serious challenge for scientists and the industry when it comes to converting fruit-based materials (such as whole fruit, juice, puree, concentrate, and pomace) into a dried form. On the one hand, achieving the desired product form with expected physical properties necessitates a deep understanding of various processing aspects, including the capabilities and efficiency of drying techniques, energy consumption considerations, optimal parameters, and more. This knowledge is essential for obtaining the appropriate type of merchandise. On the other hand, the recognition of the influence of the processing parameters on the alterations of chemical characteristics of dried fruit-based merchandises is of high importance as it may lead to negative consequences, including the degradation of natural bioactives and the formation of hazardous components, but it may also improve the composition of such commodities by increasing the content of selected constituents and antioxidant properties due to the interaction between them (Michalska et al. 2016; Michalska et al. 2019). The fruit-based matrix composition considered as a mix of different basic chemical constituents and biologically active compounds that is moderated by processing type and parameters is a key point for understanding the interactions that occur between these components affecting their availability and thus biological properties of dried merchandises (Ogawa et al. 2018; Oliveira et al. 2018; Alasalvar et al. 2023). From this perspective, the enhancement of processing technologies, particularly drying, is imperative to address the challenges associated with improving food safety and quality.

8.2 COMMONLY USED DRYING TECHNOLOGIES AND MODERN DRYING TECHNIQUES FOR DEHYDRATION OF FRUITS

Traditional fruit-based merchandises are often produced by sun drying (open sun drying, open air drying, black-cloth shade, and white-cloth shade) (Purohit et al. 2006; Mohammed et al. 2020) or by dehydration using clay ovens or smoking (Yang and Chiang 2019). Such type of production on a large scale is a low-cost process and energy-efficient approach commonly used in regions with abundant sunlight and dry climates; however, it can lead to several adverse effects on product, including high-product losses caused by inadequate drying, contamination of microorganisms, molds and fungi, encroachment of insects, etc. (Purohit et al. 2006). Additionally, due to the presence of oxygen, uncontrolled temperature and humidity during moisture removal, the drying time can be relatively long making it less efficient for a large-scale production (Raveendran et al. 2022). As a consequence, it may provoke enzymatic reactions leading to the deterioration of color being a major attribute for consumers of fruit-based products and loss of thermolabile bioactive compounds. Solar drying is a more modern approach used for dried-fruit production that also utilizes solar energy with controlled drying conditions. It involves specialized equipment like solar dryers and optimizing factors like temperature and airflow. Solar drying offers improved efficiency, making it suitable for small-scale and large-scale operations, surpassing traditional sun drying. On the other hand, quality issues linked to sensory characteristics of solar-dried fruit should be considered (Mohammed, Edna, and Siraj 2020).

When considering the drying technique suitable for fruit-based products, it is crucial to take into account factors such as production scale, efficiency, product quality, and profitability. Furthermore, in alignment with sustainable development goals, it has become essential to include considerations not only for profitability but also for energy consumption and CO_2 emissions, in order to comprehensively evaluate the overall sustainability and environmental impact of the production process. Table 8.2 compares various drying techniques applied for fruit-based product dehydration. Traditional methods like sun drying and smoking offer low-cost options but may result in significant quality loss being linked to poor retention of selected bioactive components, especially ascorbic acid (Sadler et al. 2019).

The usage of sun, solar, and air drying under uncontrolled conditions, including high temperatures, which is strongly linked to deterioration of bioactives (among others, phenolics, vitamins, etc.), is generally negatively perceived by consumers.

However, the increasing purchasers' interest in food safety and improved food quality has prompted the global industry to shift its focus from traditional sun drying

TABLE 8.2
Comparison of Drying Techniques for Fruit Products: Efficiency, Energy Consumption, and CO_2 Emission

Drying Technique		Type of Products	Type of Energy	Efficiency of Drying	Economic Values	Energy consumption	CO_2 Emission	Pros	Cons
Traditional	Sun drying	whole fruit, parts, pomace	solar	low to moderate	low	negligible	negligible	no equipment investment	drying time ↑; quality ↓
	Solar drying	whole fruit, pomace, parts, puree	solar	moderate to high	moderate to high	low	negligible	controlled drying conditions	equipment investment; quality ↓
	Air drying	whole fruit, parts, pomace	natural or ambient	moderate	low	low	negligible	simple and low-cost method	susceptible to contamination; quality ↓
	Clay ovens/ stoves	whole fruit, pomace, puree	combustion	moderate	low to moderate	moderate	medium to high	enhances flavor and aroma	drying time ↑; quality ↓
	Smoking	whole fruit	combustion	low to moderate	low	low	medium to high	adds a unique smoky flavor	limited applications; intense flavor may overpower the natural taste
Modern	Convective drying	whole fruit, pomace	electrical	high	moderate to high	high	medium to high	efficient, faster than traditional methods, controlled and uniform drying	energy consumption ↑; possible changes in the products structure (shrinkage); quality ↓
	Oven drying	whole fruit, pomace, puree	electrical	high	moderate	moderate	medium to high	controlled and uniform drying	energy consumption ↑
	Vacuum drying	whole fruit, pomace, puree, juice, concentrate, extract	electrical	high	high	high	low	preserves nutritional content	expensive equipment and operation

(Continued)

TABLE 8.2 *(Continued)*
Comparison of Drying Techniques for Fruit Products: Efficiency, Energy Consumption, and CO_2 Emission

Drying Technique	Type of Products	Type of Energy	Efficiency of Drying	Economic Values	Energy consumption	CO_2 Emission	Pros	Cons
Microwave drying	whole fruit, parts	electrical	high	moderate to high	high	low	rapid drying process	uneven overheating and potential quality loss
Freeze-drying	whole fruit, puree, juice, concentrate, extracts	electrical	high	high	high	low	retains natural color and nutrients	expensive equipment and operation; energy consumption ↑ (freezing), retention of bioactives ↑
Spray drying	fruit juice, concentrate, puree, extract	electrical	high	high	high	medium to high	easy handling and preservation	volatile compounds ↓

to more advanced techniques that are driven by the growing demand for nutritious foods. Modern techniques, such as oven, vacuum, freeze, and spray drying, offer controlled drying conditions and better nutrient retention compared to traditional methods. However, it is important to acknowledge that these techniques come with certain trade-offs, including higher energy consumption and the requirement of equipment investment (Table 8.2). When evaluating the suitability of each technique, factors, such as efficiency, economic values, and potential drawbacks, should be taken into consideration.

Fruit processing methods, including drying (traditional and modern techniques), have a significant impact on the final quality of dried products, often leading to a decrease in their nutritional properties (Weaver et al. 2014). In general, obtaining a dried fruit-based product with quality comparable to fresh fruit is challenging (Fazaeli et al. 2012; Karam et al. 2016). The nutritional value and health benefits of dried fruit-based products obtained by different drying techniques are linked to several factors, including their natural composition, nutrient density, presence of bioactive compounds (Chang et al. 2016; Sadler et al. 2019), possible formation of hazardous components *via* Maillard reaction and/or caramelization (Megías-Pérez et al. 2014; Michalska et al. 2018; Michalska-Ciechanowska et al. 2021), as well as contamination by pesticide residues or microbiological spoilage.

Convective drying is a widely used industrial method for fruit dehydration (Lewicki 2006). Boateng (2022) summarized that convective dryers are employed for the dehydration of more than 85% of food products, primarily due to their widespread adoption stemming from their inherent adaptability, controllability, and user-friendly nature. It involves exposing the fruit to hot air to remove water. The process typically consists of two phases: an initial constant drying rate phase followed by a slower drying rate phase described by an exponential function. While the first phase is efficient, convective drying may negatively impact product quality, causing significant changes in structure (e.g., shrinkage) (Figiel 2010) and alterations in chemical composition (Calín-Sánchez et al. 2013). High drying temperatures and prolonged drying times contribute to these issues (Kudra 2008), resulting in color deterioration, loss of bioactive compounds (Maskan 2001), and sensory perception (Petikirige, Karim, and Millar 2022).

Another option for fruit dehydration commonly used is oven drying that differs from convective drying as the material is placed in static or batch-type oven where heat is provided through heating elements or gas burners to remove the moisture from the material. The drying rate is lower in oven drying when compared to convective drying as the heat is not continuously supplied and the fruits are not exposed to constant flow of a hot air. This drying technique is less energy-efficient than convective drying.

By reducing the pressure during the drying process, it is possible to decrease the drying time as water evaporates more rapidly from the plant material. Vacuum drying emerges as a potential alternative to high-temperature convective and oven drying, as heat is transferred through conduction, and temperature of fruit products can be readily regulated (Lewicki 2006). Moreover, the lower temperatures involved in vacuum drying can help to mitigate the thermal degradation of biologically active constituents present in fruits. Nevertheless, this process is also time-consuming. To address these challenges, a more modern drying technique, i.e., microwave vacuum drying that combines microwave power and reduced pressure to efficiently remove water from fruits, can be applied (Wojdyło et al. 2014). By incorporating vacuum, this drying technique reduces drying time, lowers material temperature, and minimizes exposure to oxygen (Vadivambal and Jayas 2007), which improves product quality (Özcan et al. 2020). However, microwave vacuum drying has limitations when the evaporation rate exceeds the pump efficiency, necessitating adjustments in sample size or vacuum system capacity (Figiel 2010).

Among the various drying techniques, freeze-drying is commonly used to preserve bioactive components with minimal degradation, making it popular in the production of natural nutraceuticals and pharmaceuticals (Ratti 2001). Currently, a special attention was drawn to this drying technique as consumers linked it to a higher quality of fruit products. Recently, an evident surge in demand for freeze-dried fruits, notably within the bakery and breakfast cereal industries, is being observed (Global Report of Dried Fruits 2022).

While freeze-drying maintains the color and volume of fresh fruit, it is time-consuming and its application should be carefully chosen depending on the fruit product form. Michalska et al. (2017); Michalska et al. (2018); and Michalska-Ciechanowska et al. (2021) proved that when it comes to conversion of liquid products (juices and extracts) into powdered form, spray drying can be recommended for similar or even higher retention of selected bioactive compounds in powders. The recognition of potential inefficiencies in industrial-scale freeze-drying has stimulated extensive research and development endeavors aimed at exploring other drying techniques.

In the case of liquid forms of fruit, such as juices, concentrates, and purees, alternative methods of water removal can be employed to reduce production costs. Spray drying is a cost-effective option commonly used for dehydrating fruit-based products (Shishir and Chen 2017; Michalska et al. 2018; Michalska-Ciechanowska et al. 2021). Drying of these fruit forms is challenging due to their high content of low-molecular weight compounds, i.e., sugars and organic acids, which make the resulting powders hygroscopic (Bhandari 2013). To overcome this issue, carriers with higher molecular weights, such as maltodextrins, starches, and gums, are often used in powder production to increase the glass transition temperature of the products (Coimbra et al. 2021). Furthermore, Ray et al. (2016) emphasized the importance of using drying aids with Generally Recognized as Safe (GRAS) status and highlighted that drying aids should chemically remain nonreactive with biologically active substances during the drying process. It was also suggested that carriers could physically separate bioactive compounds in the mixture to prevent undesirable interactions (Ray, Raychaudhuri, and Chakraborty 2016). However, recent studies have shown that interactions between bioactives and drying aids can occur, with variations depending on the concentration and chemical composition of the carriers used during spray drying. These interactions are influenced by process parameters such as temperature and vacuum (Labuschagne 2018; Rezvankhah et al. 2020; Coimbra et al. 2021; Michalska-Ciechanowska et al. 2021). It is crucial to consider these interactions when assessing the quality of spray-dried food powders, particularly due to the formation of Maillard reaction and caramelization products (Michalska et al. 2017; Michalska et al. 2018; Michalska-Ciechanowska et al. 2021) which may have implications for human health (Tamanna and Mahmood 2015).

The complexity and varied structures of fruit-based matrices pose significant obstacles when seeking a single drying technique suitable for all fruit products. Careful consideration of the specific form and composition of fruit is essential in selecting the appropriate drying method for successful product development.

8.3 ADVANCED AND EMERGING DRYING TECHNOLOGIES TOWARD IMPROVED FRUIT-BASED PRODUCTS QUALITY

Advanced drying technologies are continuously developed to enhance the quality of food products, including fruit-based merchandises. These technologies aim to establish cost-effective innovative processes, to overcome the limitations of traditional drying techniques and offer improved product characteristics, including physical attributes and chemical properties, while taking into account the sustainable development goals, i.e., less consumption of energy and reduced environmental impact (Mujumdar and Xiao 2019). On the one hand, while advancements have been made to enhance the traditional methods of fruit drying using renewable solar energy (Leon Dharmadurai et al. 2022; Mugi et al. 2022) aligning with the latest trend of reducing energy consumption and CO_2 emissions (Leon Dharmadurai et al. 2022; Mugi et al. 2022), there are also other approaches being explored. Numerous studies were performed with a focus on the improved physical and chemical properties of dried fruit-based products (Figiel and Michalska 2017; Tontul and Topuz 2017; Majerska et al. 2019; Calín-Sánchez et al. 2020). According to Raghavan et al. (2022), the primary challenges in drying processes at present revolve around nutritional loss and energy consumption. Thus, numerous advances in modern drying technologies were applied toward improving the properties of fruit-based products, including novel processing technologies based on freeze-drying, microwave drying, and spray drying. Belwal et al. (2022) highlighted the importance of novel hybrid drying techniques, including combination of microwave vacuum drying with convective, freeze-, vacuum, fluidized bed drying, foam mat drying, and fountain drying, as well as ultrasound-assisted drying (Xu et al. 2022; Fernandes and Rodrigues 2023), electrohydrodynamic drying (Iranshahi et al. 2023) and instant controlled pressure drop drying (Rane et al. 2022). It can be observed that these technologies are constantly improved and adjusted to specific fruit matrix compositions (Karam et al. 2016; Figiel and Michalska 2017; Shishir and Chen 2017; Sabarez 2021; Boateng 2022; Richter Reis et al. 2022).

Attaining consistent and satisfactory results for all fruit products exposed to the same treatment presents a notable challenge in the realm of fruit processing as different fruit fractions (whole fruit, slices, halves, juice, concentrate, puree, pomace) that originate from diverse fruit varieties behave differently due to the matrix properties and composition, i.e., the presence and content of basic chemical components (sugars, organic acids) and numerous bioactives (pectin, fiber, phenolics, and vitamins). Thus, the adjustment of appropriate drying techniques and parameters should be linked to the final dried product's expected properties (including the form, structure, presence of bioactive compounds, nutritional values, etc.). Sheng and Wang (2023) mentioned that while developing fruit-based dried products, the selection of the whole processing chain that usually involves multiple stages (pre- and post-drying treatments),

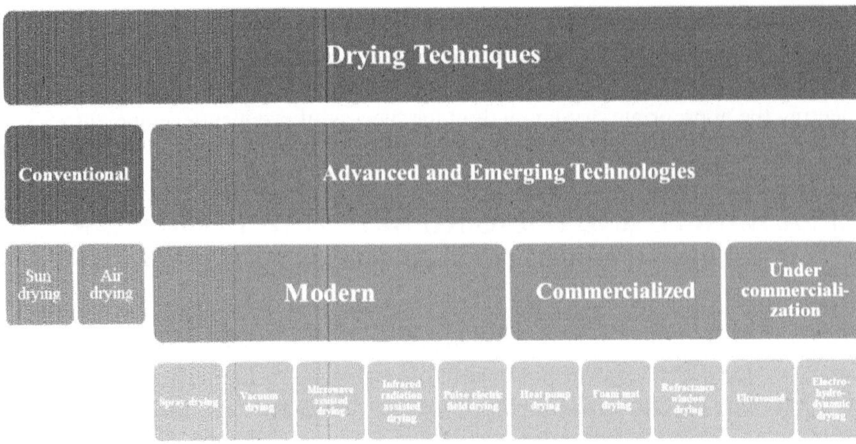

FIGURE 8.2 Conventional, advanced, and emerging technologies applied for fruit-based products (Sabarez 2021; Calín-Sánchez et al. 2020).

should comprise the recognition of specificity of the production systems, including consumers' expectations and preferences. Another issue is linked to the scale and the progress of advancement of the drying process. Considering this perspective, the commercialization and integration of advanced and emerging drying technologies hold significant implications in the context of scientific investigation and practical applications (Sabarez 2021). Some of the advanced and emerging drying technologies used for fruit-based product dehydration are presented in Figure 8.2.

Taking into account emerging global challenges connected with food safety, limited resources, and shifting climate patterns, the food industry sector is driven toward innovative solutions in drying to create a high-quality merchandises to be sustainably able to achieve a competitive advantage and generate profitable returns. The modern approach also considers processing in terms of pre-drying and post-drying treatments (Boateng 2022). The overall aim is to enhance product stability, also from a microbiological point of view, and to provide products with the appropriate characteristics. In fact, this requires a multidisciplinary approach, including characteristics from raw material to final product design, especially for specific product types (whole dried fruit, snack, powder, leather, etc.), among which an appropriate balance between retention of bioactives, formation of process contaminants influencing human health, economic issues, energy consumption, and CO_2 emissions, as well as appropriate packaging and storage (materials and conditions), should be found to create a modern way for establishing such a process chain. It is also worth mentioning that the application of any processing aids used as pre- or post-drying treatments should also be considered for their final properties (physical and chemical), as the addition of these substances will provoke alteration in composition of dried fruit-based products.

What is more, the modern approach is connected with food safety issue, not only from the microbiological point of view but also from the formation of process contaminants (described below). Taking into account clean label trends, shaping the properties of fruit-based merchandises by drying and its parameters will be a challenging approach with practical application. The latest studies indicated that process-induced modifications of chemical properties in dried fruit-based products are strongly linked to their biological properties *in vitro* (Silvan et al. 2020; Hendrysiak et al. 2023); however, more studies are needed to recognize the influence of drying on the fruit matrix, which is considered as a complex mixture. At present, the influence of only single components does not fully represent the biological properties of dried fruit-based products and does not consider interferences between different compounds induced by drying and their effect on bioactivity and bioaccessibility of the active compounds remains unknown (Oliveira et al. 2018).

8.4 QUALITY RISK ASSESSMENT OF FORMATION OF HAZARDOUS COMPOUNDS DURING FRUIT DRYING

Fruit-based dried products are an excellent source of natural bioactives, including fiber, phytochemicals (phenolics), minerals, and many others that may have a positive impact on human health (Sadler et al. 2019; Alasalvar et al. 2023). On the other hand, dried form of these merchandises could contain hazardous to human health components. These constituents could be present in raw material, added before drying or could be formed after drying (external and internal quality risk factors) (Figure 8.3).

Among internal factors, food-borne pathogens (Sheng and Wang 2023), natural/microbial and fungal toxin (Masood et al. 2015; Sheng and Wang 2023), and Maillard reaction/caramelization products (Piga et al. 2003; Megías-Pérez et al. 2014; Aktağ and Gökmen 2020; Michalska-Ciechanowska, et al. 2021) can be listed. In this group, the contaminants can originate from the presence of raw materials or can be formed during processing from components present in fruits.

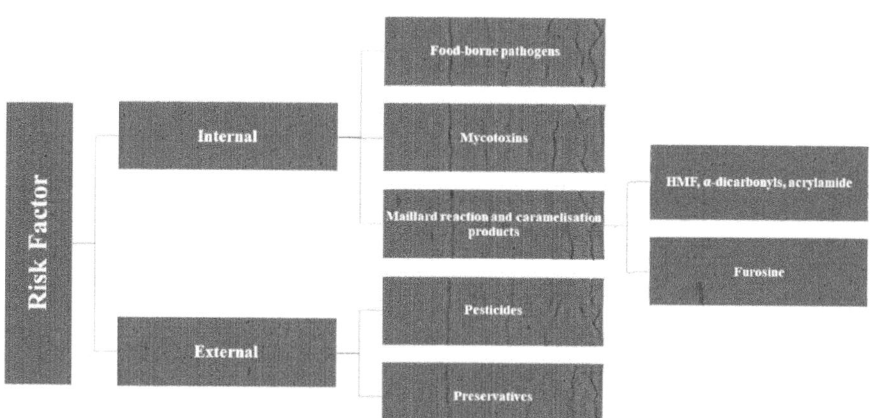

FIGURE 8.3 Quality risk factors for dried fruit-based products.

There are several ways to minimize contaminants in dried fruits that depend on the type of these hazardous substances. Table 8.3 presents the mitigation strategies that can be applied for minimalization of dried fruit adulteration. The most discussed in the literature are food-borne pathogens that might contaminate fruit during the harvest and post-harvest period (after processing, during transportation, and storage) (Alp and Bulantekin 2021). Sheng and Wang (2023) summarized pre- and

TABLE 8.3
Quality Risk Factors in Dried Fruit-Based Products and Mitigation Strategies

Quality Risks Factor	Description	Mitigation Strategies
Food-borne pathogens	Growth of microorganisms on fruit surfaces due to inadequate drying parameters Inappropriate storage Long-time presence of common food-borne pathogens, i.e., *Escherichia coli, Salmonella* spp., *Listeria monocytogenes* on dried fruits	• Blanching • Thermal treatments: microwaves, radio frequency, infrared radiation • Non-thermal treatments: cold plasma, ultrasounds, UV • Chemical treatment (addition of preservatives, e.g., ascorbic acid, citric acid, sodium metabisulfite, sulfur compounds, etc.) • Implement microbiological testing of dried fruits
Mycotoxins	Contamination of dried fruits by mycotoxins, also after removal of fungal mycelia	• Chemical treatment (ozone) • Biological treatment (probiotics, bacteriocins, bacteriophages) • Conduct regular mycotoxin testing
Formation of Maillard reaction and caramelization products	Formation of potentially hazardous components during drying and post-harvest treatments (transportation, storage): • hydroxymethyl-*L*-furfural • α-dicarbonyl compounds • acrylamide • furosine	• To use lower drying temperatures • Optimize drying time • To monitor the precursors (e.g., reducing sugars and amino acids) • Addition of inhibitors of Maillard reaction and caramelization products
Pesticide residues	Retention of pesticide residues on dried fruits, potentially above acceptable limits, impacting consumer safety	• Pre-harvest intervals for pesticide application • Washing (strong acid or alkali-oxidized water) – removal of surface-located compounds • Thermal treatments • Combined drying techniques
Preservatives and antioxidants	The addition of sulfur dioxide or potassium metabisulfite may have an adverse effect on the asthmatic population	• Alternative treatments (citric acid, etc.) • Clear label information

post-drying treatments for potential elimination of microbial and fungal contaminations for efficient hazard control of dried products. Application of selected drying techniques with consideration of raw material barrier effects (type, structure, and expected form of dried fruit-based products) can be effective; however, a combination of different dehydration techniques with pre- and post-drying treatment was found to be more successful in eliminating these risk factors (Alp and Bulantekin 2021; Sheng and Wang 2023).

During fruit processing, various compounds can be generated that are not naturally found in fresh material. These constituents arise as a result of the Maillard reaction and caramelization and are formed from the reaction between reducing sugars and amino acids, peptides, or proteins, as well as from the caramelization of sugars (Kroh 1994). The formation of these products is linked to the presence of natural precursors of this reaction being additionally accelerated by thermal-induced changes that occur during processing of fruits, including post-harvest treatments (transportation and storage). Additionally, in the case of fruits, application of sugar solution applied for preservation of these merchandises is also an additional source of carbohydrates that may interact toward formation of these compounds. Among the numerous substances formed during fruit processing *via* these reactions, some may exhibit positive effects, such as strong antioxidant properties, while others may have negative implications for human health (Tamanna and Mahmood 2015; Castillo et al. 2021; Aktag and Gökmen 2021; Michalska-Ciechanowska, et al. 2021). Previously, the attention has been focused on, among others, hydroxymethyl-*L*-furfural (HMF) found in fruit juices, concentrates (Choudhary et al. 2021), fruit leathers (Nayaka et al. 2022), in whole dried fruits (Michalska, Honke, et al. 2016; Aktağ and Gökmen 2020), as its presence was confirmed in urine of human volunteers that consumed dried fruit-based merchandises (Murkovic and Pichler 2006; Prior, Wu, and Gu 2006). Another type of fruit-based dried product in which the Maillard reaction and caramelization products may be present are powders. These products are mainly obtained from whole fruits, pomace, juices, concentrates, and extracts (after removal of sugars and organic acids). The presence of HMF was confirmed in whole plum powders (Michalska, et al. 2016), pomace powders (Michalska,, et al. 2017), powders produced from juice and extracts (Michalska et al. 2017; Michalska et al. 2018; Michalska-Ciechanowska, et al. 2021), as well as pomace extracts (Michalska-Ciechanowska, et al. 2021). It should be highlighted that before transformation of fruit juice or concentrates into powdered form, the level of HMF should be monitored as thermal treatment of juices generates its formation that influences its content in the final powdered form (Damasceno et al. 2008; Aktağ and Gökmen 2020).

Recently, Aktağ and Gökmen (2020) provided a thorough evaluation of the potential harmful components that may be introduced to human body by the consumption of fruit-based products. Besides HMF as a contributor, authors pointed out the occurrence of α-dicarbonyl compounds (including 3-deoxyglucosone; 3-DG) in dried fruits that might be responsible for diabetes, cataract, Alzheimer disease, and mortal allergenicity in young children (Hellwig et al. 2018). Authors confirmed that the content of these constituents may be significantly higher than that of HMF in selected products.

Another hazardous compound formed via Maillard reaction during fruit drying that has been classified as a potential carcinogen and neurotoxin for humans is acrylamide (IARC 1994). The presence of this compound was confirmed in dried and smoked fruits (Amrein et al. 2007; De Paola et al. 2017; Surma et al. 2018) and fruit chips (Jung et al. 2021).

Megías-Pérez et al. (2014) and Michalska et al. (2018) confirmed the presence of Maillard reaction products formed in the early stage of reaction, i.e., furosine in fruit-based dried merchandises. These studies confirmed that monitoring of Maillard reaction progress may serve as quality indicator for fruit dried commodities.

Besides formation of Maillard reaction and caramelization products from components present in fruit products, their formation may be accelerated by the addition of, e.g., carrier agents applied for drying of fruit juices and extracts. Previously, Michalska et al. (2019) proved that application of inulin may accelerate the formation of HMF. Thus, additives used for production of fruit-based products gained by drying should be monitored in terms of the formation of hazardous components for human health.

Another type of contamination of dried fruit-based products are residues of pesticides (Varela-Martínez et al. 2019). Pesticides applied on fruit can easily enter the internal structure, leading to increased residue levels (Zhao et al. 2014). Drying may raise levels of these residues, especially those having a high-thermal stability. Li et al. (2012) proved that fat-soluble pesticides have a tendency to accumulate in essential oils; thus, such components can be easier to remove by drying. Li et al. (2012) and Tang et al. (2023) presented different approaches, including drying techniques successfully used for pesticides removal. Thus, the appropriate design of drying technique adjusted to a specific type of fruit during drying may aid in removal of pesticide residues from the dried products. Nevertheless, the constant monitoring of their presence is of high importance for safe hazard control. Finally, the preservatives added before fruit drying (sulfur dioxide or potassium metabisulfite) may provoke from mild to severe effects on people with asthma, including bronchoconstriction, urticarial, and anaphylaxis (Shoaei, Heshmati, and Khorshidi 2019). In this scenario, ensuring appropriate labeling of dried fruit products holds utmost significance for consumers.

8.5 RECOMMENDATIONS AND FUTURE TRENDS

Currently, numerous researches have been made for developing new strategies for the manufacturing of dried-fruit commodities. Drying fruit-based merchandises become a challenge due to the diversity of fruit matrices submitted to dehydration that differ in terms of the structure and composition of biologically active components with a risk of presence of process contaminants. Numerous advanced drying techniques and emerging technologies have been developed to improve the physical properties of dried fruit products. Recently, due to the increased consumers' awareness about healthy lifestyle, numerous approaches were made toward superior retention of nutrients and biologically active substances during drying.

Fruits, due to the presence of sugars, organic acids, pectin, fiber, and bioactives, including vitamins and antioxidants, are a complicated matrix for processing. Thus, alterations caused by drying techniques, including pre- and post-drying treatment, require thorough analysis for food safety reasons. Currently, the challenges for the scientists rely on the establishing methods that will allow for quality assessment of fruit-based products that can be used for different processed materials (Richter Reis et al. 2022), including numerous issues of product characteristics. Considering that nowadays consumers are more aware of the quality of products and are looking for personalized products in easy-to-handle form, the introduction to the market of new dried fruit-based products requires consideration of not only physical and chemical characteristics but also a sustainable production chain within reduction of energy consumption, CO_2 emissions, and the influence of packaging and storage (including biodegradable materials) on the final product quality (Figure 8.4). Knowing the limitations of different drying technologies, to successfully achieve this goal, more research is needed to link this multidisciplinary approach, especially in terms of up-scaling the emerging drying technologies in conjunction with artificial intelligence (Raghavan et al. 2022) that enable to offer more personalized products that beyond their nutritional value will also provide the required biological properties.

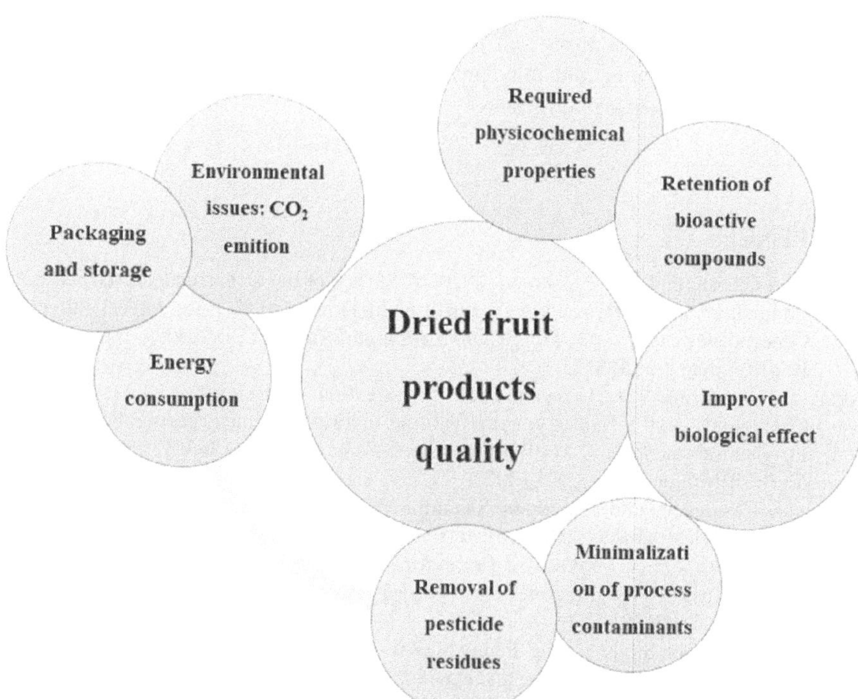

FIGURE 8.4 The factors influencing the overall dried fruit products quality.

8.6 CONCLUDING REMARKS

Taking into consideration an increasing awareness of consumers about food quality, safety, and sustainability during production of dried fruit commodities, an industry is facing a challenge due to the continuously changing expectations toward novel healthy products. Furthermore, with the growing understanding linked to the health benefits of nutrients and functional ingredients present in fruit-based products, modern consumers tend to opt for commodities that promote well-being and have a positive effect on their health. Currently, the reduction of industrial environmental footprints and increased energy consumption being recognized worldwide is of high importance. The newest trends connected with fruit processing are focused on the linkage between processing type, product, and consumer. Additionally, the matter of reducing food loss and waste has gained significant public attention as a crucial component of global initiatives aimed at addressing global hunger and enhancing food security. From this point of view, the role of drying as a technique that enables conversion of fresh, perishable merchandises into more stable and safe products is being revalorized as a new strategy for obtaining new, easy-to-handle, and nutritious fruit products that will meet modern consumers' expectations. The advancement in drying techniques coupled with novel pre- and post-drying methods may broaden the possibilities for offering consumers new merchandises, in terms of form and also chemical properties. Drying-induced alterations that occur during processing of fruit components should be recognized as the interactions and interrelations between these constituents that occur during processing and may provoke different biological responses, other than single-component behavior.

REFERENCES

Aktağ, Işıl Gürsul, and Vural Gökmen. 2020. 'A Survey of the Occurrence of α-Dicarbonyl Compounds and 5-Hydroxymethylfurfural in Dried Fruits, Fruit Juices, Puree and Concentrates'. *Journal of Food Composition and Analysis* 91: 103523. https://doi.org/10.1016/j.jfca.2020.103523

Aktag, Işıl Gürsul, and Vural Gökmen. 2021. 'Investigations on the Formation of α-Dicarbonyl Compounds and 5-Hydroxymethylfurfural in Fruit Products during Storage: New Insights into the Role of Maillard Reaction'. *Food Chemistry* 363: 130280. https://doi.org/10.1016/j.foodchem.2021.130280

Alasalvar, Cesarettin, and Fereidoon Shahidi. 2013. 'Composition, Phytochemicals, and Beneficial Health Effects of Dried Fruits: An Overview'. In *Dried Fruits: Phytochemicals and Health Effects*, edited by Cesarettin Alasalvar and Fereidoon Shahidi. Iowa: Wiley. https://www.wiley.com/en-us/Dried+Fruits%3A+Phytochemicals+and+Health+Effects-p-9780813811734

Alasalvar, Cesarettin, Sui K. Chang, Penny M. Kris-Etherton, Valerie K. Sullivan, Kristina S. Petersen, Marta Guasch-Ferré, and David J. A. Jenkins. 2023. 'Dried Fruits: Bioactives, Effects on Gut Microbiota, and Possible Health Benefits—An Update'. *Nutrients* 15 (7). https://doi.org/10.3390/nu15071611

Alp, Duygu, and Özcan Bulantekin. 2021. 'The Microbiological Quality of Various Foods Dried by Applying Different Drying Methods: A Review'. *European Food Research and Technology* 247 (6): 1333–43. https://doi.org/10.1007/s00217-021-03731-z

Amrein, Thomas M., Luca Andres, Felix Escher, and Renato Amadò. 2007. 'Occurrence of Acrylamide in Selected Foods and Mitigation Options'. *Food Additives & Contaminants* 24: 13–25. https://doi.org/10.1080/02652030701242558

Belwal, Tarun, Christian Cravotto, M. A. Prieto, Petras Rimantas Venskutonis, Maria Daglia, Hari Prasad Devkota, Alessandra Baldi, et al. 2022. 'Effects of Different Drying Techniques on the Quality and Bioactive Compounds of Plant-Based Products: A Critical Review on Current Trends'. *Drying Technology* 40 (8): 1539–61. https://doi.org/10.1080/07373937.2022.2068028

Bhandari, B. 2013. '1 - Introduction to Food Powders'. In *Handbook of Food Powders*, edited by Bhesh Bhandari, Nidhi Bansal, Min Zhang, and Pierre Schuck, 1–25. Philadelphia, PA: Woodhead Publishing. https://doi.org/10.1533/9780857098672.1

Boateng, Isaac Duah. 2022. 'Recent Processing of Fruits and Vegetables Using Emerging Thermal and Non-Thermal Technologies. A Critical Review of Their Potentialities and Limitations on Bioactives, Structure, and Drying Performance'. *Critical Reviews in Food Science and Nutrition* 0 (0): 1–35. https://doi.org/10.1080/10408398.2022.2140121

Calín-Sánchez, Ángel, Adam Figiel, Francisca Hernández, Pablo Melgarejo, Krzysztof Lech, and Ángel A. Carbonell-Barrachina. 2013. 'Chemical Composition, Antioxidant Capacity, and Sensory Quality of Pomegranate (*Punica granatum* L.) Arils and Rind as Affected by Drying Method'. *Food and Bioprocess Technology* 6 (7): 1644–54. https://doi.org/10.1007/s11947-012-0790-0

Calín-Sánchez, Ángel, Leontina Lipan, Marina Cano-Lamadrid, Abdolreza Kharaghani, Klaudia Masztalerz, Ángel A. Carbonell-Barrachina, and Adam Figiel. 2020. 'Comparison of Traditional and Novel Drying Techniques and Its Effect on Quality of Fruits, Vegetables and Aromatic Herbs'. *Foods* 9 (9): 1261. https://doi.org/10.3390/foods9091261

Cano-Lamadrid, Marina, Krzysztof Lech, Anna Michalska, Malwina Wasilewska, Adam Figiel, Aneta Wojdyło, and Ángel A. Carbonell-Barrachina. 2017. 'Influence of Osmotic Dehydration Pre-Treatment and Combined Drying Method on Physico-Chemical and Sensory Properties of Pomegranate Arils, Cultivar Mollar de Elche'. *Food Chemistry* 232: 306–15. https://doi.org/10.1016/j.foodchem.2017.04.033

Castillo, María Dolores del, Amaia Iriondo-DeHond, Maite Iriondo-DeHond, Ileana Gonzalez, Alejandra Medrano, Rosana Filip, and Jaime Uribarri. 2021. 'Healthy Eating Recommendations: Good for Reducing Dietary Contribution to the Body's Advanced Glycation/Lipoxidation End Products Pool?'. *Nutrition Research Reviews* 34 (1): 48–63. https://doi.org/10.1017/S0954422420000141

Chang, Sui Kiat, Cesarettin Alasalvar, and Fereidoon Shahidi. 2016. 'Review of Dried Fruits: Phytochemicals, Antioxidant Efficacies, and Health Benefits'. *Journal of Functional Foods* 21: 113–32. https://doi.org/10.1016/j.jff.2015.11.034

Choudhary, Ankit, Vikas Kumar, Satish Kumar, Ishrat Majid, Poonam Aggarwal, and Sheenam Suri. 2021. '5-Hydroxymethylfurfural (HMF) Formation, Occurrence and Potential Health Concerns: Recent Developments'. *Toxin Reviews* 40 (4): 545–61. https://doi.org/10.1080/15569543.2020.1756857

Coimbra, Pedro Paulo Saldanha, Flávio de Souza Neves Cardoso, and Édira Castello Branco de Andrade Gonçalves. 2021. 'Spray-Drying Wall Materials: Relationship with Bioactive Compounds'. *Critical Reviews in Food Science and Nutrition* 61 (17): 2809–26. https://doi.org/10.1080/10408398.2020.1786354

Damasceno, Leandro F., Fabiano A. N. Fernandes, Margarida M. A. Magalhães, and Edy S. Brito. 2008. 'Non-Enzymatic Browning in Clarified Cashew Apple Juice during Thermal Treatment: Kinetics and Process Control'. *Food Chemistry* 106 (1): 172–79. https://doi.org/10.1016/j.foodchem.2007.05.063

De Paola, Eleonora Laura, Giuseppe Montevecchi, Francesca Masino, Davide Garbini, Martino Barbanera, and Andrea Antonelli. 2017. 'Determination of Acrylamide in Dried Fruits and Edible Seeds Using QuEChERS Extraction and LC Separation with MS Detection'. *Food Chemistry* 217: 191–95. https://doi.org/10.1016/j.foodchem.2016.08.101

FAO. 2021. 'Tips for the Preservation of Fruits and Vegetables to Reduce Food Waste and Improve Shelf-Life | Food and Agriculture Organization of the United Nations | Food Loss Reduction CoP | Food and Agriculture Organization of the United Nations'. 2021.

Fazaeli, Mahboubeh, Zahra Emam-Djomeh, Ahmad Kalbasi Ashtari, and Mahmoud Omid. 2012. 'Effect of Spray Drying Conditions and Feed Composition on the Physical Properties of Black Mulberry Juice Powder'. *Food and Bioproducts Processing* 90 (4): 667–75. https://doi.org/10.1016/j.fbp.2012.04.006

Fernandes, Fabiano A. N., and Sueli Rodrigues. 2023. 'Ultrasound Applications in Drying of Fruits from a Sustainable Development Goals Perspective'. *Ultrasonics Sonochemistry* 96: 106430. https://doi.org/10.1016/j.ultsonch.2023.106430

Figiel, Adam. 2010. 'Drying Kinetics and Quality of Beetroots Dehydrated by Combination of Convective and Vacuum-Microwave Methods'. *Journal of Food Engineering* 98 (4): 461–70. https://doi.org/10.1016/j.jfoodeng.2010.01.029

Figiel, Adam, and Anna Michalska. 2017. 'Overall Quality of Fruits and Vegetables Products Affected by the Drying Processes with the Assistance of Vacuum-Microwaves'. *International Journal of Molecular Sciences* 18 (1): 71. https://doi.org/10.3390/ijms18010071

Global Report of Dried Fruits. 2022. 'Dry Fruits Market Size, Share & Growth - Global Report'. Report ID FBI100544. 2022. https://www.fortunebusinessinsights.com/industry-reports/dry-fruits-market-100544

Goldfein, Kara R., and Joanne L. Slavin. 2015. 'Why Sugar Is Added to Food: Food Science'. *Comprehensive Reviews in Food Science and Food Safety* 14 (5): 644–56. https://doi.org/10.1111/1541-4337.12151

Gurak, Poliana Deyse, Gicele Sbardelotto De Bona, Isabel Cristina Tessaro, and Ligia Damasceno Ferreira Marczak. 2014. 'Jaboticaba Pomace Powder Obtained as a Co-Product of Juice Extraction: A Comparative Study of Powder Obtained from Peel and Whole Fruit'. *Food Research International* 62: 786–92. https://doi.org/10.1016/j.foodres.2014.04.042

Hellwig, Michael, Sabrina Gensberger-Reigl, Thomas Henle, and Monika Pischetsrieder. 2018. 'Food-Derived 1,2-Dicarbonyl Compounds and Their Role in Diseases'. *Seminars in Cancer Biology* 49: 1–8. https://doi.org/10.1016/j.semcancer.2017.11.014

Hendrysiak, Aleksandra, Jessica Brzezowska, Nancy Nicolet, Dimitri Bocquel, Wilfried Andlauer, and Anna Michalska-Ciechanowska. 2023. 'Juice Powders from Rosehip (*Rosa canina* L.): Physical, Chemical, and Antiglycation Properties'. *Molecules* 28 (4): 1674. https://doi.org/10.3390/molecules28041674

IARC. 1994. 'Some Industrial Chemicals, Acrylamide. IARC Monographs on the Evaluation of Carcinogenic Risks to Humans'. *Geneva (Switzerland); World Health Organization*, no. 60: 389–433.

Iranshahi, Kamran, Marios Psarianos, Donato Rubinetti, Daniel I. Onwude, Oliver K. Schlüter, and Thijs Defraeye. 2023. 'Impact of Pre-Treatment Methods on the Drying Kinetics, Product Quality, and Energy Consumption of Electrohydrodynamic Drying of Biological Materials'. *Innovative Food Science & Emerging Technologies* 85: 103338. https://doi.org/10.1016/j.ifset.2023.103338

Jung, Soyoung, Yunjeong Yi, Minsu Chang, Insuk Lee, Misun Kim, Hyunjung Kim, Insil Yoo, and Yongseung Shin. 2021. 'An Improved Extraction Method for Acrylamide Determination in Fruit and Vegetable Chips through Enzyme Addition'. *Food Chemistry* 360: 129740. https://doi.org/10.1016/j.foodchem.2021.129740

Karam, Marie Céleste, Jeremy Petit, David Zimmer, Elie Baudelaire Djantou, and Joël Scher. 2016. 'Effects of Drying and Grinding in Production of Fruit and Vegetable Powders: A Review'. *Journal of Food Engineering* 188: 32–49. https://doi.org/10.1016/j.jfoodeng.2016.05.001

Kowalska, Hanna, Magdalena Trusinska, Katarzyna Rybak, Artur Wiktor, Dorota Witrowa-Rajchert, and Malgorzata Nowacka. 2023. 'Shaping the Properties of Osmo-Dehydrated Strawberries in Fruit Juice Concentrates'. *Applied Sciences* 13 (4): 2728. https://doi.org/10.3390/app13042728

Kroh, Lothar W. 1994. 'Caramelisation in Food and Beverages'. *Food Chemistry*, Interactions in Beverages, 51 (4): 373–79. https://doi.org/10.1016/0308-8146(94)90188-0

Kudra, Tadeusz. 2008. 'Energy Aspects in Food Dehydration'. In *Advances in Food Dehydration*. Boca Raton, FL: CRC Press.

Labuschagne, Philip. 2018. 'Impact of Wall Material Physicochemical Characteristics on the Stability of Encapsulated Phytochemicals: A Review'. *Food Research International* 107: 227–47. https://doi.org/10.1016/j.foodres.2018.02.026

Leon Dharmadurai, P., S. Vasanthaseelan, R. Bharathwaaj, Vijayakumar Dharmaraj, K. Gnanasekaran, D. Balaji, and Ravishankar Sathyamurthy. 2022. 'A Comparative Study on Solar Dryer Using External Reflector for Drying Grapes'. *Materials Today: Proceedings*, 2nd International Conference on Functional Material, Manufacturing and Performances (ICFMMP-2021), 50: 552–59. https://doi.org/10.1016/j.matpr.2020.11.197

Lewicki, Piotr P. 2006. 'Design of Hot Air Drying for Better Foods'. *Trends in Food Science & Technology*, EFFoST Warsaw 2004, 17 (4): 153–63. https://doi.org/10.1016/j.tifs.2005.10.012

Li, Yuncheng, Bining Jiao, Qiyang Zhao, Chengqiu Wang, Yong Gong, Yaohai Zhang, and Weijun Chen. 2012. 'Effect of Commercial Processing on Pesticide Residues in Orange Products'. *European Food Research and Technology* 234 (3): 449–56. https://doi.org/10.1007/s00217-011-1651-1

Majerska, Joanna, Anna Michalska, and Adam Figiel. 2019. 'A Review of New Directions in Managing Fruit and Vegetable Processing By-Products'. *Trends in Food Science & Technology* 88: 207–19. https://doi.org/10.1016/j.tifs.2019.03.021

Maskan, Medeni. 2001. 'Kinetics of Colour Change of Kiwifruits during Hot Air and Microwave Drying'. *Journal of Food Engineering* 48 (2): 169–75. https://doi.org/10.1016/S0260-8774(00)00154-0

Masood, Muhammad, Shahzad Zafar Iqbal, Muhammad Rafique Asi, and Noeen Malik. 2015. 'Natural Occurrence of Aflatoxins in Dry Fruits and Edible Nuts'. *Food Control* 55: 62–65. https://doi.org/10.1016/j.foodcont.2015.02.041

Masztalerz, Klaudia, Krzysztof Lech, Aneta Wojdyło, Paulina Nowicka, Anna Michalska-Ciechanowska, and Adam Figiel. 2021. 'The Impact of the Osmotic Dehydration Process and Its Parameters on the Mass Transfer and Quality of Dried Apples'. *Drying Technology* 39 (8): 1074–86. https://doi.org/10.1080/07373937.2020.1741607

Megías-Pérez, Roberto, Juliana Gamboa-Santos, Ana Cristina Soria, Mar Villamiel, and Antonia Montilla. 2014. 'Survey of Quality Indicators in Commercial Dehydrated Fruits'. *Food Chemistry* 150: 41–48. https://doi.org/10.1016/j.foodchem.2013.10.141

Michalska, Anna, Aneta Wojdyło, Grzegorz P. Łysiak, and Adam Figiel. 2017. 'Chemical Composition and Antioxidant Properties of Powders Obtained from Different Plum Juice Formulations'. *International Journal of Molecular Sciences* 18 (1): 176. https://doi.org/10.3390/ijms18010176

Michalska, Anna, Aneta Wojdyło, Grzegorz P. Łysiak, Krzysztof Lech, and Adam Figiel. 2017. 'Functional Relationships between Phytochemicals and Drying Conditions during the Processing of Blackcurrant Pomace into Powders'. *Advanced Powder Technology* 28 (5): 1340–48. https://doi.org/10.1016/j.apt.2017.03.002

Michalska, Anna, Aneta Wojdyło, Jessica Brzezowska, Joanna Majerska, and Ewa Ciska. 2019. 'The Influence of Inulin on the Retention of Polyphenolic Compounds during the Drying of Blackcurrant Juice'. *Molecules* 24 (22). https://doi.org/10.3390/molecules24224167

Michalska, Anna, Aneta Wojdyło, Joanna Honke, Ewa Ciska, and Wilfried Andlauer. 2018. 'Drying-Induced Physico-Chemical Changes in Cranberry Products'. *Food Chemistry* 240: 448–55. https://doi.org/10.1016/j.foodchem.2017.07.050

Michalska, Anna, Aneta Wojdyło, Krzysztof Lech, Grzegorz P. Łysiak, and Adam Figiel. 2016. 'Physicochemical Properties of Whole Fruit Plum Powders Obtained Using Different Drying Technologies'. *Food Chemistry* 207: 223–32. https://doi.org/10.1016/j.foodchem.2016.03.075

Michalska, Anna, Joanna Honke, Grzegorz Łysiak, and Wilfried Andlauer. 2016. 'Effect of Drying Parameters on the Formation of Early and Intermediate Stage Products of the Maillard Reaction in Different Plum (*Prunus domestica* L.) Cultivars'. *LWT - Food Science and Technology* 65: 932–38. https://doi.org/10.1016/j.lwt.2015.09.015

Michalska-Ciechanowska, Anna, Aleksandra Hendrysiak, Jessica Brzezowska, Aneta Wojdyło, and Agnieszka Gajewicz-Skretna. 2021a. 'How Do the Different Types of Carrier and Drying Techniques Affect the Changes in Physico-Chemical Properties of Powders from Chokeberry Pomace Extracts?'. *Foods* 10 (8): 1864. https://doi.org/10.3390/foods10081864

Michalska-Ciechanowska, Anna, Jessica Brzezowska, Aneta Wojdyło, Agnieszka Gajewicz-Skretna, Ewa Ciska, and Joanna Majerska. 2021b. 'Chemometric Contribution for Deeper Understanding of Thermally-Induced Changes of Polyphenolics and the Formation of Hydroxymethyl-L-Furfural in Chokeberry Powders'. *Food Chemistry* 342: 128335. https://doi.org/10.1016/j.foodchem.2020.128335

Miletić, Nemanja, Branko Popović, Olga Mitrović, Miodrag Kandić, and Aleksandar Leposavić. 2014. 'Phenolic Compounds and Antioxidant Capacity of Dried and Candied Fruits Commonly Consumed in Serbia'. *Czech Journal of Food Sciences* 32 (4): 360–98. https://doi.org/10.17221/166/2013-CJFS

Mohammed, Ssemwanga, Makule Edna, and Kayondo Siraj. 2020. 'The Effect of Traditional and Improved Solar Drying Methods on the Sensory Quality and Nutritional Composition of Fruits: A Case of Mangoes and Pineapples'. *Heliyon* 6 (6): e04163. https://doi.org/10.1016/j.heliyon.2020.e04163

Mugi, Vishnuvardhan Reddy, Pritam Das, Ramakrishna Balijepalli, and Chandramohan Vp. 2022. 'A Review of Natural Energy Storage Materials Used in Solar Dryers for Food Drying Applications'. *Journal of Energy Storage* 49: 104198. https://doi.org/10.1016/j.est.2022.104198

Mujumdar, Arun S., and Hong-Wei Xiao. 2019. *Advanced Drying Technologies for Foods*. Boca Raton, FL: CRC Press.

Murkovic, Michael, and Nicole Pichler. 2006. 'Analysis of 5-Hydroxymethylfurfual in Coffee, Dried Fruits and Urine'. *Molecular Nutrition & Food Research* 50 (9): 842–46. https://doi.org/10.1002/mnfr.200500262

Nayaka, V. S. K., R. B. Tiwari, C. K. Narayana, K. Ranjitha, Azeez Shamina, C. Vasugi, R. Venugopalan, S. Bhuvaneswari, and O. J. Sujayasree. 2022. 'Comparative Effect of Different Sugars Instigating Non-Enzymatic Browning and Maillard Reaction Products in Guava Fruit Leather'. *Journal of Horticultural Sciences* 17 (1): 174–83.

Nizamlioglu, Nizam Mustafa, Sulhattin Yasar, and Yasemin Bulut. 2022. 'Chemical versus Infrared Spectroscopic Measurements of Quality Attributes of Sun or Oven Dried Fruit Leathers from Apple, Plum and Apple-Plum Mixture'. *LWT* 153: 112420. https://doi.org/10.1016/j.lwt.2021.112420

Ogawa, Yukiharu, Natthawuddhi Donlao, Sukanya Thuengtung, Jinhu Tian, Yidi Cai, Florencio C Reginio, Sunantha Ketnawa, Nami Yamamoto, and Masatsugu Tamura. 2018. 'Impact of Food Structure and Cell Matrix on Digestibility of Plant-Based Food'. *Current Opinion in Food Science*, Food Chemistry and Biochemistry * Food Bioprocessing, 19: 36–41. https://doi.org/10.1016/j.cofs.2018.01.003

Oliveira, Ana, Ana L Amaro, and Manuela Pintado. 2018. 'Impact of Food Matrix Components on Nutritional and Functional Properties of Fruit-Based Products'. *Current Opinion in Food Science*, Foodomics Technologies 2018 * Innovations in Food Science, 22: 153–59. https://doi.org/10.1016/j.cofs.2018.04.002

Özcan, Mehmet Musa, Fahad Al Juhaimi, Isam A. Mohamed Ahmed, Nurhan Uslu, Elfadil E. Babiker, and Kashif Ghafoor. 2020. 'Effect of Microwave and Oven Drying Processes on Antioxidant Activity, Total Phenol and Phenolic Compounds of Kiwi and Pepino Fruits'. *Journal of Food Science and Technology* 57 (1): 233–42. https://doi.org/10.1007/s13197-019-04052-6

Petikirige, Jasmine, Azharul Karim, and Graeme Millar. 2022. 'Effect of Drying Techniques on Quality and Sensory Properties of Tropical Fruits'. *International Journal of Food Science & Technology* 57 (11): 6963–79. https://doi.org/10.1111/ijfs.16043

Piga, Antonio, Alessandra Del Caro, and Giampaola Corda. 2003. 'From Plums to Prunes: Influence of Drying Parameters on Polyphenols and Antioxidant Activity'. *Journal of Agricultural and Food Chemistry* 51 (12): 3675–81. https://doi.org/10.1021/jf021207+

Prior, Ronald L., Xianli Wu, and Liwei Gu. 2006. 'Identification and Urinary Excretion of Metabolites of 5-(Hydroxymethyl)-2-Furfural in Human Subjects Following Consumption of Dried Plums or Dried Plum Juice'. *Journal of Agricultural and Food Chemistry* 54 (10): 3744–49. https://doi.org/10.1021/jf0601113

Purohit, Pallav, Atul Kumar, and Tara Chandra Kandpal. 2006. 'Solar Drying vs. Open Sun Drying: A Framework for Financial Evaluation'. *Solar Energy* 80 (12): 1568–79. https://doi.org/10.1016/j.solener.2005.12.009

Raghavan, Vijaya, Alex Martynenko, and Shivanand S. Shirkole. 2022. 'Role of Drying in Food Quality, Security, and Sustainability'. *Drying Technology* 40 (8): 1499. https://doi.org/10.1080/07373937.2022.2089819

Rane, Riddhi, Thankamani Marar, Sachin K Sonawane, and Ashish Dabade. 2022. 'A Review on Instant Controlled Pressure Drop Technology (DIC) Associated with Drying Technology and Effect on Quality Characteristics'. *Food Chemistry Advances* 1: 100114. https://doi.org/10.1016/j.focha.2022.100114

Ratti, C. 2001. 'Hot Air and Freeze-Drying of High-Value Foods: A Review'. *Journal of Food Engineering* 49 (4): 311–19. https://doi.org/10.1016/S0260-8774(00)00228-4

Raveendran, Dhivya, Madhura Bhagwat, Duggonahally Veeresh Chidanand, Sugumar Anandakumar, and Chikkaballapur Krishnappa Sunil. 2022. 'Highlight on Drying Fruit Slices with Better Retention of Bioactive Compounds'. *Journal of Food Process Engineering* 45 (8): e14048. https://doi.org/10.1111/jfpe.14048

Ray, Sohini, Utpal Raychaudhuri, and Runu Chakraborty. 2016. 'An Overview of Encapsulation of Active Compounds Used in Food Products by Drying Technology'. *Food Bioscience* 13: 76–83. https://doi.org/10.1016/j.fbio.2015.12.009

Rezvankhah, Amir, Zahra Emam-Djomeh, and Gholamreza Askari. 2020. 'Encapsulation and Delivery of Bioactive Compounds Using Spray and Freeze-Drying Techniques: A Review'. *Drying Technology* 38 (1–2): 235–58. https://doi.org/10.1080/07373937.2019.1653906

Richter Reis, Felipe, Caroline Marques, Ana Carolina Sales de Moraes, and Maria Lucia Masson. 2022. 'Trends in Quality Assessment and Drying Methods Used for Fruits and Vegetables'. *Food Control* 142: 109254. https://doi.org/10.1016/j.foodcont.2022.109254

Sabarez, Henry. 2021. '3.04 - Advanced Drying Technologies of Relevance in the Food Industry'. In *Innovative Food Processing Technologies*, edited by Kai Knoerzer and Kasiviswanathan Muthukumarappan, 64–81. Oxford: Elsevier. https://doi.org/10.1016/B978-0-08-100596-5.23042-4

Sadler, Michele Jeanne, Sigrid Gibson, Kevin Whelan, Marie-Ann Ha, Julie Lovegrove, and Jennette Higgs. 2019. 'Dried Fruit and Public Health – What Does the Evidence Tell Us?'. *International Journal of Food Sciences and Nutrition* 70 (6): 675–87. https://doi.org/10.1080/09637486.2019.1568398

Samborska, Katarzyna, Sadegh Poozesh, Alicja Barańska, Mariia Sobulska, Aleksandra Jedlińska, Cordin Arpagaus, Narjes Malekjani, and Seid Mahdi Jafari. 2022. 'Innovations in Spray Drying Process for Food and Pharma Industries'. *Journal of Food Engineering* 321: 110960. https://doi.org/10.1016/j.jfoodeng.2022.110960

Samborska, Kinga, Lovisa Eliasson, Agata Marzec, Jolanta Kowalska, Dariusz Piotrowski, Andrzej Lenart, and Hanna Kowalska. 2019. 'The Effect of Adding Berry Fruit Juice Concentrates and By-Product Extract to Sugar Solution on Osmotic Dehydration and Sensory Properties of Apples'. *Journal of Food Science and Technology* 56 (4): 1927–38. https://doi.org/10.1007/s13197-019-03658-0

Sheng, Lina, and Luxin Wang. 2023. 'Approaches for a More Microbiologically and Chemically Safe Dried Fruit Supply Chain'. *Current Opinion in Biotechnology* 80: 102912. https://doi.org/10.1016/j.copbio.2023.102912

Shishir, Mohammad Rezaul Islam, and Wei Chen. 2017. 'Trends of Spray Drying: A Critical Review on Drying of Fruit and Vegetable Juices'. *Trends in Food Science & Technology* 65: 49–67. https://doi.org/10.1016/j.tifs.2017.05.006

Shoaei, F., A. Heshmati, and M. Khorshidi. 2019. 'Risk Assessment of Sulphite Intake through Dried Fruit Consumption in Hamadan, Iran'. *Journal of Food Quality and Hazards Control*. https://doi.org/10.18502/jfqhc.6.3.1386

Silvan, Jose Manuel, Anna Michalska-Ciechanowska, and Adolfo J. Martinez-Rodriguez. 2020. 'Modulation of Antibacterial, Antioxidant, and Anti-Inflammatory Properties by Drying of *Prunus Domestica* L. Plum Juice Extracts'. *Microorganisms* 8 (1): 119. https://doi.org/10.3390/microorganisms8010119

Surma, Magdalena, Anna Sadowska-Rociek, and Ewa Cieślik. 2018. 'Assessment of Thermal Processing Contaminant Levels in Dried and Smoked Fruits'. *European Food Research and Technology* 244 (9): 1533–43. https://doi.org/10.1007/s00217-018-3067-7

Tamanna, Nahid, and Niaz Mahmood. 2015. 'Food Processing and Maillard Reaction Products: Effect on Human Health and Nutrition'. *International Journal of Food Science* 2015: e526762. https://doi.org/10.1155/2015/526762

Tang, T., M. Zhang, R. Ju, A.S. Mujumdar, and D. Yu. 2023. 'Novel Drying and Pretreatment Methods for Control of Pesticide Residues in Fruits and Vegetables: A Review'. *Drying Technology* 41 (2): 151–71. https://doi.org/10.1080/07373937.2022.2041029

Thongcharoenpipat, Chanoknate, and Ram Yamsaengsung. 2023. 'Microwave-Assisted Vacuum Frying of Durian Chips: Impact of Ripening Level on the Drying Rate, Physio-Chemical Characteristics, and Acceptability'. *Food and Bioproducts Processing* 138: 40–52. https://doi.org/10.1016/j.fbp.2023.01.001

Tontul, Ismail, and Ayhan Topuz. 2017. 'Spray-Drying of Fruit and Vegetable Juices: Effect of Drying Conditions on the Product Yield and Physical Properties'. *Trends in Food Science & Technology* 63: 91–102. https://doi.org/10.1016/j.tifs.2017.03.009

Vadivambal, R., and D. S. Jayas. 2007. 'Changes in Quality of Microwave-Treated Agricultural Products—A Review'. *Biosystems Engineering* 98 (1): 1–16. https://doi.org/10.1016/j.biosystemseng.2007.06.006

Varela-Martínez, Diana Angélica, Miguel Ángel González-Curbelo, Javier González-Sálamo, and Javier Hernández-Borges. 2019. 'Analysis of Multiclass Pesticides in Dried Fruits Using QuEChERS-Gas Chromatography Tandem Mass Spectrometry'. *Food Chemistry* 297: 124961. https://doi.org/10.1016/j.foodchem.2019.124961

Weaver, Connie M., Johanna Dwyer, Victor L. Fulgoni, Janet C. King, Gilbert A. Leveille, Ruth S. MacDonald, Jose Ordovas, and David Schnakenberg. 2014. 'Processed Foods: Contributions to Nutrition'. *The American Journal of Clinical Nutrition* 99 (6): 1525–42. https://doi.org/10.3945/ajcn.114.089284

Wojdyło, Aneta, Adam Figiel, Krzysztof Lech, Paulina Nowicka, and Jan Oszmiański. 2014. 'Effect of Convective and Vacuum–Microwave Drying on the Bioactive Compounds, Color, and Antioxidant Capacity of Sour Cherries'. *Food and Bioprocess Technology* 7 (3): 829–41. https://doi.org/10.1007/s11947-013-1130-8

Xu, Baoguo, Essodézam Sylvain Tiliwa, Weiqiang Yan, S. M. Roknul Azam, Benxi Wei, Cunshan Zhou, Haile Ma, and Bhesh Bhandari. 2022. 'Recent Development in High Quality Drying of Fruits and Vegetables Assisted by Ultrasound: A Review'. *Food Research International* 152: 110744. https://doi.org/10.1016/j.foodres.2021.110744

Yang, Kai-Min, and Po-Yuan Chiang. 2019. 'Effects of Smoking Process on the Aroma Characteristics and Sensory Qualities of Dried Longan'. *Food Chemistry* 287: 133–38. https://doi.org/10.1016/j.foodchem.2019.02.017

Zhao, Xiangsheng, Weijun Kong, Jianhe Wei, and Meihua Yang. 2014. 'Gas Chromatography with Flame Photometric Detection of 31 Organophosphorus Pesticide Residues in Alpinia Oxyphylla Dried Fruits'. *Food Chemistry* 162: 270–76. https://doi.org/10.1016/j.foodchem.2014.04.060

Zhu, Jinxing, Yuan Liu, Chuanhe Zhu, and Min Wei. 2022. 'Effects of Different Drying Methods on the Physical Properties and Sensory Characteristics of Apple Chip Snacks'. *LWT* 154: 112829. https://doi.org/10.1016/j.lwt.2021.112829

9 Techno-Economic Evaluation for Cost-Effective Drying of Different Fruit Products

Viplav Hari Pise
Institute of Chemical Technology Mumbai
Maharashtra, India

9.1 INTRODUCTION

Fruits, the seed-bearing part of the plant developed from flowers, are consumed for taste, health benefits, and energy sources. The production of fruits across the globe and in India in 2020 was reported to be about 887.027 million tonnes and 105.971 million tonnes (FAO 2022). There is a huge gap between the recommended fruit consumption (approximately 400 g per day) and against the actual. This could be due to several reasons like uneven distribution, seasonal production, transport requirements, low shelf-life, and so on. Considering the overall statistics of production, consumption rate (recommended and actual), and overall distribution, a huge surplus quantity is available for different levels of processing at several locations in significant quantities. The consumption and demand for fruits are continuously increasing, which also demands processing to overcome perishability issues and seasonal availability (Bhatkar et al. 2022b). The classification of fruits as pre-processed, semi- or partially-processed, processed, and ultra-processed is discussed in detail by Bhatkar et al. (2022a). There are several ways of processing fruits in the form of fresh cut, dehydrated pieces, fruit powders, concentrates, pulps, and purees, which are in high demand in the health and wellness sector (Bhatkar et al. 2022a). This processing is required to preserve the surplus produce within the shelf-life for resource utilization and to serve seasonal fruit reserve requirements. Presently, fruit processing is focused on developing novel processed products with safety to meet consumer requirements, preservation of nutritional characteristics, and meeting quality parameters (Tylewicz et al. 2019; Giannakourou and Tsironi 2021).

Drying is one of the most widely used and oldest methods for preserving any agricultural produce, including fruits. It is also an extensively researched process with several technologies meeting maturity; however, innovative approaches are still required in the domain of drying to meet new challenges, global competitiveness, energy efficiency, environmental impacts, and meeting desired product qualities.

DOI: 10.1201/9781003402916-9

Drying is an industrial preservation method implemented for agricultural products to minimize degradation due to biochemical reactions (Simal et al. 2005). The process can simply be stated as the removal of water to reduce the moisture content and lower the water activity; however, it is a highly complex operation considering the tissue structures of the fruits and the removal of water from within the cell structures. Considering the associated complexity, to implement a dryer, the balance between technological advancement and product value addition must be well balanced for feasibility. This section addresses the background for any techno-economic feasibility study (TEFS), including various parameters centred, cost considerations, projection of incomes, cost-benefit analysis, and profitability indices around the optimization of processes.

9.2 NEEDS FOR TECHNO-ECONOMIC EVALUATION OF THE DRYING PROCESS

A process is defined considering the steps of unit operations involved and the objective to be achieved. There is scope for improvement in every unit operation at any given time. However, better technology is inherently associated with higher capital and operational costs. It is, therefore, pertinent to carry out a TEFS on various processes stepwise, considering the value addition at each step. The process's technical, operational, and economic aspects are highly desired to ensure the acceptability of the overall process. The detailed TEFS ensures the proper dryer selection, safeguards the investment, and prevents tangible and intangible losses (Subramanian and Taghizadeh-Hesary 2021). This study is required to balance the technological advances, product quality, capital investment associated, operational expenses, and profits (Figure 9.1).

The risk of failure in implementing the equipment depends on intrinsic and extrinsic factors. *Intrinsic factors* like the optimum performance of the equipment, operating parameters, application energy and its mode, loading capacity, and mass transfer rate determine the quality of the products contributing. *Extrinsic factors* like ease of operation, utility and workforce requirements, batch duration, capital cost, and annual operational days ensure the unit's profitability. The TEFS for any process determines the risks associated with units and highlights the areas of concern to be

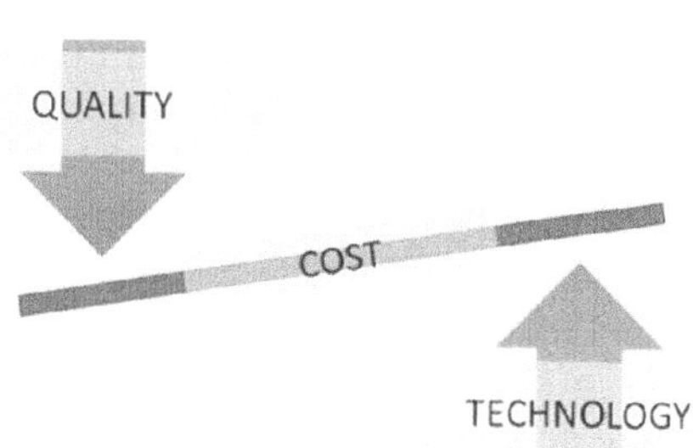

FIGURE 9.1 Techno-economic feasibility study (TEFS).

worked upon to meet the project's economic viability and financial feasibility. TEFS helps to establish a new processing system, merging advancements or modifications of an existing system, contingency and mitigation plans, and optimizing the time frame for erection, execution, and operation for investment recovery (Subramanian and Taghizadeh-Hesary 2021).

Drying is a post-harvest processing operation through which significant value-added products from fruits and vegetables can be obtained. This not only reduces the harvest losses but also helps the better distribution of produce, lowering the transport expenses by a reduction in mass and volume and overcoming the seasonal limitation of fruits. In the agricultural domain, producing dehydrated fruits and vegetables has captured a good market and become a good source of foreign exchange earnings and a significant domestic and international market. Having said so, dehydration is an energy-intensive process; hence, to keep it economically feasible, optimization needs to be undertaken. Optimization and efficient energy utilization require advanced technology, which is inherently associated with high capital investment as operational criticalities. Hence, a techno-economic evaluation of the entire process is required before the finalization of the scheme.

9.3 COST CONTRIBUTING FACTORS FOR FRUIT DRYING

Fresh fruits are typically associated with high moisture content ranging from about 75% to 90%. The most widely used commercial technology for the dehydration of fruits, as of now, is convective drying. The process involves unit operations of sorting, washing, peeling, cutting, blanching, sulphiting, and drying. Various dehydration parameters to preserve final product functional properties are studied and reported (Simal et al. 2005; Akbarian, Ghasemkhani, and Moayedi 2013; Salehi 2020;). The dehydration of fruits is reported to be highly influenced by the maturity and season of harvest. Similarly, the key challenges encountered for agricultural products can be narrowed down to the irregular shape of the agricultural produces (grains, legumes, fruits and vegetables, classified under durable and perishable harvests), irregular matrix of the agricultural produce with complex tissues building blocks, significant volumes for processing, and short times for processing.

The plants' organs are organized considering three tissue layers: the epidermis, the ground tissue, and the vascular system. The chemical structures of the primary cell walls of the different parts of plants and their progenitors differ for different plant species. They vary in the complex glycans that interlace and cross-link the cellulose microfibrils to form a robust framework, in the nature of the gel matrix surrounding this framework, and in the types of aromatic substances and structural proteins that covalently cross-link the primary and secondary walls and lock cells into shape (Carpita 1996). The chemistry of structural elements like polysaccharides, aromatic substances, and proteins of different plant parts is unique and results in synthesizing and assembling dynamic and functional cell walls (Megías, Molist, and Pombal 2022).

The seeds and fruits are developed from the ovule (post-fertilization of the egg cell by the microspore) of the flower ovary and ovary walls, respectively. The histological organization of the ovary is similar to a leaf with an inner and an outer epidermal layer and parenchyma with vascular bundles in between. The development of these layers becomes the pericarp, which is actually the fruit without the seed. The pericarp consists of the exocarp, mesocarp, and endocarp. The exocarp is the outer layer of the fruit, and the endocarp covers the seed. The tissue between these two layers is the mesocarp, made up of storing parenchyma or sclerenchyma. In this structure, the mesocarps can be said to have a uniform tissue structure, but the separation of the same is difficult, which leads to the drying of combined exocarp, mesocarp, and endocarp tissues having different internal mass transfer rates and moisture holdings. Such complex structures of the tissues further lead to the need for complex drying mechanisms adding to the costs associated with the drying of fruits.

Various types of dryers have been researched, developed, and applied to overcome such internal mass transfer of water molecules. However, during the drying operations, considerations for challenges like inhomogeneous drying (over- or under-dried parts), the efficiency of the process, and the quality of the product (colour, texture, brittleness, taste, and physical and chemical attributes) add to the complexity and costs associated with the drying. The same cannot be ignored, and the selection of an appropriate dryer is crucial, as any minor changes in the feed or performance could adversely affect the final dried product characteristics. Reported drying methods and operating conditions for different fruits are given in Table 9.1.

9.4 COST-IMPACT OF DIFFERENT MODES OF DRYING

Drying, by definition, is '*the separation operation that converts solids, semi-solids or liquids to solid products by evaporation of associated liquid into the vapour phase through the application of heat*'. Being associated with the latent heat of vaporization, it is one of the most energy-intensive and inherently inefficient unit operations similar to distillation. Drying is a mandatory unit operation for one or several reasons, such as ease of handling, preservation, storage, transport cost reduction, achieving desired product quality, and so on. It is a critical and systematic process to be conducted with proper understanding. Drying is a part of life, widely associated with sun-based drying. The desiccating effect of air is the driving force for mass transfer, and solar energy acts as the activation energy. In recent years, technology has been developed for controlling the exposure and enhancing the utilization of solar energy through various configurations of solar dryers (Chavan et al. 2020).

Drying at the industrial scale is carried out by vaporization of liquids through the supply of activation energy and provision of the diffusion pathway to transfer vapours. The heat supply can be through *convection, conduction* (in direct or indirect contact mode), or *radiation* through various frequencies (microwave, far-infrared, infrared, ultra-violet, and radiofrequency energy). The selection and quantification of these energies are dependent on the size, shape, and structure of the objects and are regulated by maintaining the chamber temperatures from freezing to hot conditions, i.e., −50°C to 100°C, using refrigerated systems and heaters (fuelled by electricity, natural gas, coal/briquettes, firewood, and so on), respectively. Regulating these parameters, the energy of activation for vaporization and the inherently slow process of drying make this unit operation one of the most energy-intensive as well as cost-intensive (Mujumdar 2014).

TABLE 9.1
Reported Drying Methods and Operating Conditions for Fruits (Bhatkar et al., 2022a, 2022b)

Fruit	Method	Operating Conditions	References
Banana	HPP treatment and Osmotic dehydration	HPP: 200 MPa, 5 min at 26°C, 60° Brix at 40°C for 5 h or 40° Brix at 70°C and drying at 70°C	(da Júnior et al. 2018)
Banana (powder)	Spray drying	Inlet temperature = 150°C, 30% (w/w) maltodextrin	(Wong, Teoh, and Putri 2018)
Watermelon	Osmotic pretreatment and vacuum IR drying	Osmotic dehydration: 150 min, 50°C, 50° Brix with 3% salt. IR: 180 min, 55°C, 0.005 mmHg	(Chakraborty and Mondal 2017)
Watermelon	Spray drying (powder)	Inlet temperature = 130°C	(Quek, Chok, and Swedlund 2007)
Apple	Osmotic pretreatment and HAD	Pretreatment: 60° Brix sorbitol, 8 h, 60°C. Hot air-drying: 80°C, 1 m/s MW - 180 W, 70°C IR: 6 kW, 150°C, 45 s	(Shewale, Rajoriya, and Hebbar 2019)
Orange	Osmotic pretreatment and MW drying	Pretreatment = 7% (w/w) 90 min, 900 W power 50°C, 334 W power, at 12.31 kPa	(Bozkir 2020)
Orange (powder)	Spray drying Spouted bed dryer	Inlet temperature = 120°C, soluble solid to maltodextrin (Feed concentration) = 5:2 @ 60°C, air flow rate = 1.7 m^3/min	(Barbosa, Brandão, and Teixeira 2017)
Mango	Ultrasound-assisted hot air drying	45–60 W power, 52–55°C with the frequency of 3 min/30 min	(Méndez-Calderón, Ocampo-Castaño, and Orrego 2018)
Pineapple	Mixed-mode solar dryer	Solar dryer with a flat plate solar collector and drying chamber	(Rani and Tripathy 2021)
Pear	Ultrasound-assisted infrared drying	Ultrasound at 60 W power and infrared at 220 W Drying at 35°C, 1 m/s	(Liu et al. 2019)
Pear (powder)	Hot air drying with rice bran fibre	Rice bran fibre and maltodextrin at 3, 5, and 7% (w/w). Drying: 65°C, 2 m/s	(Jiang, Ameer, and Eun 2020)
Papaya	Osmotic drying	37°C, 60 °Brix for 4.25 h	(Jain et al. 2011)

The diffusion of vapours is considered in two steps: internal diffusion (movement of vapours in the interiors of the matrix) and external diffusion (removal of vapours from the surface), which basically govern the drying regimes (Mujumdar and Devahastin 2000; Pai et al. 2021). The external mass transfer conditions can be controlled by circulation draft (natural or forced), vacuum (regulating partial pressures), and relative humidity for providing moisture gradient.

From a research point of view, the drying process mainly depends on the product size (microns to tens of centimetres), product porosity (0% to 99%), drying time (0.25 sec to 5 months), production capacity (0.1 kg/h to 100 tonnes/h), product speed (stationary to 2000 m/min) or the residence time, drying temperature (below triple point to above critical point of water), operating pressure (fraction of millibar to 25 atm), and mode of heat transfer (Mujumdar 2014). Thus, it can be seen that the drying process needs a great deal of attention. The right choice of the most suitable dryer must provide a good return on investment (RoI) and be techno-economically feasible.

Typically, the drying unit operation falls under two methods – *Natural and Artificial Drying. The natural drying method* is undertaken by direct or indirect exposure to the sun, especially when the relative humidity is below 70% and sufficient space is available for spreading out agricultural produces. *The artificial drying method* is a part of the progressive mechanization of agriculture. It supports the drying operation even in humid and subtropical zones with unfavourable conditions.

The construction of a mechanical dryer includes the chamber/body of the dryer, the heating element or activation energy source, and the ventilators. Further, the construction can be categorized as *static dryers* or *continuous* dryers. Static dryers are inexpensive, low maintenance, and adaptable to small and marginal collection centres. Continuous dryers are high-flow dryers with complex construction, complementary equipment, planning, and organization. These are more favourable or, moreover, required for high-volume harvest processing (Gunathilake et al. 2018).

Different types of dryers have been designed and implemented for agricultural produces, considering their processing volumes. They may be categorized as *batch-type, batch-type with continuous flow, batch-type fluidized bed, or continuous fluidized dryer.* These types may further be subjected to minor modifications to meet specific requirements. The construction of the drying chambers can also lead to one of the major cost-impacting factors. A few classifications are listed as below:

- *Tray Dryers* – the simplest dryer in which trays are used for spreading out the product to be dried uniformly in thin layers.
- *Tunnel Dryers* – a developed tray dryer version with trolley trays moving in the tunnel.
- *Pneumatic Dryers* – products to be dried are rapidly conveyed in the air stream.
- *Rotary Dryers* – horizontal, inclined cylinders used for handling the material. Either the cylinder rotates or the paddles/screws rotate, and the cylinders are stationary.
- *Trough Dryers* – trough-shaped belt used for transport of material while air is blown through the bed of material.
- *Bin Dryer* – bins with perforated bottoms are used for drying the products.
- *Belt Dryer* – horizontal mesh/solid belt for material transport with drying medium.

Similarly, to meet the performance implementation, different techniques lead to several types of dryers also leading to increase in the capital cost associated with the system:

- *Freeze Dryer* – freezing the feed product at a very low temperature of (–)80°C to (–)50°C solidifying the water, and then removing it in the form of water vapour by sublimation. However, it leads to damage to the fibres, increasing the product's brittleness and is also ecologically unfriendly.

- *Spray Dryer* – method developed for obtaining dried powders from slurry by rapid evaporation.
- *Flash Dryer* – flash drying is similar to spray drying in which super-heated fluid/slurry is sprayed or exposed to atmospheric conditions in the agitated state so that the moisture content is released rapidly giving a dehydrated product.
- *Drum Dryer* – method developed for processing pureed raw materials. The liquid slurries are dried at relatively low temperatures over high-capacity rotating drums and collected in the form of sheets.
- *Vacuum Dryer* – process developed by changing operating conditions to lower pressures. This leads to faster removal of moisture from the product at atmospheric temperatures.
- *Microwave-assisted Dryer* – Microwaves can be successfully adopted for drying food crops with the advantage of less drying time. Microwaves can penetrate moist food materials, vibrate water molecules, and generate heat. However, it is associated with the disadvantages of non-homogeneous drying of the material. This can be overcome by clubbing with a fluidized bed to get homogeneous heating, which again adds to the construction cost. It is also reported that microwave and fluidized bed have resulted in considerable time and energy saving compared to conventional fluidized bed drying.

The selection of dryer is a critical process for contributing to the overall growth and development through the manufacturing activity. The process takes into consideration criticalities and principles associated with the process. It focuses on preservation of agricultural produces while retaining their quality through a sustainable and economically viable approach. Parameters that can be considered for the selection of dryer and drying conditions are graphically shown in Figure 9.2.

Recent industrial-scale agricultural cultivation and significant advances in research and technology have led to the exploration and implementation of new technologies for the dehydration of agricultural produces as well as processed foods. These are crucially developed to meet the well-controlled dehydration and product requirements as demanded by the market. Such advances have led to further different types of dryers, including:

- *Multi-stage Dryer* – method developed based on the heat and mass transfer characteristics along with the drying curve showcased by the material. Keen observation of drying curves and heat and mass transfer rates indicates distinct mechanisms for optimized drying rates in different conditions. Ideally, each mechanism needs different operating conditions and different techniques for drying. This calls for the need for a multi-mode/multi-stage dryer.
- *Super-heated Steam Dryer* – the replacement of hot air with super-heated steam is a revolutionary technique considering drastic changes in operating conditions. The equipment, as such, needs auxiliary devices for meeting the heating and cooling of the product, and the chamber can either be pressurized or under vacuum. The concept behind using super-heated steam is to increase moisture mobility in the feed material to increase the drying rate.

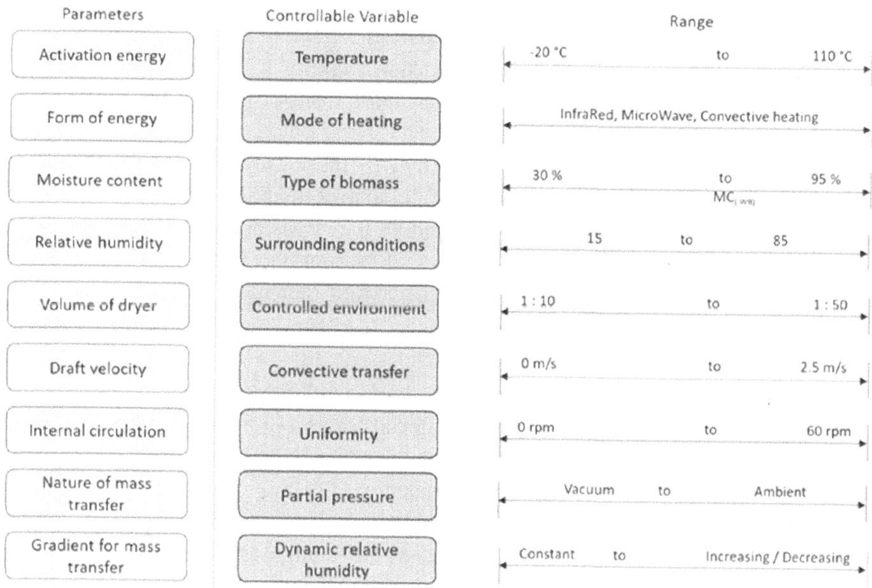

FIGURE 9.2 Parameters that can be considered for the selection of dryers.

- *Contact-sorption Dryer* – feed material is dehydrated by contacting with heated particulate inert material. The particle-particle contact induces heat transfer and hence mass transfer facilitating the drying operation. The heated, highly hygroscopic inert material generates a high chemical potential gradient. The drying rate is further enhanced in this method by the introduction of sorbent material along with hot air.
- *Spouted-bed Dryer* – these are modified fluidized beds for coarse particles that are large and heavier. The design is such that the motion is rather regular than random. These designs are further modified to have the rotating jet spout-bed dryers.
- *Heat Pump Dryer* – heat pumps are known to be efficient for drying operations. It provides an advantage of the recovery of heat for the exhaust and control over temperature and mainly humidity of the drying medium. It can be stated to be a refrigeration system in reverse operation. The refrigeration cycle in this process is used for dehumidification of air and heating the air for sensible heating to the desired temperature. A well-designed heat pump supplies heating and cooling as the drying process requires. The process is considered more efficient as both the latent heat and sensible heat are recovered from the exhaust air improving the thermal efficiency.
- *Fluidized-bed Dryer* – the passing drying medium through the permeable support resting the feed material to be dried is utilized not just for drying but also for fluidization. The superficial velocity of the drying medium, mostly heated air, is maintained such that the drying material is at the onset of fluidization and below the terminal velocity. This method is considered

a gentle, has a high degree of efficiency, and gives a uniform product. The thin boundary layer surrounding the particles to be dried due to rapid mixing ensures rapid heat and mass transfer, making it favourable for heat-sensitive materials. The rapid heat and mass transfer prevent the particles from getting overheated.

- *Heat Pump-assisted Fluidized-bed Dryer* – these are similar to the fluidized-bed systems but add a heat pump to control the humidity of the drying medium. These are still more efficient as they consider the advantages of rapid heat and mass transfer due to rapid mixing and a thin layer over the particles to be dried, as well as the improved thermal efficiencies due to heat recovery.

Drying is a major operation in agricultural and food industries. It is often used as a primary operation for preserving food materials or as a secondary process in some manufacturing operations. Drying is a higher energy-consuming process. As seen above, any degree of product quality or efficiency of drying can be met with a well-designed dryer; however, it is associated with a designing and capital cost, which need to be optimized using TEFS. Considering the complexity of the heat and mass transfer and associated physical and structural changes, it is ideal in such a scenario to have the flexibility for operating parameters and for handling different feed products (Delele, Weigler, and Mellmann 2015). However, operational flexibility further leads to complex designs and indirectly leads to higher capital as well as operating costs, which again call for a balance of quality and technology through TEFS.

9.5 TECHNO-ECONOMIC FEASIBILITY STUDY (TEFS)

9.5.1 Description of Process

TEFS for any process determines the risks associated with units and highlights the areas of concern to be worked upon to meet the project's economic viability and financial feasibility. Techno-economic analysis, by definition, is the process of assessing the value of a given technology to guide investment and resource allocation. Techno-economic evaluation is carried out by comparing the Levelized cost of processing (Kobos et al. 2020).

Generally, TEFS considers all relevant factors for implementing the project, including technical, economic, legal, scheduling, and operational aspects. This study defines the proposed system's technical elements, operational factors, and financial analysis for the dehydration process. Based on these aspects, to determine the feasibility of dehydration operations, case studies for drying fruits can be carried out before the implementation of the project and are projected in the later section. A generic platform is provided here to guide how the TEFS is conducted. The platform can be utilized for any crop, with the valuation of the project and operational costs to determine the TEFS and later generate a Detailed Project Report (DPR).

Preparation of TEFS begins with the ideation of a problem statement. Need for process and challenges associated with the process need to be clearly defined. It is then expected to overcome the processing challenges, finalize technical parameters,

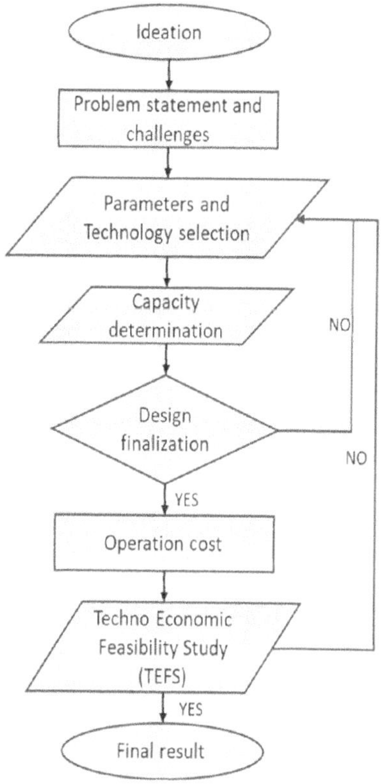

FIGURE 9.3 Equipment selection process based on techno-economic feasibility study (TEFS).

and select the most suitable technology considering the advancements in the sector of dehydration. The availability of the resources, marketable products, duration of harvest, the time for dehydration, and the capacity need to be taken into account to get the cost estimation of the unit. The operation costs, along with the other associated costs, are also considered. Finally, the TEFS is projected based on the value-addition seen through the process on the selected product. Equipment selection process based on TEFS has been graphically depicted in Figure 9.3.

9.5.2 Technological Consideration

Considering the different dryers stated above, parameters and principles for drying for specific applications can be broadly classified with a mode of heating, temperature and pressure of operation, method of material handling, and continuous/batch operation considered. With this classification and different product applications, more than 200 types of commercial dryers are reported in the literature.

To narrow down the selection, technical, operational, and economic aspects of the process are highly desired to ensure the acceptability of the overall drying process.

The selection of dryer is critical and plays a vital role in preserving the produce, retaining quality, ensuring sustainability of the processand economic viability, and contributing to overall growth and development through the manufacturing activity. The detailed TEFS ensures the proper dryer selection, safeguards the investment, and prevents tangible and intangible losses (Subramanian and Taghizadeh-Hesary 2021). The TEFS for any process determines the risks associated with units and highlights the areas of concern to be worked upon to meet the project's economic viability and financial feasibility. TEFS helps to establish a new processing system, merging advancements or modifications of an existing system, contingency and mitigation plans, and optimizing the time frame for erection, execution, and operation for investment recovery (Subramanian and Taghizadeh-Hesary 2021). The profitability aspect of the proposed drying projects needs to be evaluated through TEFS.

In this chapter, the technical elements, operational factors, and financial analysis for the dehydration process for the proposed system shall be defined for determined dehydration parameters of fruits.

9.5.3 Identification of Specific Process

Owing to the several advances in the process of drying and the development of various dryers for application-specific drying requirements, the finalization of specific process and selection of dryer is one of the most complex and poorly understood areas. Several classifications based on criteria, available information, and material to be dried are reported for the preparation of equipment trees based on bench tests and basic flowsheet parameters. Based on the theory and practical experiences, algorithms and expert systems for the selection are emphasized (Kemp 1999). The selection of dryers can be stated to be governed by equipment, material, or flowsheet-based factors. This can be further executed through a five-stage algorithm, as stated in Figure 9.4.

The selection sequence begins with stating the requirement and defining the issues related to the process. On a broader scale, considering the needs, the stated nine parameters need to be finalized and a crude design of the dryer can be made. Further intricacies are looked into for the suitability and selection of the shortlisted option is confirmed and further subject to fine-tuning/refinements. Considering the raw material specifications and the product characterization requirement of specific features are looked into and implemented. The finalized parameters are combined in a system, constructed as a dryer, and taken up for industrial scale drying.

9.6 COST-EFFECTIVE APPROACHES FOR DRYING OF DIFFERENT FRUIT PRODUCTS

9.6.1 Technology Advancement and Associated Costs of Dehydration

Drying at the molecular level involves activating water molecules, mobilizing water molecules within the matrix, and transferring molecules outside the matrix. These three activities are carried out sequentially and at times simultaneously in dryers by controlling parameters such as the temperature of drying for the activation energy,

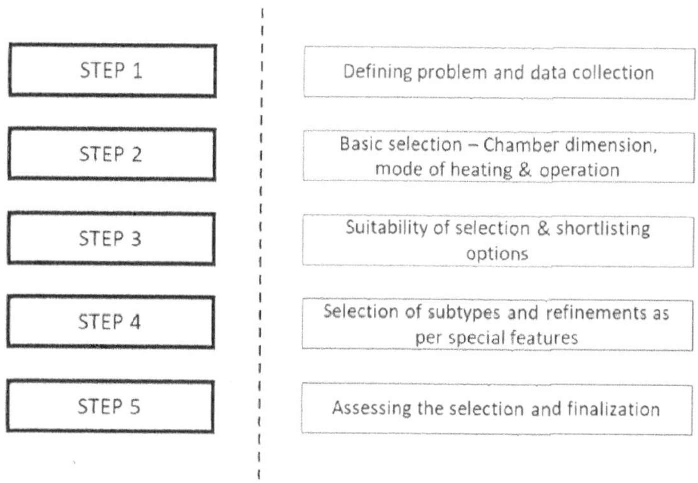

FIGURE 9.4 Dryer selection primary algorithm.

mode of heating for internal diffusion and draft (natural or forced), and surface area exposure for external mass transfer (Khan et al. 2017; Mujumdar 2014). The technical performances of the dryer, including drying rate (DR), capacity, energy consumption, and efficiency, can be regulated, as required, by controlling each of these parameters. Different drying techniques and combinations thereof have evolved, considering the products to be dried, and desired responses were like colour, shrinkage, bulk density, porosity, phytochemical and volatile retention, anti-oxidants, sugars, proteins, and sensory attributes. An ideal dried product should reconstitute all properties of the original product on rehydration (Khaing Hnin et al. 2019).

In natural drying process, mass transfer occurs by desiccating effect of air, and activation energy for the evaporation of water molecules is obtained by exposure to the sun. Considering the cost aspects, natural drying is the least expensive drying method, as only space is required for the uniform spreading of the fruits to be dried needs to be taken into account. Exposure to the sun can be controlled and enhanced by using different configurations of solar dryers. This assists the process but adds to the cost of the design and fabrication of the dryers compared to natural drying. As stated earlier, DR of the fruits under consideration highly depends upon size, shape, structure, moisture content, and the tissue structure of the fruit matrix. To drive out the fruit's moisture from the core, temperature or the activation energy needs to be regulated. Suitable heating mechanisms with direct or indirect heat, such as conduction, convection, or radiation mode using microwave (MW), far-infrared, infrared, ultra-violet, and radiofrequency (RF) energy, can be taken into consideration. This is the point where the cost of the dryer starts to increase. The cost of energy utility and the efficiency of utilization of the energy start impacting the capital cost as well as the operation cost and time. The selection of the heating source (combustion furnace, electrical heating element, IR lamp, magnetron, RF generator), the required ducting transfer to the chamber, and controlling mechanism further start adding up to the capital cost.

With the finalization of the source for activation energy along with the transfer and controlling mechanism, the second step associated with the drying operation is the transfer of these water molecules from the object to be dried to the surroundings. These external mass transfer conditions are controlled by circulation draft (natural or forced), vacuum (regulating partial pressures), and relative humidity for providing moisture gradient (I. C. Kemp and Bahu 1995; Ian C. Kemp 1999; Mujumdar 2014).

Further, the product quality is closely related to dehydration parameters, as seen at the micro-structural level (Ramos et al. 2004; Savitha et al., 2022; Mayor et al., 2005). The rate at which the moisture is removed must be well-balanced to control the quality of the dried product (Pise et al., 2022). The high-temperature drying and the more significant variation between the internal diffusion and evaporation rate can be catastrophic, causing permanent physical and chemical changes (Tanko et al. 2005). During drying operation, the dehydration rate can be governed by either the internal mass diffusion or the external mass transfer. High temperature leads to faster water removal, unbalanced internal moisture diffusion, and external mass transfer. However, due to the quicker mass transfer rate, the internal stresses at the cellular level are seen to increase, resulting in distortion of cellular structures. The control over dehydration by knowing the relative rates of internal diffusion and external mass transfer rate can be achieved by novel approach defined based on the drying (Chavan and Thorat 2020; Pai et al. 2021). For ideal drying conditions, the controlled activation energy for balanced internal mass diffusion along with regulated relative humidity for balanced external mass transfer, low temperature, and high surface areas are required, which results in higher cost for the drying system. The most suitable dryer offers optimized conditions for higher capacity operation, better quality products, good efficiency, lower cost, and minimal environmental impact (Mujumdar 2014; Khaing Hnin et al. 2019). Thus, it can be seen that the drying process needs a great deal of attention. The right choice of the most suitable dryer must provide a good RoI and be techno-economically feasible.

9.6.2 Cost-effective Approach for Drying

Drying of fruits itself leads to an array of dryers considering the uniqueness of each fruit owing to its water content, low molecular weight sugar and acids, nutrients, minerals, and morphological characteristics (Akbarian et al. 2013; Bhatkar et al. 2022a, 2022b). Based on the parameters mentioned in the above section for drying and its applications, these dryers can be broadly classified based on the modes of heat transfer, raw-material handling capacities, particle shapes, structures and sizes, time for dehydration, utility requirement, energy efficiency, and capital and operating cost (Mujumdar and Law 2010; Mujumdar 2014). Inappropriate dryers can make the process energy inefficient due to thermal energy consumption if the TEFS is not worked out properly.

The feasibility of the project is typically determined and verified in multiple aspects. A few elements for conducting the studies relevant to dehydration of fruit projects include technical, economic, scheduling, legal/ethical issues, resource planning, marketing, real estate, and cultural and comprehensive feasibility (Subramanian and Taghizadeh-Hesary 2021). All these must be considered to ensure that the proposed project is technically feasible overall. As far as fruit dehydration is concerned, the following three are the major areas, and the same are tried to be discussed as a part of this study,

i. *Engineering design/technical performance,*
ii. *Operational/functional performance, and*
iii. *Economic analysis.*

The evaluation of the *technical performance* mainly involves the determination of utility consumption and efficiency of the dryer. The latent heat of vaporization and moisture extraction rate (MER) are the key points to be considered for utility consumption (mainly heat utilized for evaporating water). The energy (utility) consumption can be determined based on the unit's thermal efficiency and the capacity of processing. This utility could be different for different dryers. For solar-based dryers – radiation is received (with no additional cost); for vacuum dryers – electricity is for the vacuum pump and the heating element; for heat pump dryers – electricity is for dehumidification; for hot air dryers – electricity for the heating element or the units of LPG/briquettes/coal consumed; and so on. Additional power requirements for the generation of the draft shall be accounted for separately. The total energy consumed shall be the sum of utility consumed and power for the draft.

To determine the dryer's technical performance, *Specific heat energy consumption* and *Specific power consumption* can be defined as the energy utilized for unit associated water (bound/unbound) removal in *kJ/kg*.

$$Q_s = {Q_h}/{M_w} = {F \cdot P \cdot T}/{M_w} \qquad \text{(Eq. 9.1)}$$

Similarly, the total energy required is the sum of heat energy and utility supplied

$$Q_t = Q_h + F \cdot P \cdot T \qquad \text{(Eq. 9.2)}$$

Details of technical performances considered in techno-economic feasibility studies are given in Table 9.2.

As discussed above, after ideation and based on the definition of problem statements and challenges, preliminary data need to be gathered or generated through experimentation to determine the *technical performance*. Evaluation is initiated by the selection of fruits to be dried, listing the initial and desired moisture content, required drying time, mode of heating, and mass transfer mechanism. Drying time is closely linked with the parameters such as the loading density, draft velocity, inlet and outlet air temperature, relative humidity, and airflow rate. The material thickness, density, porosity, and other properties also significantly influence drying time. These are to be monitored and mapped critically, as they affect the efficiency of the scaled-up dryer. The DR curve shall be obtained for bench-scale, pilot level, demo unit, and, if possible, plant scale drying unit to monitor performance (Genskow 1994).

Operational performance defined for the economic model shall be specific to the fruit under consideration. The loading density, DR, and operating conditions from

TABLE 9.2
Details of Technical Performances Considered in Techno-Economic Feasibility Studies

Sr. No.	Description	Notation	Unit	Remarks
A	*Technical Performance*			
1	Initial moisture content	M_i	%	The moisture content of the fresh fruits at the time of harvest
2	Desired moisture content	M_f	%	Safe moisture content limit for better shelf-life. Typically, <15%
3	Drying time	t_d	*h*	Ideal batch time considered for achieving the desired moisture content
4	Loading capacity	B	*kg*	Ideal batch size considered for calculating the MER and SMER
		B_{fr}	*kg/h*	Flowrate consideration for continuous process, as per dehydration time
5	Drying rate	DR	*kg/h*	Produced dried product per unit of time by the drying system
6	Water evaporated	M_w	*kg*	Amount of water removed by the drying systems for a batch
7	Moisture extraction rate	MER	*kg/h*	Water removal rate through the unit during the drying process
8	Airflow rate	M_a		Draft of air required for carrying the necessary dehydration
9	Latent heat of vaporization	L_v	*kJ/kg*	Heat required to convert liquid (at its boiling point) to vapour
10	Inlet temperature	T_{Ai}	*°C*	The inlet temperature of the air draft used for dehydration
11	Outlet temperature	T_{Ao}	*°C*	The outlet temperature of the air draft used for dehydration
12	Heat utilized to evaporate water	Q_e	*kJ*	Based on the flow rate, specific heat and temperature of inlet and outlet air
13	Energy (Utility) consumption	Q_h	*kJ*	The energy supplied will vary depending on the type of dryer, mode of heating, and drying conditions
	LPG	$Q_{h(LPG)}$	*kg/h*	
	Biomass	$Q_{h(Biomass)}$	*kg/h*	
	Electricity	$Q_{h(Electricity)}$	*Units*	
14	Energy efficiency of Dryer	$\eta_D = Q_e/Q_h$	%	Ratio of energy utilized for drying to the energy supplied to the unit
15	Heat energy requirement	Q_h	*kJ*	
16	Power consumption (blower)		*kW*	Required for providing the draft for moisture removal
17	Power factor of the motor	F		A typical value is 0.75

the technical performance lead to the dryer chamber selection. Harvest load & feasible time/duration from fresh cut to dehydration provide the required capacity of the dryer. The harvest area under cultivation gives the raw material availability or drying load on the system during harvesting season. Details of operational performances considered in techno-economic feasibility studies are given in Table 9.3.

TABLE 9.3
Details of Operational Performances Considered in Techno-Economic Feasibility Studies

Sr. No.	Description	Notation	Unit	Remarks
B	*Operational performance*			
1	Satisfactory annual harvest	Y_a	*ton/ha/year*	Max./optimum/min. yield per unit area. Potential quantity for dying annually.
2	No. of harvest	N	*per year*	The frequency of the processing load based on tenure of the crop and harvest season.
3	Yield per harvest	Y_h	*ton/ha*	For multiple harvest annually, potential peak load for dehydration per harvest
4	Harvest seasons	t_h	*day*	No. of harvesting days at maturity indicating minimum duration of drying operation.
5	Area under consideration	A_{cap}	*ha*	Area required under cultivation to meet the supply of the raw material as per the processing capacity of dryer for smooth and uninterrupted feed.
6	Processing requirement per day	C_{pro}	*ton/day*	With the peak processing load determined, to minimize the losses capacity to be processed every day needs to be finalized.
7	Initial moisture content	M_i	*%*	The moisture content of the fresh harvest.
8	Final moisture content	M_f	*%*	Desired moisture content post-dehydration for preservation/processing.
9	Drying batch time	T_d	*h*	Ideal batch time long with loading and unloading time considered for achieving the desired moisture content (based on the drying rate curve and scale-up).
10	Loading density	L_d	*kg/m²*	Loading on trays/conveyors/packed columns/fluidized state based on the experimental readings to workout chamber volume and dryer details
11	Operating condition	-	Continuous/ Batch	Continuous – Capacity processing per hour based on the required residence time. Batch – Capacity of dryer based on drying time and number of feasible batches.
12	Fresh biomass cost	C_{fresh}	*USD/kg*	Raw material costing considered for value addition through drying.
13	Dried biomass cost	C_{dried}	*USD/kg*	Product processing cost – Raw material cost is distributed over the reduced weight + processing cost per unit weight.

Further information shall be determined based on the details like optimum harvest per unit area, harvesting period, the season of harvesting, technical performance of the dryer, and processing requirement for a season. The number of drying batches per day (residence time in case of continuous operation) and dependent utility requirement will give the operating costs for the process. The number of operational days, the feasibility of 24-h operation, and multi-products with different harvest seasons should be opted to optimize the dryer's capacity and distribution of the associated capital costs.

The *economic evaluation* can be carried out based on annualized cost method and life cycle saving. In the case of annualized cost method, the total associated drying expenses are distributed over the amount of product dried over the year. In the case of conventional energy utilization, the drying cost keeps increasing with the increase in utility costs. However, in the case of solar dryers, the variation in the cost of drying is negligible over the operational life. Hence, to carry out the economic analysis of solar-based dehydration, the savings over the dryer's life are taken into account. In either case, a short pay-back period is more attractive to drive the buyers to the dehydrating process (Chavan and Thorat 2021). The economic evaluation of the drying system considers factors like initial cost (C_I) or the capital cost of the drying unit. Details like foundation preparation, structure erection, and associated auxiliary costs are also included in capital costs. Repair and maintenance cost ($C_{R\&M}$), typically 5% of the initial cost annually and depreciation cost (C_D) at a depreciation rate of about 10% to 12% annually give the drying system's actual cost value. Scrap value (C_S), about 10% of the initial cost, is taken into account to associate the depreciation rate and the life of the equipment. Details of economic performances considered in techno-economic feasibility studies are given in Table 9.4.

TABLE 9.4
Details of Economic Performances Considered in Techno-Economic Feasibility Studies

Sr. No.	Description	Notation	Unit	Remarks
C	*Economic Performance*			
1	Capacity of the dryer	C_{cap}		Defined based on the technical performance and operating performance
2	Area covered	A_{cap}	*ha/yr.*	Area required under cultivation to meet the supply of the raw material
3	Material losses anticipated	M_{loss}	*%*	Waste material/parts that removed during pre-processing and losses during drying Can be in range of 2–20% of the fresh weight for fruits
4	Product output	M_{prod}	*tons*	Dried products processed/obtained through the unit
5	Operation days	OD_{pm}	*per month*	Typically considered to be 20–25 days
		OD_{py}	*per year*	Typically considered to be 265–300 days depending on the harvest

(Continued)

TABLE 9.4 *(Continued)*
Details of Economic Performances Considered in Techno-Economic Feasibility Studies

Sr. No.	Description	Notation	Unit	Remarks
6	Production capacities	$C_{prod(pd)}$	*per day*	Annual product output distributed over the operational periods
		$C_{prod(pm)}$	*per month*	
		C_{prod}	*per year*	
7	Initial cost for dryer	C_e	*USD (INR)*	The equipment cost includes the drying chamber, utilities, and Auxiliary units. Hence, the Initial cost of dryer Ci = Ce +
8	Installation and commissioning cost	Ci&c	*USD (INR)*	Typically, 10% of the equipment cost
9	Design life	N	*Yr.*	Typically considered to be 5–15 years
10	Scrap value	C_S	*USD*	Can be considered from 0% to 25% of the Initial cost
11	Depreciation cost	C_D	*USD*	Cost actual cash value of the unit with the decided depreciation rate and design life
12	Repair and Maintenance cost	$C_{R\&M}$	*USD*	It can be considered in the range of 2–10% based on the type of dryer
13	Labour cost	C_{Lcost}	*USD*	Based on the wages, number of shifts and number of workers required for the operation of the drying system
14	Utility requirement	C_U	*USD*	Total energy consumed for a selected type of dryer over the operational period
15	Total operating costs	C_{TO}	*USD*	Sum of utility cost and labour cost
16	Net raw material cost	C_{fresh} or C_{raw}	*USD*	The raw material purchase cost for the processing
17	Net-dried product cost	C_{dried} or C_{prod}	*USD*	Typically, this is the sum of raw material cost, processing cost, and profit margin associated with distributed over the dried product quantity
18	Net earnings/ Net cash flow	A_t	*USD*	Difference between the selling price of the product and raw material cost with processing cost
19	Discount rate considered	I	*%*	It is the key variable that determines the NPV of the investment. May vary from 7%, 10%, 14%, 21%, or 28%
20	Pay-back period	C_N	*Yrs.*	Time required to recover the Investment
21	Net present value	NPV	*USD*	Provides the economic viability of the process. (a positive NPV indicates an economically viable investment or project)
22	Internal rate of returns	IRR	*%*	IRR is the discount rate that makes the NPV of all cash flows from a particular investment equal zero (higher IRR indicates the more desirable project)
23	Benefit-cost ratio	BCR		It is the ratio of total discounted benefits to total discounted costs. A project with a ratio greater than 1 has more significant benefits and is favourable

Considering the C_I, C_D, C_S, and operational year of the unit, the net present value (NPV) is calculated. This, along with the $C_{R\&M}$ and cash flow, helps analyse the life-cycle cost analysis of the dryer system (Barnwal and Tiwari 2008). Further, sensitivity analysis can be performed by varying one factor at a time to observe the impact of each parameter fluctuation in the range of a 10% increase or decrease (Pise and Thorat 2023).

9.7 TECHNO-ECONOMIC EVALUATION (PROFITABILITY INDICES) WITH CASE STUDY

The cost of the dryer depends on the technical performance and the drying requirement, including DR or MER (also considered as the throughput rate), specific moisture removal rate (SMER, also referred to as moisture removal per unit energy consumption), and has been used in determining the dryer efficiency (η_D) (Obeng-Akrofi et al. 2021).

$$\text{DR} = (\text{M}_\text{I} - \text{M}_\text{F})/\text{t} \qquad \text{(Eq. 9.3)}$$

where DR – Drying rate (kg/h), M_I – Initial moisture content (% wet basis), and M_F – Final moisture content (% wet basis)

$$\text{MER} = \left[\text{W}_\text{I}\left\{1 - \left(\frac{(1-\text{M}_\text{i})}{(1-\text{M}_\text{F})}\right)\right\}\right]/\text{t} \qquad \text{(Eq. 9.4)}$$

where MER – Moisture extraction rate (kg water/h) and W_I – Initial weight of material to be dried (kg),

$$\text{SMER} = \left[\text{W}_\text{I} \times \left(\frac{(\text{M}_\text{I} - \text{M}_\text{F})}{(100-\text{M}_\text{F})}\right)\right]/\text{Q}_\text{H} \qquad \text{(Eq. 9.5)}$$

where SMER – Specific moisture extraction rate (kg water/kWh) and Q_H – Total energy consumed (kWh)

$$\eta_\text{D} = \frac{\text{Q}_\text{E}}{\text{Q}_\text{T}} \times 100 = \frac{\text{M}_\text{a}(\text{C}_\text{i}.\text{T}_\text{i} - \text{C}_\text{o}.\text{T}_\text{o}).\ \text{t}_\text{d}}{\text{Q}_\text{H} + \text{f}.\text{P}_\text{E}\ .\ \text{t}_\text{d}} \times 100 \qquad \text{(Eq. 9.6)}$$

where η_D – Efficiency of the dryer (%), M_a – Mass flow rate of air (kg/h), C_i & C_o – Specific heat capacity of inlet & outlet air (kJ/kg), T_i & T_o – Inlet & Outlet temperatures (°C), P_E – Power consumption (kW), t_d – Dehydration time (h), and f – power factor of electrical motor.

With this saving per unit in hand, the saving over the operation period (days/months/years) can be determined to calculate the NPV.

Based on the cash flow for the period, the Internal Rate of Returns (IRR), benefit-to-cost ratio (BCR), and RoI are determined.

$$\text{C}_\text{D} = (\text{C}_\text{I} - \text{C}_\text{S})/\text{N} \qquad \text{(Eq. 9.7)}$$

where N is the design life of a dryer (yrs.) (typically considered as 15 yrs.).

The economic performance factors determine the Pay-back period (CN), the time required to recover investment cost, and the Capital Recovery Factor (CRF). The product of CRF and CN gives the annualized uniform cost for a drying system. In the case of solar dehydration, the savings per drying unit (S_{unit}) is determined by taking into account the price of fresh produce (C_{fp}), cost of dried products (C_{dp}), cost of drying per unit dried product (C_s), and comparing with the selling price of the similar marketed product (C_b).

$$S_{unit} = \left(C_b - \left(\left(C_{fp} \times \frac{M_f}{M_d} \right) + C_s \right) \right) \quad \text{(Eq. 9.8)}$$

where M_f and M_d are mass of fresh and dried products, respectively.

The annualized cost of the equipment is determined considering a discounted rate of 5.4% and a life span (n) of 10 years. The CRF is determined using Eq. 9.9, and the annualized CAPEX is determined as the product of CRF and capital cost (Zayed et al. 2020). A depreciation rate (d) of 10%, an interest rate of 7.65% and an inflation rate of 7.1% are considered for this project. The project's profit is the difference between annual revenue and production costs. The net profit is subjected to a 30% corporate tax rate in India post-depreciation deduction and interest payment for the working capital loan. The annual cash flow is determined as the sum of profit after tax and the depreciation of the equipment considered.

$$CRF = \frac{d(1+d)^n}{\left[(1+d)^n - 1\right]} \quad \text{(Eq. 9.9)}$$

Profitability indices considered for the determination of profitability are NPV, IRR, Pay-back Time, and RoI.

9.7.1 Net Present Value (NPV)

The NPV of the project is determined based on the value of money in future, which is worth lesser than the present value over time. It is the cash flow occurring in the future at a discounted rate reflecting the reduced value of the current time. An investment is considered feasible if NPV > 1 also indicates the returns more than the opportunity cost of funds.

$$NVP = \sum_{t=0}^{n} \frac{(A_t)}{(1+i)^t} = \sum_{1}^{n} \frac{\text{Net Cash Flow in the nth year}}{(1+d)^n} \quad \text{(Eq. 9.10)}$$

where A_t – cash flow for the period t or at a particular time, t is the project lifetime, and i is the annual minimal acceptable interest rate/discounted rate.

The NPV depends on the cash flow and the interval of time between the cash flows. As seen in the equation, it is the value of all future cash flows over the entire life of an investment discount to the present.

9.7.2 Internal Rate of Returns (IRR)

IRR is used to determine the profitability of potential Investment. IRR is determined by considering discounted rate (*r*) that makes the NPV zero. The maximum interest rate is affordable when the entire investment is borrowed to meet the breakeven of the investment in the project's lifetime. The project's investment is considered acceptable, subjected to IRR > *r* (Salazar T. et al. 2021).

$$NPV = \sum_{t=0}^{n} \frac{(A_t)}{(1+IRR)^t} = 0 \qquad \text{(Eq. 9.11)}$$

9.7.3 Return on Investment (RoI)

The feasibility of the investment is determined by considering the ratio of gains to cost, known as RoI, evaluated in terms of percentage, per period and rate of returns on money invested.

$$\text{Return on Investment (ROI)} = \frac{\text{Annual Profit}}{\text{Total investment costs}} \times 100 \qquad \text{(Eq. 9.12)}$$

Benefit-to-Cost ratio is another term for projecting the economic analysis. It projects the relationship between relative cost and benefit of a proposed project, in qualitative or monetary terms.

$$B_{CR} = \sum_{t=0}^{n} \frac{(B_t)}{(1+i)^t} \Bigg/ \sum_{t=0}^{n} \frac{(C_t)}{(1+i)^t} \qquad \text{(Eq. 9.13)}$$

where B_t is the benefit of the project in *t*th year, C_t is the cost of the project in *t*th year, and *i* is the discount rate.

9.7.4 Pay-Back Back Period

The project's desirability is also determined based on the breakeven achievable for the investment, i.e., the pay-back time for the project. As an indication, the project's pay-back period should be less than the defined project lifetime.

$$\text{Payback period} = \frac{\text{Initial investment}}{\text{Annualised cash inflow}} = \frac{C_I}{A_t} \qquad \text{(Eq. 9.14)}$$

The mentioned method of TEFS is elaborated for hot air dryers and products from four fruits to provide ease of understanding. The selected dryer is a typical hot air dryer powered by liquefied petroleum gas (LPG) fired air heater of 33.3 kW. Forced draft is generated by radial fan at pressure of 250 kPa and 3600 m^3/h. The drying cabinet considered has dimensions of 4.0 m × 2.0 m × 1.0 m divided into four equal sections which can be mounted by four trays. It provided a control over the inlet temperature using a thermostat (Soysal and Öztekin 2001). Using this dryer, TEFS for dehydration of apples, pineapples, papaya, and banana is projected herewith. TEFS evaluation of hot air dryer for dehydration of apples, pineapples, papaya and banana has been given in Table 9.5.

TABLE 9.5
TEFS Evaluation of Hot Air Dryer for Dehydration of Apples, Pineapples, Papaya, and Banana

			Selected Fruits			
Description	**Notation**	**Unit**	**Apple**	**Pineapple**	**Papaya**	**Banana**
Technical Performance						
Initial moisture content	Mi	%	89.00%	87.50%	80.00%	80.00%
Desired moisture content	Mf	%	7.50%	7.50%	7.50%	7.50%
Drying time	T	H	9	6.5	12	5
Loading capacity	B	Kg	100	100	100	100
Drying rate	DR	kg/h	11.11	15.38	8.33	20.00
Water evaporated	Mw	Kg	88.11	86.49	78.38	78.38
Moisture extraction rate	MER	kg/h	9.79	13.31	6.53	15.68
Air flowrate		kg/h	19500	19500	19500	19500
Specific heat of air		kJ/kg.K	1	1	1	1
Inlet temperature		C	25	25	25	25
Outlet temperature		C	46	46	46	46
Latent heat of vaporization		kJ/kg	2260	2260	2260	2260
Heat utilized to evaporate water	Qe		608624	604959	586635	586635
Energy (Utility) consumption	Qh		1007584	1417678	1387989	1061403
LPG		kg/kg db	27.380	38.200	37.400	28.600
Efficiency		kg/h				
Low heating value of LPG		kJ/kg				
Energy efficiency of dryer	Eta = Qe/Qh	%	60.40%	42.67%	42.27%	55.27%

(Continued)

TABLE 9.5 *(Continued)*
TEFS Evaluation of Hot Air Dryer for Dehydration of Apples, Pineapples, Papaya, and Banana

			Selected Fruits			
Description	**Notation**	**Unit**	**Apple**	**Pineapple**	**Papaya**	**Banana**
Specific heat energy requirement	Qh	kJ	1007584	1417678	1387989	1061403
Power consumption (Blower)	1	Hp	0.745	0.745	0.745	0.745
Power factor of motor	0.8	F	0.80	0.80	0.80	0.80
Specific heat energy consumption	Qs = Qh/Mw	kJ/kg	11436	16392	17709	13542
Specific power consumption	Qp = f.P.t/Mw	kJ/kg	219	161	328	137
Total energy requirement	Qt = Qh + f. P. t	kJ	1007803	1417840	1388317	1061540
Operational performance						
Satisfactory annual harvest .	Y	tons/ha/year	12	70	74	100
No. of harvest	N	per year	1	1	2	1
Yield per harvest	Yi	tons/ha	12	70	37	100
Harvest seasons	D	Days	60	90	180	240
Peak processing requirement		kg/day	0.20	0.78	0.21	0.42
Feasible number of batches			2.00	3.00	2.00	4.00
Drying time		H	9	6.5	12	5
Area per unit		m^2	27.36	27.36	27.36	27.36
Loading density		kg/m^2	3.655	3.655	3.655	3.655
Fresh fruit cost		*USD/kg*	*0.6*	*0.3*	*0.2*	*0.22*
Dried fruit cost		*USD/kg*	*3.75*	*2.8*	*2*	*2*
Economic Performance						
Sizing/Capacity of dryer (batch)		Kg	100	100	100	100
Area covered		ha/yr.	2.80	0.82	0.78	0.72
Material losses anticipated		%	15.00%	18.00%	10.00%	22.00%
Material processed per year		tons/yr.	33.6	57.6	57.6	72
Evaporated water		kg/yr.	29604.3	49816.2	45145.9	56432.4

(Continued)

TABLE 9.5 *(Continued)*

TEFS Evaluation of Hot Air Dryer for Dehydration of Apples, Pineapples, Papaya, and Banana

			Selected Fruits			
Description	**Notation**	**Unit**	**Apple**	**Pineapple**	**Papaya**	**Banana**
Throughput of dried biomass		kg/yr.	3396	6383	11209	12143
Operation days		per month	14	16	24	15
		per year	168	192	288	180
Production		per day	20.216	33.243	38.919	67.459
		per month	283.027	531.892	934.054	1011.892
Initial cost for dryer	Ci	USD	2500	2500	2500	2500
Design life		Yrs.	7.5	7.5	7.5	7.5
Depreciation cost		USD	333.33	333.33	333.33	333.33
Repair and maintenance cost	10%	USD	250	250	250	250
Labour cost	`@ 20 USD/day	USD	3360	3840	5760	3600
Utility requirement						
Annual LPG consumption		Kg	6216	11520.0	15840.0	19800
Annual electricity consumption			3024	3744	6912	3600
LPG	`@USD 0.5872/kg	USD	3650.04	6764.54	9301.25	11626.56
Power	`@USD 0.0674/kWh	USD	203.82	252.35	465.87	242.64
Total operating costs			7797.19	11440.22	16110.45	16052.53
Net dried product cost			12736.22	17871.57	22417.30	24285.41
Net earnings/Net cash flow	At		4939.03	6431.34	6306.85	8232.87
Discount rate considered	I	%	10.00%	10.00%	10.00%	10.00%
Net present value		USD	4490.03	5846.68	5733.50	7484.43
Fresh raw material			2037.8	1914.8	2241.7	2671.4
Net earning			2901.24	4516.53	4065.12	5561.48
Pay-back period			0.862	0.554	0.615	0.450
IRR			23%	23%	11%	20%
Investment			−10297.19	−15855.03	−20852.18	−21223.93
Cash flow 1 year			2901.24	4516.53	4065.12	5561.48
Cash flow 8 year			2901.24	4516.53	4065.12	5561.48

9.8 CONCLUDING REMARKS

Drying of fruits is an important unit operation. Different drying techniques and combinations thereof have evolved, considering the products to be dried and meeting the desired responses like colour, shrinkage, bulk density, porosity, phytochemical and volatile retention, anti-oxidants, sugars, proteins, and sensory attributes. An ideal dried product should reconstitute all properties of the original product on rehydration. Controlled drying is crucial in obtaining the quality product to meet the vast market and consumer demand. Research advances have led to the development of many dryers for imparting the necessary energy of activation for water molecules. The energy is mainly required for the two sequentially and parallelly happening processes such as mobilizing water molecules within the matrix and transferring molecules outside the matrix while preserving the desired characteristics. Selection of the most appropriate dryer for dehydration is equally important, hence the need for techno-economic feasibility study.

TEFS was carried out considering the technical performance, and economic evaluation of the dryer. In this chapter, the details of TEFS specific to dehydration of apple, pineapple, papaya, and banana were discussed. Need for TEFS, methodologies for conducting it, and parameters considered the same were elaborated. Guidelines for determination of suitable dryer capacity, determination of area under cultivation, and carrying out economic evaluation accordingly were properly summarized using a case study for hot air dryer for dehydration.

REFERENCES

Akbarian, Mina, Nila Ghasemkhani, and Fatemeh Moayedi. 2013. 'Osmotic Dehydration of Fruits in Food Industrial : A Review'. *International Journal of Biosciences* 6655 (December): 1–16.

Barbosa, Joana, S. Brandão, and Paula Teixeira. 2017. 'Spray Drying Conditions for Orange Juice Incorporated with Lactic Acid Bacteria'. *International Journal of Food Science and Technology* 52 (9): 1951–58. https://doi.org/10.1111/ijfs.13472

Barnwal, P., and G. N. Tiwari. 2008. 'Life Cycle Cost Analysis of a Hybrid Photovoltaic/Thermal Greenhouse Dryer'. *The Open Environmental Journal* 2 (1): 39–46. DOI: 10.2174/1876325100802010039

Bhatkar, Nikita S., Shivanand S. Shirkole, Charles Brennan, and Bhaskar N. Thorat. 2022a. 'Review Pre-Processed Fruits as Raw Materials : Part I – Different Forms, Process Conditions and Applications'. *International Journal of Food Science and Technology* 57 (8): 1–18. https://doi.org/10.1111/ijfs.15891

______. 2022b. 'Review Pre-Processed Fruits as Raw Materials : Part II – Process Conditions, Demand and Safety Aspects'. *International Journal of Food Science and Technology* 57 (8): 1–18. https://doi.org/10.1111/ijfs.15887

Bozkir, Hamza. 2020. 'Effects of Hot Air, Vacuum Infrared, and Vacuum Microwave Dryers on the Drying Kinetics and Quality Characteristics of Orange Slices'. *Journal of Food Process Engineering* 43 (10): 1–12. https://doi.org/10.1111/jfpe.13485

Carpita, Nicholas C. 1996. 'Structure and Biogenesis of the Cell Walls of Grasses'. *Annual Review of Plant Physiology and Plant Molecular Biology* 47 (1): 445–76. https://doi.org/10.1146/annurev.arplant.47.1.445

Chakraborty, Rajat, and Pijus Mondal. 2017. 'Effects of Intermittent CO_2 Convection Under Far-Infrared Radiation on Vacuum Drying of Pre-Osmodehydrated Watermelon'. *Journal of the Science of Food and Agriculture* 97 (11): 3822–30. https://doi.org/10.1002/jsfa.8246

Chavan, Anand, and Bhaskar Thorat. 2020. 'Mathematical Analysis of Solar Conduction Dryer Using Reaction Engineering Approach'. *International Journal of Chemical Reactor Engineering* 18 (5–6). https://doi.org/10.1515/ijcre-2019-0220

Chavan, Anand, and Bhaskar N. Thorat. 2021. 'Techno-Economic Comparison of Selected Solar Dryers: A Case Study'. *Drying Technology* 40 (10): 2105–15. https://doi.org/10.1080/07373937.2021.1919141

Chavan, Anand, Vivek Vitankar, Arun Mujumdar, and Bhaskar Thorat. 2020. 'Natural Convection and Direct Type (NCDT) Solar Dryers: A Review'. *Drying Technology* 22: 1–22. https://doi.org/10.1080/07373937.2020.1753065

Delele, M. A., F. Weigler, and J. Mellmann. 2015. 'Advances in the Application of a Rotary Dryer for Drying of Agricultural Products: A Review'. *Drying Technology* 33 (5): 541–58. https://doi.org/10.1080/07373937.2014.958498

FAO, UN. 2022. *World Food and Agriculture – STATISTICAL YEARBOOK.*

Genskow, Larry R. 1994. 'Dryer Scale-up Methodology for the Process Industries'. *Drying Technology* 12 (1–2): 47–58. https://doi.org/10.1080/07373939408959949

Giannakourou, Maria C., and Theofania N. Tsironi. 2021. 'Application of Processing and Packaging Hurdles for Fresh-Cut Fruits and Vegetables Preservation'. *Foods* 10: 1–23.

Gunathilake, D. D. M. C., D. P. Senanayaka, G. Adiletta, and Wiji Senadeera. 2018. 'Chapter 14 – Drying of Agricultural Crops'. In *Advances in Agricultural Machinery and Technology*, edited by G. Chen, First, 331–67. CRC Press Taylor & Francis Group.

Jain, S. K., R. C. Verma, L. K. Murdia, H. K. Jain, and G. P. Sharma. 2011. 'Optimization of Process Parameters for Osmotic Dehydration of Papaya Cubes'. *Journal of Food Science and Technology* 48 (2): 211–17. https://doi.org/10.1007/s13197-010-0161-7

Jiang, Guihun, Kashif Ameer, and Jong Bang Eun. 2020. 'Encapsulation of Hot Air-Dried Asian Pear Powders Using Rice Bran Dietary Fiber'. *Food Bioscience* 38 (December 2019): 100742. https://doi.org/10.1016/j.fbio.2020.100742

Kemp, I. C., and R. E. Bahu. 1995. 'A New Algorithm for Dryer Selection'. *Drying Technology* 13 (5–7): 1563–78. https://doi.org/10.1080/07373939508917039

Kemp, Ian C. 1999. 'Progress in Dryer Selection Techniques'. *Drying Technology* 17 (7–8): 1667–80. https://doi.org/10.1080/07373939908917644

Khaing Hnin, Kay, Min Zhang, Arun S. Mujumdar, and Yuelin Zhu. 2019. 'Emerging Food Drying Technologies with Energy-Saving Characteristics: A Review'. *Drying Technology* 37 (12): 1465–80. https://doi.org/10.1080/07373937.2018.1510417

Khan, Md Imran H., R. Mark Wellard, Szilvia Anett Nagy, M. U. H. Joardder, and M. A. Karim. 2017. 'Experimental Investigation of Bound and Free Water Transport Process during Drying of Hygroscopic Food Material'. *International Journal of Thermal Sciences* 117: 266–73. https://doi.org/10.1016/j.ijthermalsci.2017.04.006

Kobos, Peter H, Thomas E Drennen, Alexander Sasha Outkin, Erik K Webb, M Scott, and Steven Wiryadinata. 2020. 'Techno-Economic Analysis: Best Practices and Assessment Tools'. *Sandia Report.*

Liu, Yunhong, Changying Sun, Yuqing Lei, Huichun Yu, Huihan Xi, and Xu Duan. 2019. 'Contact Ultrasound Strengthened Far-Infrared Radiation Drying on Pear Slices: Effects on Drying Characteristics, Microstructure, and Quality Attributes'. *Drying Technology* 37 (6): 745–58. https://doi.org/10.1080/07373937.2018.1458317

Mayor, L., M. A. Silva, and A. M. Sereno. 2005. 'Microstructural Changes during Drying of Apple Slices'. *Drying Technology* 23 (9–11): 2261–76. https://doi.org/10.1080/07373930500212776

Megías, M., P. Molist, and M. A. Pombal. 2022. 'Atlas of Plant and Animal Histology'. 2022.

Méndez-Calderón, Erika K., Juan C. Ocampo-Castaño, and Carlos E. Orrego. 2018. 'Optimization of Convective Drying Assisted by Ultrasound for Mango Tommy (*Mangifera Indica* L.)'. *Journal of Food Process Engineering* 41 (1). https://doi.org/10.1111/jfpe.12634

Mujumdar, Arun S., and Chung Lim Law. 2010. 'Drying Technology: Trends and Applications in Postharvest Processing'. *Food and Bioprocess Technology* 3 (6): 843–52. https://doi.org/10.1007/s11947-010-0353-1

Mujumdar, Arun S., and Sakamon Devahastin. 2000. 'Chapter 1 – Fundamental Principles of Drying'. In *Exergex*, 1st ed., 1–22. Brossard: Exergex Corporation.

Mujumdar, Arun Sadashiv. 2014. 'Section I – Fundamental Aspects'. In *Handbook of Industrial Drying*, edited by Arun Sadashiv Mujumdar, Forth, 32–155. CRC Press Taylor & Francis Group.

Obeng-Akrofi, George, Joseph O. Akowuah, Dirk E. Maier, and Ahmad Addo. 2021. 'Techno-Economic Analysis of a Crossflow Column Dryer for Maize Drying in Ghana'. *Agriculture (Switzerland)* 11 (6): 1–15. https://doi.org/10.3390/agriculture11060568

Pai, Kalash R., Voruganti Sindhuja, Palghat A. Ramachandran, and Bhaskar N. Thorat. 2021. 'Mass Transfer 'Regime' Approach to Drying'. *Industrial and Engineering Chemistry Research* 60 (26): 9613–23. https://doi.org/10.1021/acs.iecr.1c01680

Pise, Viplav Hari, and Bhaskar Narayan Thorat. 2023. 'Techno-Economics Feasibility of Supercritical CO_2 Ambrette Seed Extraction for Optimised Parameters Obtained Using RSM'. *Journal of Applied Research on Medicinal and Aromatic Plants*, 1–9. https://doi.org/10.1016/j.jarmap.2023.100490

Pise, Viplav, Shivanand Shirkole, and Bhaskar N. Thorat. 2022. 'Visualization of Oil Cells and Preservation during Drying of Betel Leaf (Piper Betle) Using Hot-Stage Microscopy'. *Drying Technology*. https://doi.org/10.1080/07373937.2022.2048848

Quek, Siew Young, Ngan King Chok, and Peter Swedlund. 2007. 'The Physicochemical Properties of Spray-Dried Watermelon Powders'. *Chemical Engineering and Processing: Process Intensification* 46 (5): 386–92. https://doi.org/10.1016/j.cep.2006.06.020

Ramos, Inês N., Cristina L. M. Silva, Alberto M. Sereno, and José M. Aguilera. 2004. 'Quantification of Microstructural Changes during First Stage Air Drying of Grape Tissue'. *Journal of Food Engineering* 62 (2): 159–64. https://doi.org/10.1016/S0260-8774(03)00227-9

Rani, Poonam, and P. P. Tripathy. 2021. 'Drying Characteristics, Energetic and Exergetic Investigation during Mixed-Mode Solar Drying of Pineapple Slices at Varied Air Mass Flow Rates'. *Renewable Energy* 167: 508–19. https://doi.org/10.1016/j.renene.2020.11.107

Salazar T, María Belén, M. Fernanda San Martín-González, Hua Cai, and Jen Yi Huang. 2021. 'Economic and Environmental Performance of Instantaneous Water Heating System for Craft Beer Production'. *Food and Bioproducts Processing* 127: 472–81. https://doi.org/10.1016/j.fbp.2021.04.006

Salehi, Fakhreddin. 2020. 'Recent Applications and Potential of Infrared Dryer Systems for Drying Various Agricultural Products: A Review Recent Applications and Potential of Infrared Dryer Systems for Drying Various Agricultural Products: A Review'. *International Journal of Fruit Science* 20 (3): 586–602. https://doi.org/10.1080/15538362.2019.1616243

Savitha, S., Snehasis Chakraborty, and Bhaskar N. Thorat. 2022. 'Microstructural Changes in Blanched, Dehydrated, and Rehydrated Onion'. *Drying Technology* 40 (12): 2550–67. https://doi.org/10.1080/07373937.2022.2078347

Shewale, Sandhya R., Deependra Rajoriya, and H. Umesh Hebbar. 2019. 'Low Humidity Air Drying of Apple Slices: Effect of EMR Pretreatment on Mass Transfer Parameters, Energy Efficiency and Quality'. *Innovative Food Science and Emerging Technologies* 55 (May): 1–10. https://doi.org/10.1016/j.ifset.2019.05.006

Júnior, Silva, Aluízio Freire da, Wilton Pereira da Silva, Juarez Everton deFarias Aires, Kalina Lígia, and C. A. Farias Aires. 2018. 'Numerical Approach to Describe Complementary Drying of Banana Slices Osmotically Dehydrated'. *Heat and Mass Transfer/Waerme – Und Stoffuebertragung* 54 (2): 267–76. https://doi.org/10.1007/s00231-017-2120-6

Simal, S., A. Femenia, M. C. Garau, and C. Rossell. 2005. 'Use of Exponential, Page's and Diffusional Models to Simulate the Drying Kinetics of Kiwi Fruit'. *Journal of Food Engineering*, 66: 323–28. https://doi.org/10.1016/j.jfoodeng.2004.03.025

Soysal, Y., and S. Öztekin. 2001. 'Technical and Economic Performance of a Tray Dryer for Medicinal and Aromatic Plants'. *Journal of Agricultural and Engineering Research* 79 (1): 73–79. https://doi.org/10.1006/jaer.2000.0668

Subramanian, M., and Farhad Taghizadeh-Hesary. 2021. 'Techno-Economic Feasibility Study Method in Startup Financing'. In *Investment in Startup and Small Business Finances*, 107–36. World Scientific.

Tanko, Hakibu, Danielle Julie Carrier, Lijun Duan, and Ed Clausen. 2005. 'Pre- and Post-Harvest Processing of Medicinal Plants'. *Plant Genetic Resources* 3 (2): 304–13. https://doi.org/10.1079/pgr200569

Tylewicz, Urszula, Cinzia Mannozzi, Santina Romani, Juan Manuel Castagnini, Kinga Samborska, Pietro Rocculi, and Marco Dalla Rosa. 2019. 'Chemical and Physicochemical Properties of Semi-Dried Organic Strawberries Enriched with Bilberry Juice-Based Solution'. *LWT – Food Science and Technology* 114 (July): 108377. https://doi.org/10.1016/j.lwt.2019.108377

Wong, Chen Wai, Chin Ying Teoh, and Clarisa Eka Putri. 2018. 'Effect of Enzymatic Processing, Inlet Temperature, and Maltodextrin Concentration on the Rheological and Physicochemical Properties of Spray-Dried Banana (*Musa acuminata*) Powder'. *Journal of Food Processing and Preservation* 42 (2): 1–9. https://doi.org/10.1111/jfpp.13451

Zayed, Mohamed E, Jun Zhao, Wenjia Li, Ammar H Elsheikh, Zhennan Zhao, and A. Khalil. 2020. 'Performance Prediction and Techno-Economic Analysis of Solar Dish/Stirling System for Electricity Generation'. *Applied Thermal Engineering* 164 (September 2019): 114427. https://doi.org/10.1016/j.applthermaleng.2019.114427

10 Future Prospects and Global Market Demand for Dried Fruit Products

Macherla Sangeetha, Parul Shukla, Keshram Dulait, and Shivanand S. Shirkole
Institute of Chemical Technology Mumbai, ICT-IOC Bhubaneswar, Odisha, India

10.1 INTRODUCTION

Drying is a preservation technique which works for the reduction of the moisture content as well as water activity of the product up to a desired level, which increases the shelf life of the product and prevents microbial spoilage due to bacteria, fungi, yeast, and mould. For making the dried fruit products, moisture content needs to be reduced from the fruits by means of either thermal or non-thermal drying techniques, which evaporate most of the free water present and end up making the concentrated/dried product, which has almost all the nutrients which are available in raw fruits. Dried fruits consist of a large variety of phytochemicals, such as flavonoids, phytoestrogens, carotenoids, phenolic acid, etc., and are equivalently nutritious to raw/ripen fresh fruits in less serving count. It consists of a peculiar combination of flavour and aroma with vital macro- and micronutrients. According to some, dried fruits are crucial for human health since they offer excellent nutrition and health advantages (Kiat et al., 2016).

Drying is proven to be an effective technique to reduce post-harvest losses, as in case of surplus production, the excess fruits can be dried and stored for later use in off-seasons. Likewise, most fruits are only seldom accessible but are in high demand all year round. In this case, surplus fruits are dried whole, cut into halves, or in slices. The most commonly available dried fruit products are berries, prunes, grapes, dates, plums, apricots, papayas, cherries, apples, peaches, figs, etc. "Conventional" or "traditional" dried fruits include apples, apricots, dates, figs, peaches, pears, prunes, and raisins. Before drying, certain fruits, like blueberries, cranberries, cherries, strawberries, and mangoes, are typically infused with sugar solutions or fruit juice concentrate.

Dried fruits emerged as a substitute for processed food or sugary snacks and have come out as one of the best alternatives to unhealthy snacking globally. It also has the benefit of being a fruit-based product, which makes it wholesome and nourishing, easy to store and distribute, and always available, even in places where it cannot normally grow. Studies have also shown that traditional dried fruit comes under the category of low- to moderate-GI foods and is good for diabetic people. Different types of fruits and their mostly grown countries are given in Table 10.1.

DOI: 10.1201/9781003402916-10

TABLE 10.1
Types of Fruits and Their Mostly Grown Countries

Sr. No.	Fruits	Mostly Grown Countries
1	Apples	China, USA, Poland, Turkey, India
2	Apricots	Turkey, Iran, Uzbekistan, Algeria, Italy
3	Banana	India, China, Indonesia, Brazil, Ecuador
4	Blueberries	USA, Peru, Canada, Mexico, Spain
5	Cherries	Turkey, USA, Russia, Iran, Uzbekistan
6	Coconut	India, Indonesia, Philippines, Brazil, Sri Lanka
7	Cranberries	USA, Canada, Chile, Belarus, Azerbaijan
8	Dates	Egypt, Saudi Arabia, Iran, Algeria, Iraq
9	Figs	Turkey, Egypt, Morocco, Algeria, Iran
10	Grapes	China, Italy, Spain, France, USA
11	Kiwi	China, New Zealand, Italy, Greece, Iran
12	Mango	India, Indonesia, China, Pakistan, Mexico
13	Olives	Spain, Italy, Morocco, Turkey, Greece
14	Papaya	India, Dominican Republic, Brazil, Mexico, Indonesia
15	Peach	China, Spain, Italy, Turkey, Greece
16	Pear	China, Italy, USA, Argentina, Turkey
17	Pineapple	Costa Rica, Indonesia, Philippines, Brazil, China
18	Plum	China, European Union, Romania, USA, Serbia
19	Strawberries	China, USA, Mexico, Turkey, Egypt
20	Watermelon	China, Turkey, India, Brazil, Algeria

According to statistics from 2014 on the production of commercially significant dried fruits worldwide, raisins and currants come in first place in terms of production with 1,394,566 metric tonnes (MT), followed by dates (756,000 MT), prunes (216,853 MT), cranberries (152,000 MT), figs (135,744 MT), and apricots (88,129 MT). The production of other dried fruits, like berries, citrus fruits, and cherries, is often not well documented in statistics (Omolola et al., 2015; Fruit et al., 2018).

Several steps which are crucial in making the dried fruits are as follows:

1. Sorting of fruits – from the given lot, the sorting or grading of fruits is done for the selection of fruits which are going to be used for drying. The selection criteria can be the degree of ripening, colour, firmness, size, sugar content, etc.
2. Washing & cleaning – thorough cleaning of the selected fruits is done carefully, to remove any dirt, debris, or the pesticide residues which come along with the fruits. The washing of fruits can be done using chlorinated water to reduce the microbial load which is followed by the pat drying of the fruits.
3. Preparation of fruits to get them dried – after washing, the removal of unwanted parts of the fruits, such as stems, leaves, seeds, or any inedible part of the fruits, is done. As per the requirement of drying and type of

fruits, the operation is performed, such as peeling, slicing, pulping, etc., in an organised manner to ensure uniform drying.

4. Pre-treatment – though it is an optional step but can give some significant benefits such as enhancement of flavour, colour, texture, etc. Some pre-treatment includes blanching, i.e., dipping the diced or sliced fruits into the boiling water and then immersing them into the chilled water, dipping the pieces into lemon juice of ascorbic acid, or soaking in the sugar syrup solution, which adds sweetness to the final product. The pre-treatment also prevents the product from enzymatic browning and oxidative damage.
5. Drying – there are several methods of drying which can be used to make the final product. The most commonly used drying techniques are as follows:
    a. Sun drying – it is a conventional and inexpensive method of drying which has been used for decades. The sun drying can be done in an open atmosphere or using modern solar dryers, which use sunlight as a source of heat.
    b. Hot air oven – it is a method of drying in which the sliced or diced pieces of fruits are placed on the trays which are placed inside a hot air oven in which hot air is blown continuously, which evaporates the free water available. The source of heat is hot air, and the mode of heat transfer is convection.
6. Drying time – it varies depending on the type and thickness of the fruit slices, the method of drying used, and the amount of moisture present in the fruits. It is essential to check the drying time after regular intervals to get the optimum and desired texture and final moisture content in the final product.
7. Storage – the dried fruits can be stored in airtight containers or re-sealable bags. Storage of the product should be done in such a manner that it should be away from the sunlight and moisture. If stored properly, the dried fruit products can be stored for months to years.

There are several dried fruits which are available in the market, and their demands are increasing day by day. These products are convenient to use, easy to carry, portable, can be eaten before or after meal snack, healthy, and tasty, which give them an additional advantage.

Dried fruit products can be used as an ingredient to make some other food product or added to some other food category for the purpose of value addition. It can be used in breakfast cereals, dairy products, bakery goods, snacks, bars, fruit leathers, confectionaries, etc.

It is advised to consume dried fruits daily for optimal benefit from the important nutrients, antioxidants, and phytochemicals they contain. This occurs because of the fact that fruits contain important vitamins, minerals, and phytochemicals in addition to basic nutrients that have a substantial positive impact on health.

Dried fruits have low-fat content (0.32–0.93 g/100 g), making them important sources of dietary fibre and potassium. According to sources, 40 g (on a per-serving basis) of dried fruits provide more than 9% dietary fibre and 3.8–9.9% potassium. Prunes are one of the best-dried fruit sources of vitamin K. Five prunes provide roughly 28 g of vitamin K, which is 23% of the daily intake for men and 31% for women (Carughi et al., 2020). Dried fruit consumption alters the diversity of the gut bacteria by increasing the relative number of healthy microorganisms while decreasing the overall presence of harmful microbes. Consuming dried fruit may also reduce the prevalence of obesity, many types of cancer, type 2 diabetes, cardiovascular disease (CVD), and other chronic diseases (Alasalvar et al., 2020).

The primary flavonoids identified in dried fruits include anthocyanidins, dihydrochalcones, flavanols, flavones, and flavan-3-ols. Dietary phytoestrogens present in dried fruits have generated a lot of research due to their potential preventive effects against a variety of disease conditions, including cancer, CVD, osteoporosis, and menopausal symptoms. Frequent consumption of dried fruit has been linked to improvements in lipid profiles, hunger and satiety control, CVD and cardiometabolic syndrome (such as inflammation, endothelial function, and BP), glucose and insulin homeostasis, bone health, and cancer (Mossine et al., 2019).

There are several methods which can be used to determine some important phytochemicals from the dried fruits. Total phenolics and anthocyanins can be determined by using non-specific methods such as pH differential, and specific methods include ultraviolet-visible spectroscopy (UV-VIS), high performance liquid chromatography (HPLC), and mass spectrometry (MS).

According to several research studies, dried fruits had higher bioactive components and antioxidant activity than those of their comparable fresh counterparts. There is evidence that the phenolic compounds in dried fruits are bio-accessible. According to a recent study (Scrob et al., 2022), dates and cranberries had the highest levels of phenolic bio-accessibility, while dates and prunes had the lowest, from the group of six dried fruits (dates, raisins, coconuts, cranberries, prunes, and bananas).

The main sugars which are present in all dried fruits are fructose and glucose, except for dates and peaches, in which the abundant sugar is sucrose. Dried fruits have a high sugar content; hence, it is expected that they will have a high glycaemic index (70 and above), resulting in elevated insulin responses. But according to recent studies, dried fruits have low (55 and under) to moderate (56–69) glycaemic and insulin indices, and their glycaemic and insulin responses are similar to those of fresh fruits. The presence of phenolics and fibre may have an impact on how the body reacts to sugar, changing the glycaemic response.

Market research includes analysing market trends, growth patterns, and the effects of many factors on market value and market dynamics, such as macroeconomic and microeconomic conditions, and geopolitical and demographic factors. Reported different phenolic compounds in various dried fruits are graphically depicted in Figure 10.1.

10.2 PRESENT GLOBAL MARKET SCENARIO FOR DRIED FRUIT PRODUCTS

The demand for ready-to-eat snacks has increased, which has had a beneficial impact on the market for dried fruits and nuts. Nowadays, most consumers want their food to satisfy complex health, wellness, and sustainability standards. Healthy eating

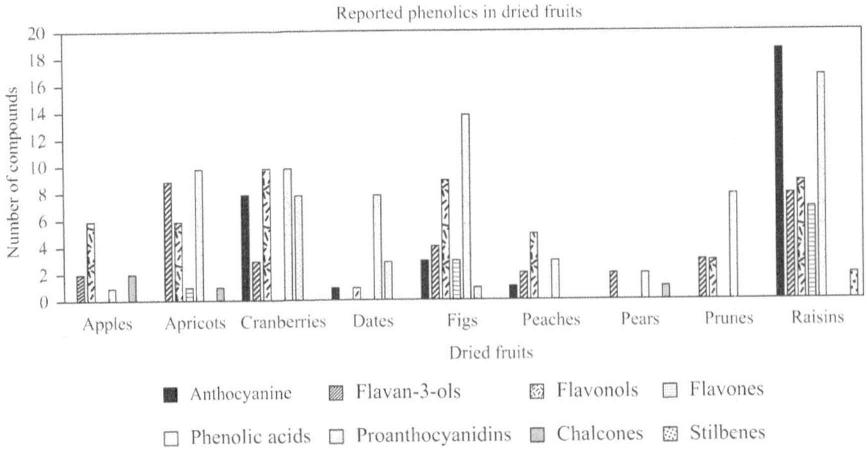

FIGURE 10.1 Reported phenolics in dried fruits (Alasalvar et al., 2020).

has thus become the norm. The popularity of healthy snacks among consumers has increased as a result of growing awareness of obesity and diseases linked to lifestyle. According to the Product Complexity Index (PCI), dried fruit is ranked 835th position. Due to the rise in global consumption, the market for dried fruits and nuts is expanding rapidly.

The global market size of dried fruits was \$7.86 billion in 2022 and it has reached \$8.68 billion in 2023 at a compound annual growth rate (CAGR) of 10.5%. The possibility of a global economic recovery following the COVID-19 pandemic was hampered by the Russia-Ukraine conflict. The dispute between these two countries has disrupted supply chains, increased commodity prices, and imposed economic sanctions on a number of other countries, leading to an inflation of goods and services on many global markets. The market size of dried fruits is expected to reach \$11.62 billion in 2027 at a CAGR of 7.6%. Dried fruit is the 660th most traded good in the world with a total trade of \$2.98 billion.

In the next few years, it is expected that the market for dried fruits will rise due to the increasing number of healthier consumers. Due to the growing consumer demand for dried fruits that are made from ethically sourced, environmentally friendly, and clean-label ingredients, producers have been forced to invest in processing and the development of unique products. Consequently, the market's growth is constrained by the expensive production and cultivation of dried fruit. Production of dried fruits worldwide during 2021–2022 is graphically shown in Figure 10.2.

In 2022, Asia-Pacific was the first largest region in the dried fruit market and the second was North America. Values in the dried fruits market are "factory gate" values, which are the prices at which goods are sold by their creators or producers, either directly to end users or to other entities. In a survey done in January 2021 under the direction of biotechnology start-up Gelesis, it was discovered that 51% of Americans wanted to reduce weight and 60% of them wanted to feel healthier. As a result, due to a rise in consumers who want to be fit and healthy, the demand for

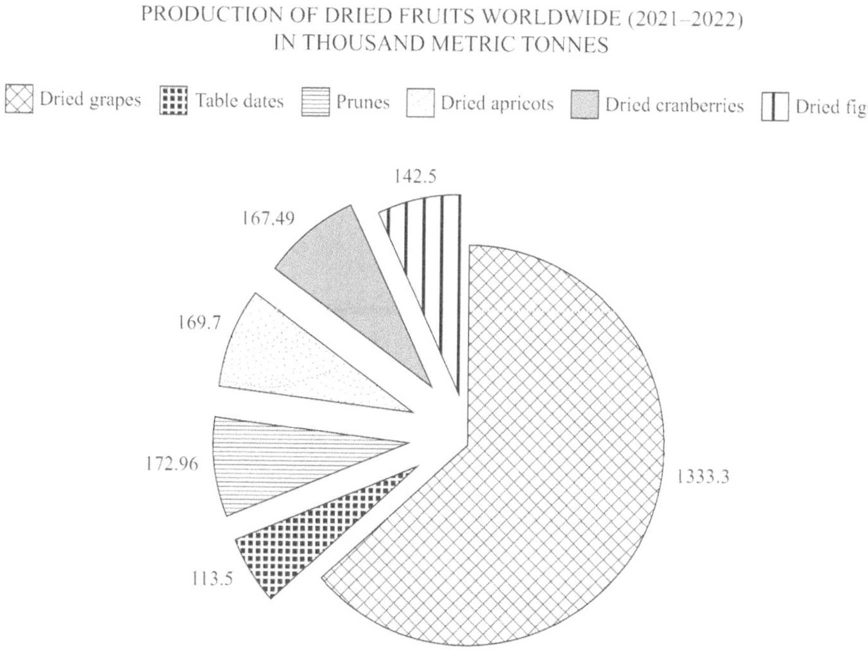

FIGURE 10.2 Production of dried fruits worldwide during 2021–2022 (in thousand metric tons).

these kinds of products is increasing in the market. In the years 2021–2022, dried fig production worldwide peaked at 0.142 million metric tonnes, while dried grape production peaked at 1.33 million metric tonnes.

The major exporters of dried fruits in 2021 were Turkey ($399M), Thailand ($290M), the United States ($254M), China ($247M), and Germany ($233M) and the major importers of dried fruits in 2021 were China ($367M), United States ($318M), Germany ($263M), France ($163M), and the United Kingdom ($139M). The top 15 exporters and importers of dried fruits are graphically shown in Figure 10.3. Whereas quantity-wise export of top dried fruits with respective trade values during 2021 is given in Table 10.2.

10.2.1 Segmentation of the Dried Fruits Market

The market for dried fruits is segmented based on four attributes, which are type, category, application, and geography.

1. The types of dried fruits are raisins, apricots, berries, figs, dates, prunes, etc.
2. Based on category, the dried fruits are divided into organic and conventional.

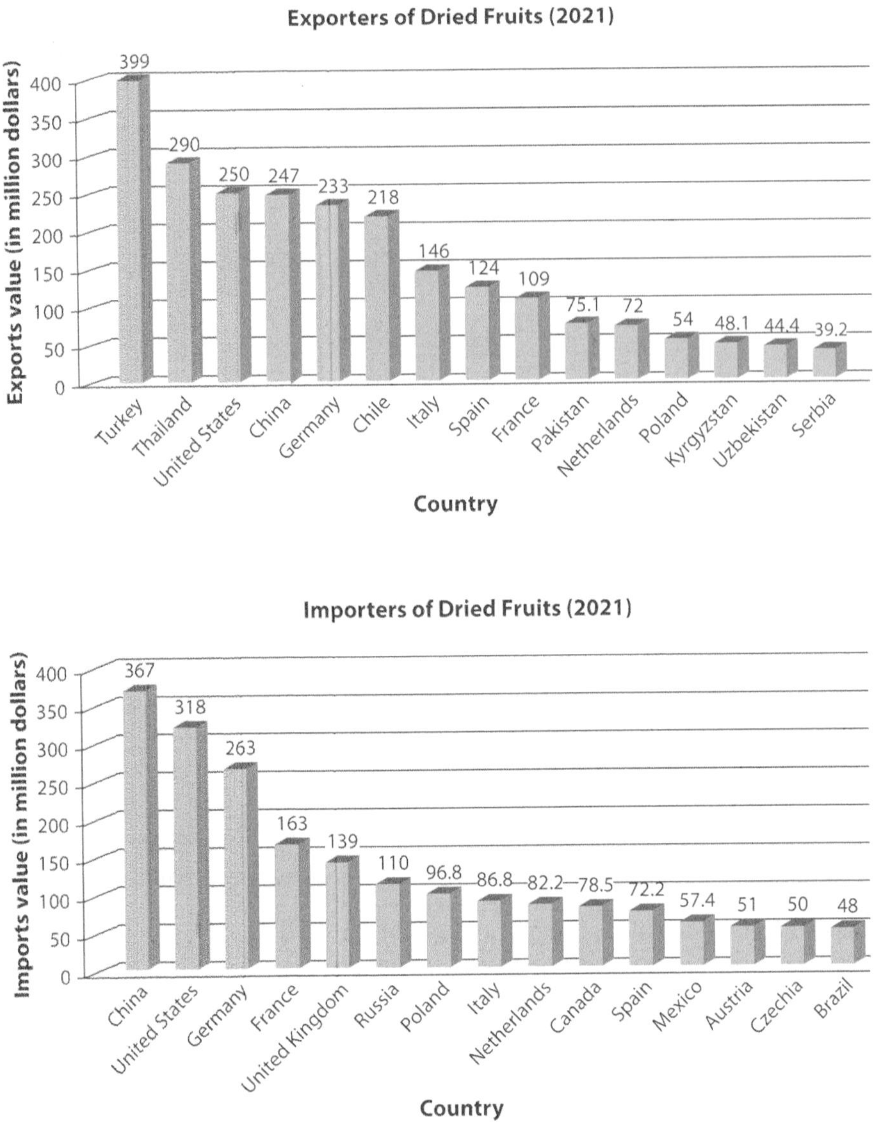

FIGURE 10.3 Top 15 exporters and importers of dried fruits during 2021 (Source – OEC).

3. Dried fruits can be added to bakery, dairy, snacks, confectionery, and cereal products.
4. Segmentation based on regional wise –
   North America (United States, Canada, and Mexico), South America (Brazil and Argentina), Europe (Germany, France, Italy, Spain, UK, and Russia), Asia Pacific (China, India, Japan, and Australia), Middle East & Africa (South Africa and UAE)

TABLE 10.2
Quantity-wise Export of Top Dried Fruits with Respective Trade Values during 2021

Dried Fruits	Quantity (kg)	Trade Value 1000 USD
Dried grapes	707000000	1464902.81
Prunes	182000000	609528.90
Dried Apricots	144000000	451766.11
Dried Apple	47100000	154578.49

10.2.2 Major Companies in Dried Fruits Market

- National Raisin Company, United States
- Kiantama Oy, Finland
- Traina Foods, United States
- Red River Foods Incorporated, United States
- Dole Food Company Inc, United States
- Sunbeam Foods Pty Ltd, Australia
- Lion Raisins Inc, United States
- Sun-Maid Growers of California, United States
- Bergin Fruit and Nut Company Inc, United States
- Paradise Fruits, Germany

10.3 PRESENT REGULATIONS FOR THE IMPORT/EXPORT OF DRIED FRUITS

The import and export of any product throughout the world depend on the regulations imposed by the regulatory bodies. It is worth noting that regulations are not the same everywhere in the world, they vary from one country to another. Based on the specific standards or toxicity of the product, each country has its own standards or limits of toxicity. For instance, the Aflatoxin, total maximum limit of dried fruits in the United States is 20 ppb, whereas in the European Union (EU) is 4–15 ppb, which clearly defines the variation of standards among the countries. In the case of India, an IEC Certificate or Import-Export Code is required before commencing any dry fruit export or import operation. The Director General of Foreign Trade (DGFT) issues this 10-digit code, which is valid for life. The U.S. Dried Fruits and Nut Association, Australian Dried Fruits Association (Australia), NDFTA – The Nut and Dried Fruit Trade Association Limited (United Kingdom), and Japan Dried Fruits Importers Association are Dried Fruits Associations for respective countries. There are even specific Associations and Commissions for respective dried fruits, such as American Pistachio Growers Association, California Prune Board, California Raisin Administration Commission, California Walnut Commission, U.S. Cranberry Marketing Committee, U.S. Highbush Blueberry Council, and U.S. Pecan Growers Council. The Plant Protection Act, Food Sanitation Act, and Customs Act are the primary regulations that govern the importation of dried fruits in Japan.

In order to export/import the dried fruits, they have to be devoid of mould, rotten contaminants, live or dead insects, and insect parts. Regulations governing food safety have received greater scrutiny because concerns about food contamination have grown. These regulations had a direct impact on the food trade. Aflatoxins are considered to be the most toxic mycotoxin due to their mutagenic and carcinogenic potencies. *Aspergillus parasiticus* and *Aspergillus flavus* produce aflatoxin B1 and B2, while *Aspergillus parasiticus* and *Aspergillus nomius* produce aflatoxin G1 and G2.

The author studied the Aflatoxin B1 content in Apricots and Figs of 20 samples each. It was reported that out of 20 samples of Figs, 12 samples were non-contaminated and 8 samples were contaminated with an aflatoxin toxin range of 14.43–6.72 ppb (Zahra et al., 2018). In the case of 20 samples of Apricots, it was reported that there were 19 non-contaminated samples and only 1 contaminated sample. The amount of aflatoxin B1 found in the sample of Apricots was 13.2 ppb, which is more than the USFDA's allowable level, of 10 ppb. They observed that Aflatoxins B2, G1, and G2 were absent in both of the dried fruits. A study by Naeem et al., (2022) concluded that Iranian dates were found to contain AFB1 and AFT at concentrations between 0.6–6.6 µg/kg and 0.9–8.1 µg/kg, respectively.

The Rapid Alert System for Food and Feeds recorded 399 notifications involving aflatoxins in 2021. Dried fruit intended for direct human consumption should have aflatoxin B1 maximum levels must be less than 2 g/kg, and total aflatoxins (AFT) (B1, B2, G1, and G2) maximum levels must be less than 4 g/kg, 5 g/kg for AFB1, and 10 g/kg for AFT in Turkey, the United States, Canada, and Australia.

The EU unveiled the European Green Deal in 2020 with the intention of transforming the European economy into one that is carbon neutral by 2050. By 2030, the action plan calls for a reduction in pesticide use of 50% and an increase in the proportion of agricultural land used for organic farming of 25%. Different quality parameters with their maximum limits for dried fruits for export in different countries are given in Table 10.3.

TABLE 10.3
Different Quality Parameters with Their Maximum Limit in Dried Fruits for Export in Different Countries

Dried Fruit	Country	Parameter	Maximum Limit (µg/kg)
Dried fruits	European Union (EU)	Aflatoxin B1	2
(Other than		Total aflatoxin (AFT)	4
dried figs)	India, European Union	Aflatoxin B1	6
Dried figs		Total aflatoxin (AFT)	10
All dried fruits	Pakistan		
Dried vine fruit	EU	Total aflatoxin (AFT)	20
(Currants, raisins, dried figs, sultanas)	India	Ochratoxin	8
Other		Ochratoxin	2
All dried fruits		Inorganic bromide	30

10.4 ISSUES ASSOCIATED WITH THE GLOBAL MARKET DEMAND

We know that food products which are dried have reduced water activity, and they have an edge over other perishable food products, which have more chances to get spoiled by microbial contamination due to high moisture content/water activity. Well, this is true to some extent; reduced water activity makes the dried food products shelf stable, but it is still not sterile. In the case of fruits, which are high in organic acids and have a low pH due to their acidic nature are prone to spoilage due to the growth of yeast and fungi, which can easily grow even at low moisture content if not processed or stored in hygienic conditions. Hence, if the dried fruits are not processed or stored in aseptic or hygienic environment, the product can be deteriorated easily, which affects the global market demand for the product. To avoid this, GHP, GMP, GSP, and HACCP practices need to be followed strictly.

According to Alexander Todorov's 2013 case study, Bulgaria, a European Union member subject to EU laws, has a sizable market for dried fruits and nuts (Fruit et al., 2013). A GfK (Growth from Knowledge) survey report in late 2011 showed that 87% of consumers consider the price as an important purchasing factor, and for 41%, the price was the most important reason for shopping. Affordable prices were stated as the main reason for shopping at certain retailers, followed by good quality products, which affect the global market demand (Fruit et al., 2013). If the prices are less and the product is easily affordable, the market demand for dried fruits will obviously increase, which will both promote the import as well as manufacturing of the dried fruits.

Another factor which affects the market demand is the occurrence of any occasion and less availability of fresh fruits due to off-seasons. As a traditional component of the regional cuisine, dried fruits and nuts are consumed in the biggest amounts during the Christmas and Easter holidays. During the winter months, when fresh fruits are scarce in the market, dried fruit sales also increase. Along with all these stated reasons, lack of promotion and advertisement also affects the overall product demand. There are people who are not much available on online platforms and they get to know about various trends, technologies, or new products from the local advertisements through print media, hoarding, telecommunications, etc. Hence, the lack of advertisements for these products does not make people aware of dried fruits, which can be substituted as a healthy snacking habit in place of processed food. Due to this, demand is not huge for the products, which are not well advertised. To raise consumer demand and spread awareness of their products, businesses should employ a variety of communication methods.

Another major issue which hugely affects the global market demand is the risk of infestation and contamination from live insects and mites. A massive infestation of mites has been observed in dried fruits which are imported from the Mediterranean region of Central Europe. It was infested by the *Carpoglpyhus lactis L. (Acarina: Carpoglyphidae)*. About 180 samples from supermarkets were used in a small-scale investigation by Hubert et al. (2011); 13% of these samples were contaminated; the contamination levels ranged from 0 to 660 mites per g of dried fruit. Dried apricots, figs, plums, and raisins all had higher levels of contamination. To study it even more, the dried fruits are kept under certain environmental conditions in different packaging materials, and all the dry fruit packaging materials that were evaluated, including polypropylene and aluminium foils, were found to be open to mite entry. This suggests that mites can migrate from package to package in supermarkets. It is well known that mites are both allergen makers and the carriers of fungi that cause mycotoxin. These results suggest that dried fruit has a higher chance of *C. lactis* infection. This comes out to be a huge problem with respect to the global market demand for the fruits-related products.

Another major and one of the most critical issues associated with the global market demand is the "Prevalent standards and non-compliances by exporting countries." Every country may have its own standards, or it may follow the general standards which are set by the CODEX for food products. If the importing country follows the CODEX standards, then it is easy to maintain the standards, which are uniform for all, but if the country follows its own national standards, then it is a little difficult to comply with all the standards of the exporting country. It leads to less or limited import of the dried fruits, or the import can be done only from the limited countries who can comply with all the standards, which creates a gap between the demand and supply of the dried fruit products. Different marketing standards exist in the segment of dried fruits. The EU general marketing standard applies to the countries who are the part of the EU. The general marketing standard of the EU forbids adhering to the minimum standards for quality and maturity, imposes a maximum tolerance, and mandates that products bear the entire name of their country of origin on their labels (Report et al., 2020).

If we talk about the compounds which are present in fruit products, we get to know that several undesirable compounds which do not deactivate even after the processing, can also cause health issues which affect the demand for the product. The most suitable example of this is Amygdalin, commonly known as laetrile, which is a natural substance that can be found in the seeds of apricots, bitter almonds, apples, peaches, and plums but is not found in the fruit itself. Consuming a lot of the seeds or pits, however, might be dangerous as enzymes in the intestines can convert amygdalin into cyanide, which can result in cyanide poisoning (FAO).

Similarly, there are some toxins, which are the secondary metabolites of the microorganisms who are contaminating the product and affecting the overall acceptance of the dried fruits. From these toxins, the presence of mycotoxins is the most prevalent because high-acid fruit products have more chances to get infected with fungi, and mycotoxin is the secondary metabolite of filamentous fungi. Since these substances can cause severe toxicity in the body even at low doses, their presence in the food chain raises serious concerns for human health. The mycotoxin-contaminated dried fruits not only create the health concerns among humans but also result in huge economic losses for exporting countries. The most common types of mycotoxin which can be found in dried fruits are aflatoxins (raisins), and ochratoxin A, which is commonly found in vine fruits such as currants, raisins, and sultanas, patulin and Alternaria toxins. (Fernández-cruz et al., 2010).

One of the main areas of concern is aflatoxins. They are most prevalent in dried figs. When dried fruits are kept in warm, humid environments, the fungi *Aspergillus flavus* and *Aspergillus parasiticus* can easily grow on them and create a group of poisons known as aflatoxin. As a potent liver carcinogen, aflatoxin is closely monitored to assure the safety of food. In accordance with Commission Regulation 1881/2006, the maximum aflatoxin levels for the entire EU, including Germany, are specified. This is also applied to imported products. Maximum permitted level of aflatoxin in dried fruit in European countries is 4 micrograms per kg, which is used for direct consumption and ochratoxin A is 10 micrograms per kg in dried vine fruits.

Several ochratoxin-producing fungal sources are *Aspergillus carbonarius* and others which have been isolated from grapes or dried vine fruits including *A. niger* and *A. tubingensis*. It has also been observed several times that commercially sold dried fruits can be contaminated with mycotoxins at high level which exceeds the number as per mentioned in the regulations in some countries (Trucksess & Scott, 2008).

Another most prevalent problem in food sector is the contamination of food due to microorganisms. Even at the reduced water activity, the growth of microorganisms can take place. It can lead to growth of some pathogens, yeasts, and moulds during the storage, transportation, and supply chain. It can even cause the serious health problems if there are enough viable spores and pathogens remain in the product. These microorganisms can even come from when the raw food is contaminated by the fungi or other pathogens such as *E. coli* O157:H7, *Listeria monocytogenes*, *Staphylococcus aureus*, and *Salmonella spp.* (Alp & Bulantekin, 2021).

It is also true that colour has a significant impact on how well-liked a product is, and dried fruits are no exception. Colour can influence people's perception of flavour of many dried fruit products, which affects the desirability of the product and hence affects the overall demand for dried fruits (Omolola et al., 2015).

The typical U.S. label does not adhere to EU labelling regulations. European Parliament and Council Directive 2000/13/EC lays forth general rules for pre-packaged foods sold in the EU, including rules for labelling, packaging, and advertising. It is applicable to foods designed for supply to restaurants, hospitals, and other mass caterers, as well as foods sold to the final customer. Additionally, according to the EU labelling regulations, dried fruits that were preserved with sulphur in excess of 10 mg SO_2/kg must bear a label which gives information about the product (Fruit et al., 2013). Countries who are not able to comply with the EU labelling standards can trade among the countries who are the part of the EU.

Countries who are members of the EU should be in complete compliance with the tariffs and regulations imposed by the EU, which happened in the case of Bulgaria. Depending on the nation of origin, there are different quotas or direct import taxes for dried fruits and nuts under EU customs, rules, and tariffs (European Commission, 2018; Report et al., 2020).

The EU requires phytosanitary certificates for the import of a number of goods, including dried fruit products. When imported from specific nations, these goods are also subject to toxin inspections at customs. These checks must be requested by importers at various customs locations, and products are detained by officials until they get the results. The EU is authorised to formally recognise a third country's pre-export controls (PEC) on feed and food prior to export under Article 23 of the EU regulation on official food and feed controls (Regulation (EC) No 882/2004), provided that it complies with EU standards. The EU may only approve a third country's pre-export inspection system if the checks conducted there are sufficient to reduce the need for import checks once the goods arrive in the EU. Countries which are the part of the EU face less competition from the nations like India, Iran, Gambia, China, etc. due to their inability to meet the EU sanitary/safety import regulations.

Another reason for the increase in demand is the rise in incomes of certain consumer groups, which increases their purchasing power, and they give their contribution to the market to grow high-value processed food. Along with this, there is a perception among the consumers regarding the food safety of products from certain countries. Because of the U.S.'s well-established food safety regulatory system, Bulgarian customers have faith in American goods, which raises demand and, as a result, encourages the import of food items from these nations.

Market entry strategy uses by different exporting or importing countries also affects the demand for the products. For example, the main benefit of adopting direct distribution is that it reduces costs and retains the products' freshness. Because in this case, the products need not to go through different channels of distribution and the storage time is reduced by manifolds.

The consumer and producer behaviour patterns affect the price, demand, and consumption trends.

10.5 FUTURE PROSPECTS AND GLOBAL MARKET DEMAND/CHALLENGES

Global Dried Fruits Market Size is 4.7 million USD in 2022. Over the forecast period, the dried fruit market is anticipated to expand at a CAGR of 5.2%, reaching a valuation of 5.10 million USD in 2027. The size of the dried fruit market is anticipated to reach USD 5.10 million, expanding at a CAGR of 5.2% from 2020 to 2025. Due to their numerous health benefits, including managing blood pressure, which is one of the key concerns of the elderly population, dried grapes account for a significant portion of the global dried fruit industry (GARFIELD, 1993) – Market data forecast 2022.

The consumption of dried fruits has drastically increased recently due to a rise in health awareness, especially among the young population. This has also led to studies and the identification and quantification of various groups of phytochemicals, which have significantly aided this market segment's growth. An increase in research and development (R&D) coupled with creative methods to employ dried fruits in various applications is also supporting the market growth that has been observed recently, due to awareness about various health benefits of dried fruits. They are widely utilised to improve flavours in dairy and bakery items, contributing to the market's overall growth. The rising use of dried fruits in bakery and confectionary products to enhance flavour, texture, and nutritional composition is aiding the market growth. These factors will surely help to increase the demand in the coming future, which will also help to reduce post-harvest losses.

The growth of the middle class and rising earnings are driving up the demand for quick snacks like raisins and nuts. Additionally, the rising demand for healthy breakfast and snack cereals like cornflakes, oats, muesli, etc. that contain a variety of dried fruits to boost their nutritional value is driving the market's expansion. Additionally, the expansion of the retail sector, including convenience stores, supermarkets, and hypermarkets, is enhancing consumer access to a variety of dried fruit items, supporting the expansion of the sector. The rise of e-commerce platforms, which provide a wide range of dried fruit goods with ease of convenience in both developing and developed countries, is another factor in the growth of the market.

Dried fruit products as an individual segment are eventually growing its market value due to the stated reasons. Along with this, in association with other food segments such as desserts and dairy, snacks, cereals, bars, bakery, and confectionary and its demand is increasing and will keep on increasing as these products are majorly consumed by the mass population.

On top of that, the leading companies' innovative and environmentally friendly packaging practices are supporting the expansion of the dried fruit industry. The bakery, snack, dairy, and confectionary industries are driving market expansion with the trend of adding dried fruits to boost the nutritional content of a variety of delicacies. Furthermore, top corporations are investing more money in dried fruit packaging to make the product more appealing, which is promoting market expansion. Furthermore, the emergence of sustainable packaging that offers better portion control and ingredient transparency due to rising environmental concerns among customers is giving the sector growth another momentum.

In addition, the introduction of portable, healthy foods that can be consumed anywhere is boosting the dried fruit industry. This trend will undoubtedly continue as working professionals don't have time to prepare meals and prefer to eat healthy snacks. The market is anticipated to continue to grow in the projected period of 2023–2028, expanding at a CAGR of 4.9%, driven by the rising demand for food products that improve immunity as a result of the coronavirus outbreak.

The dried fruit market can be divided into conventional and organic products based on various types. The industry can be divided into many product categories, such as figs, dates, raisins, apricots, and more. The market is divided among other categories like bakery and confectionery, cereals, snacks and bars, dairy and desserts, and others. The industry is segmented, among others, into supermarkets and hypermarkets, convenience stores, and Internet-based distribution networks. North America, Europe, Asia Pacific, Latin America, the Middle East, and Africa are the main regional markets for dried fruits.

Germany is the second-largest consumer of dried fruits and nuts, despite not producing much of them (Giles et al., 2013). Although German customers often favour domestically produced goods, the demand for various foreign quality foods, such as nuts and dried fruits, has steadily increased in recent years EDA (2018).

Considering that just 43.5% of respondents from a survey claimed that, they consume dried fruit once or twice a month, it appears that people eat it infrequently. It means that even today, people consume dried fruit less frequently than fresh alternatives, but people are getting aware of the various benefits, which include health benefits as well as convenience in consuming the dried fruits, which will increase the demand in future (Jesionkowska et al., 2008).

Along with all the stated reasons, the establishment of an efficient food safety regulatory system including GHP, GMP, HACCP, GSP (Good Storage Practices), etc. and the compliance of all the international standards and fair play a crucial role in the growth of this sector in coming time.

10.6 CONCLUDING REMARKS

The demand and growth of dried fruits sector is expected to increase on a very positive scale as people are getting aware of the health benefits of the dried fruit products and the convenience to use. However, latest technologies and proper processing and handling practices need to be implemented effectively in order to ensure the exports/imports standards by various countries. Turkey, followed by Thailand, is among the top leading exporters, followed by China, and the top importing countries are China and the United States.

REFERENCES

Alasalvar, C., Chang, S. K., & Shahidi, F. (2020). Dried fruits. *Health Benefits of Nuts and Dried Fruits, C*, 335–368. https://doi.org/10.1201/9781315173337-16

Alasalvar, C., Salvadó, J. S., & Ros, E. (2020). Bioactives and health benefits of nuts and dried fruits. *Food Chemistry, 314*(December 2019), 126192. https://doi.org/10.1016/j.foodchem.2020.126192

Alp, D., & Bulantekin, Ö. (2021). The microbiological quality of various foods dried by applying different drying methods: A review. *European Food Research and Technology, 247*(6), 1333–1343. https://doi.org/10.1007/s00217-021-03731-z

Carughi, A., Gallaher, D., & Mandalari, G. (2020). Bioavailability of Nutrients and Phytochemicals from Dried Fruits. In Health Benefits of Nuts and Dried Fruits (pp. 369–396). CRC Press.

EDA. (2018). Dried fruits in Germany. *Export Development Association, May.*

European Commission. (2018). COMMISSION IMPLEMENTING REGULATION (EU) 2018/1602 of 11 October 2018 amending Annex I to Council Regulation (EEC) No 2658/87 on the tariff and statistical nomenclature and on the Common Customs Tariff THE. *Official Journal of the European Union, March 2011*, 1–960. https://eur-lex.europa.eu/legal-content/EN/TXT/PDF/?uri=CELEX:32018R1602&from=EN

Fernández-cruz, M. L., Mansilla, M. L., & Tadeo, J. L. (2010). *Mycotoxins in Fruits and Their Processed Products: Analysis, Occurrence and Health Implications,* 113–122. https://doi.org/10.1016/j.jare.2010.03.002

Fruit, D., Nuts, T., Brief, P., Henney, M., Todorov, A., & Assistant, A. M. (2013). *Bulgaria Product Brief – Dried Fruits and Nuts.*

Fruit, D., Nuts, T., Slette, J., & Figurska, J. (2018). *Dried Fruit and Nut Sector.*

GARFIELD, J. (1993). Research report. *Teaching Statistics, 15*(1), 30–31. https://doi.org/10.1111/j.1467-9639.1993.tb00256.x

Hubert, J., Erban, T., Nesvorna, M., & Stejskal, V. (2011). Emerging risk of infestation and contamination of dried fruits by mites in the Czech Republic. *Food Additives and Contaminants – Part A, 28*(9), 1129–1135. https://doi.org/10.1080/19440049.2011.584911

Jesionkowska, K., Sijtsema, S., Simoneaux, R., Konopacka, D., & Plocharski, W. (2008). Preferences and consumption of dried fruit and dried fruit products among Dutch, French and Polish consumers. *Journal of Fruit and Ornamental Plant Research, 16*(January), 261–274.

Kiat, S., Alasalvar, C., & Shahidi, F. (2016). *Review of Dried Fruits: Phytochemicals, Antioxidant Efficacies, and Health Benefits. 21*, 113–132.

Mossine, A. V., Tanzey, S. S., Brooks, A. F., Makaravage, K. J., Ichiishi, N., Miller, J. M., ... & Scott, P. J. (2019). One-pot synthesis of high molar activity 6-[18 F] fluoro-l-DOPA by Cu-mediated fluorination of a BPin precursor. Organic & biomolecular chemistry, 17(38), 8701–8705. https://doi.org/10.1039/C9OB01758E

Naeem, I., Ismail, A., Rehman, A. U., Ismail, Z., Saima, S., Naz, A., ... & Aslam, R. (2022). Prevalence of aflatoxins in selected dry fruits, impact of storage conditions on contamination levels and associated health risks on Pakistani consumers. International journal of environmental research and public health, 19(6), 3404. https://doi.org/10.3390/ijerph19063404

Omolola, A. O., Jideani, A. I. O., & Kapila, P. F. (2015). *Critical Reviews in Food Science and Nutrition Operation Quality Properties of Fruits as Affected by Drying Operation. February*, 37–41.

Scrob, T., Covaci, E., Hosu, A., Tanaselia, C., Casoni, D., Torok, A. I., ... & Cimpoiu, C. (2022). Effect of in vitro simulated gastrointestinal digestion on some nutritional characteristics of several dried fruits. *Food Chemistry*, 385, 132713. https://doi.org/10.1016/j.foodchem.2022.132713

Trucksess, M. W., & Scott, P. M. (2008). *Mycotoxins in Botanicals and Dried Fruits: A Review. 0049.* https://doi.org/10.1080/02652030701567459

USDA (2020). *Report Name: Product Brief Dried Fruits and Nuts,* Retrieved from: https://apps.fas.usda.gov/newgainapi/api/Report/DownloadReportByFileName?fileName=Product%20Brief%20Dried%20Fruits%20and%20Nuts_Berlin_Germany_01-07-2020

Zahra, N., Khan, M., Mehmood, Z., Saeed, M. K., Kalim, I., Ahmad, I., & Malik, K. A. (2018). Determination of aflatoxins in spices and dried fruits. *Journal of Scientific Research*, *10*(3), 315–321. https://doi.org/10.3329/jsr.v10i3.37075

Index

Note: Page numbers in **bold** and *italics* refer to tables and figures, respectively.

Q

R

S

T

U

V

W

X

Z